Colloidal and Morphological Behavior of Block and Graft Copolymers

Colloidal and Morphological Behavior of Block and Graft Copolymers

Proceedings of an American Chemical Society Symposium
held at Chicago, Illinois, September 13-18, 1970

Edited by Gunther E. Molau
Polymer Science Group
Physical Research Laboratory
The Dow Chemical Company
Midland, Michigan

PLENUM PRESS • NEW YORK—LONDON • 1971

Library of Congress Catalog Card Number 72-148414

ISBN-13: 978-1-4684-1910-8 e-ISBN-13: 978-1-4684-1908-5
DOI: 10.1007/978-1-4684-1908-5

Softcover reprint of the hardcover 1st edition 1971

A Division of Plenum Publishing Corporation
227 West 17th Street, New York, N.Y. 10011

United Kingdom edition published by Plenum Press, London
A Division of Plenum Publishing Company, Ltd.
Davis House (4th Floor), 8 Scrubs Lane, Harlesden, NW10 6SE, England

CONTRIBUTORS

A. Banderet, Ecole Supérieure de Chimie, Mulhouse, France
S. J. Borchert, Department of Chemical Engineering, University of Wisconsin, Madison, Wisconsin
R. A. Botham, Polymer Research Section, Materials Research, The National Cash Register Company, Dayton, Ohio
E. B. Bradford, The Dow Chemical Company, Midland, Michigan
C. W. Childers, Research and Development Department, Phillips Petroleum Company, Bartlesville, Oklahoma
E. H. Cirlin, Science Center, North American Rockwell Corporation, Thousand Oaks, California
E. Clark, Research and Development Department, Phillips Petroleum Company, Bartlesville, Oklahoma
S. L. Cooper, Department of Chemical Engineering, University of Wisconsin, Madison, Wisconsin
R. G. Crystal, Rochester Research Laboratories, Xerox Corporation, Rochester, New York
E. F. Cuddihy, Jet Propulsion Laboratory, California Institute of Technology, Pasadena, California
W. G. Davies, Research Department, Midland Silicones Limited, Barry, Glamorgan, United Kingdom
G. M. Estes, Department of Chemical Engineering, University of Wisconsin, Madison, Wisconsin
F. W. G. Fearon, Polymer Research Laboratories, Dow Corning Corporation, Midland, Michigan
A. K. Fritzsche, Physical Chemistry Branch, General Electric Research and Development Center, Schenectady, New York
D. E. Green, Institute for Enzyme Research, University of Wisconsin, Madison, Wisconsin
J. T. Gruver, Research and Development Department, Phillips Petroleum Company, Bartlesville, Oklahoma
T. Hashimoto, Department of Polymer Chemistry, Faculty of Engineering, Kyoto University, Kyoto, Japan
M. G. Huguet, The B. F. Goodrich Research Center, Brecksville, Ohio
H. Inagaki, Institute for Chemical Research, Kyoto University, Kyoto, Japan

T. Inoue, Department of Polymer Chemistry, Faculty of Engineering, Kyoto University, Kyoto, Japan
D. P. Jones, Research Department, Midland Silicones Limited, Barry, Glamorgan, United Kingdom
D. H. Kaelble, Science Center, North American Rockwell Corporation, Thousand Oaks, California
K. Kato, Development Research Laboratories, Toray Company, Limited, Ohtsu, Japan
H. Kawai, Department of Polymer Chemistry, Faculty of Engineering, Kyoto University, Kyoto, Japan
T. Kotaka, Institute for Chemical Research, Kyoto University, Kyoto, Japan
G. Kraus, Research and Development Department, Phillips Petroleum Company, Bartlesville, Oklahoma
S. Krause, Department of Chemistry, Rensselaer Polytechnic Institute, Troy, New York
W. J. MacKnight, Chemistry Department, University of Massachusetts, Amherst, Massachusetts
M. Matsuo, Central Research Laboratory, The Japanese Geon Company, Limited, Kawasaki, Japan
J. Moacanin, Jet Propulsion Laboratory, California Institute of Technology, Pasadena, California
G. E. Molau, Polymer Science, The Dow Chemical Company, Midland, Michigan
M. Moritani, Department of Polymer Chemistry, Faculty of Engineering, Kyoto University, Kyoto, Japan
M. Motegi, Department of Polymer Chemistry, Faculty of Engineering, Kyoto University, Kyoto, Japan
H. Ohnuma, Institute for Chemical Research, Kyoto University, Kyoto, Japan
T. R. Paxton, The B. F. Goodrich Research Center, Brecksville, Ohio
J. Periard, Ecole Supérieure de Chimie, Mulhouse, France
F. P. Price, Physical Chemistry Branch, General Electric Research and Development Center, Schenectady, New York
G. Riess, Ecole Supérieure de Chimie, Mulhouse, France
J. C. Saam, Polymer Research Laboratories, Dow Corning Corporation, Midland, Michigan
S. Sagaye, Central Research Laboratory, The Japanese Goen Company, Limited, Kawasaki, Japan
W. Schoene, Polymer Institut, Universität Karlsruhe, Karlsruhe, Germany
R. W. Seymour, Department of Chemical Engineering, University of Wisconsin, Madison, Wisconsin
C. P. Shank, Polymer Research Section, Materials Research, The National Cash Register Company, Dayton, Ohio
M. Shen, Department of Chemical Engineering, University of California, Berkeley, California
H. Stutz, Polymer Institut, Universität Karlsruhe, Karlsruhe, Germany

C. Thies, Polymer Research Section, Materials Research, The National Cash Register Company, Dayton, Ohio
H. Y. Tom, Jet Propulsion Laboratory, California Institute of Technology, Pasadena, California
L. A. Utracki, Shawinigan Research, Gulf Oil Canada Limited, Quebec, Canada
G. Vanderkooi, Institute for Enzyme Research, University of Wisconsin, Madison, Wisconsin
B. Vollmert, Polymer Institut, Universität Karlsruhe, Karlsruhe, Germany
D. E. Walmsley, Jet Propulsion Laboratory, California Institute of Technology, Pasadena, California

PREFACE

The molecules of block and graft copolymers are molecules of a higher order; they consist of homopolymer subchains which are interconnected by chemical valence bonds. This structural complexity is manifested in the unusual behavior of block and graft copolymers both in solution and in bulk. Many types of interactions are possible in block and graft copolymers in the solid state. Polymer subchains of one molecule can interact with other polymer subchains which may belong to the same molecule or to different molecules. Since polymer chains of chemically different composition are usually incompatible, thermodynamically unfavorable as well as thermodynamically favorable interactions exist in the solid state. In solutions of block and graft copolymers, the situation becomes even more complex, because interactions between the solvent molecules and the various subchains of the copolymer molecules occur in addition to the interactions between the polymer chains. This multitude of interactions gives rise to a wide spectrum of colloidal and morphological properties which have no parallel in less complex polymer systems such as homopolymers or random copolymers.

Research on the colloidal and morphological behavior of block and graft copolymers is a relatively new field of endeavor. It started in 1954, when F. M. Merrett fractionated mixtures of grafted natural rubber with the corresponding homopolymers and observed that colloidal sols were formed at certain points during his fractional precipitations. His further experiments with these sols showed that the same samples of grafted rubber could be obtained in the form of two different modifications, which he called "hard form" and "soft form." Merrett's colloidal sols and his hard and soft morphological modifications seem to be the first experiments reported in the literature which demonstrate that graft copolymers can have colloidal properties and that these properties can lead to the formation of different types of morphologies.

In the years following Merrett's discovery, further progress in this special area was slow. A few isolated papers appeared in intervals of several years. Studies in France revealed that block copolymers of styrene and ethylene oxide can form liquid crystalline solutions. Work on high-impact polystyrenes and related systems in the United States showed that graft or block copolymers have an emulsifying effect in immiscible polymer solutions. In the last two to three years, the interest of the research community in the colloidal and morphological aspects of block copolymers has increased dramatically since the discovery was made that certain triblock copolymers of butadiene or isoprene with styrene are "thermoplastic rubbers" which act like crosslinked rubbers at room temperature but can be molded like regular thermoplastic materials at elevated temperatures. In addition to this discovery, the research interest was stimulated by the finding that butadiene or isoprene units in block copolymers or in rubber-modified polymer systems can be stained with osmium tetroxide, thus allowing direct observation of the stained portions of the polymer chains in the transmission electron microscope.

The Polymer Division of the American Chemical Society has sponsored a symposium which has focused, for the first time, on the collodial and morphological aspects of research on block and graft copolymers. Under the title "Colloidal and Morphological Behavior of Block and Graft Copolymers," this symposium was held as a part of the 160th National Meeting of the American Chemical Society in Chicago, September 13-18, 1970. The purpose of the symposium was to present papers which are representative of the work currently being done on colloidal and morphological aspects of block and graft copolymers. At the same time, an attempt was made to bring people of different research areas together who could be expected to have mutual interests. These different but interrelated research areas are concerned with membranes, ionomers, urethane elastomers, and block and graft copolymers.

The symposium consisted of five half-day sessions. In the first session, the peculiar morphology of block copolymers was illustrated in papers on electron microscopy of block copolymers of styrene with butadiene or isoprene. A theoretical treatment followed which explained why the various types of morphological structures are formed. Block copolymers containing siloxane chains, which were newcomers in the field, were the subject of papers presented in the second session. The third session started with a special feature, a lecture on biological membranes. This lecture was given in order to make people interested in block copolymer morphology aware that biological membranes have structures which are, in many ways, analogues to structures observed with block copolymers. An interchange of ideas and knowledge between membrane technology and block copolymer morphology is desirable in the light of the current industrial interests in membranes, e.g., for desalination of water. Other papers in the third session presented work on ionomers and on certain urethane elastomers, again materials which are analogous to block copolymers because they too have "physical crosslinks."

The fourth and fifth sessions covered papers on the emulsifying effect of block and graft copolymers, on colloidal aspects of high-impact polymers, and on the problem of microphase separation in block copolymers. One paper which presented work on the adsorption of block copolymer molecules on solid surfaces gives rise to the hope that better fractionation methods will be found based on the adsorption of block copolymers, e.g., in adsorption chromatography.

The symposium has covered a considerable portion of the work currently being carried out on colloidal and morphological behavior of block and graft copolymers and many researchers working in this field have participated. Previous published work in the same area is covered by a recent survey article also entitled "Colloidal and Morphological Behavior of Block and Graft Copolymers." This article has appeared as a part of another Plenum book which covers the proceedings of a recent symposium on general aspects of block copolymers (Proceedings of the Symposium on Block Polymers at the meeting of the American Chemical Society in New York City, September, 1969, published by Plenum Press, New York, 1970).

I would like to thank all authors for their fine cooperation in expediting the publication of this book and I am grateful to Dr. R. F. Boyer, Dr. J. K. Rieke, and Dr. D. R. Weyenberg for their presiding over three of the sessions.

September 26, 1970 Gunther E. Molau

CONTENTS

MICROMORPHOLOGY-PROPERTY RELATIONSHIPS IN GRAFT AND BLOCK COPOLYMERS

M. Matsuo* and S. Sagaye

The Japanese Geon Co., Ltd., Central Research Laboratory

1-2-1 Yako, Kawasaki, Japan

INTRODUCTION

Graft and block copolymers composed of plastic and rubber components have attracted much attention because of their interesting combination of physical properties. ABS plastics which possess the rigidity of plastics and the toughness of rubbers at the same time are good examples. The properties of ABS have been realized by incorporating fine rubber particles in plastic matrix, i.e., by designing the phase structure. The importance of phase structure was first recognized when Kato(1) demonstrated microstructure of the ABS plastic clearly under the electron microscope by using his osmium tetroxide (OsO_4) staining and hardening procedure. With the aid of Kato's method, it has now become possible to study the micromorphologies in plastic-rubber systems and to correlate them with their physical properties. This method of investigation has been quite new in this field and supplements the traditional method of setting up a hypothesis, carrying out the experiments and modifying the hypothesis.

An accumulation of the data on the morphology-property relationship may lead us to a new field, POLYMEROGRAPHY,(2,3) proposed after the name of Metallography. In POLYMEROGRAPHY, attempts are being made to systematize the phase structure - property relationships mainly by the use of electron microscopy. Moreover, it is intended to develop a new property by creating

*Present address: Bell Telephone Laboratories, Incorporated, Murray Hill, New Jersey 07974

a new phase-structure based on the above systematization. This type of investigation in polymers, however, has had only a short history and unfortunately many results are often kept secret by those hoping to exploit the commercial development. In the present paper, several examples of POLYMEROGRAPHY which, we believe, are still in germination step will be discussed.

ABS POLYMERS AS AN IMPACT MODIFIER OF PVC

In order to improve the impact properties of rigid PVC without sacrificing its transparency, specially prepared ABS (or MBS) graft copolymers are mixed with PVC. To retain the transparency, it is basically required to match the refractive index of the modifier with that of PVC and at the same time, to reduce its particle size to less than ca. 0.1μ in diameter. However, such a reduction of particle size usually causes a decrease of impact strength. Consequently, many research works have concentrated on the grafting technique to improve the impact strength. As a result, one of the interesting patents described that the addition of a large amount of water soluble salt in the grafting process resulted in an improvement of impact strength.[4]

The ABS modifiers for the present study have been prepared by carrying out the emulsion polymerization of acrylonitrile (AN) and styrene (ST) monomers in the presence of polybutadiene rubber latex, with or without addition of salt according to the above patent. When 10 parts of modifier, prepared in the presence of salt, are blended with 100 parts of PVC, the Charpy impact strength of 15 $Kg.cm/cm^2$ was obtained. This is much higher than the case where no salt was added, 8 $Kg.cm/cm^2$. Transparency of the former did not decrease to a noticeable degree. The electron micrograph of an ultrathin section of the ABS-modified PVC (with the addition of salt) is shown in Fig. 1. It is interesting to note that the modifier particles form a cluster of 0.1 to 0.3μ in diameter, and that a part of the grafted AN-ST copolymer seems to be included in the cluster. On the contrary the ABS modifier without salt exhibited a single particle dispersion in the PVC matrix. It is thus reasonable to consider that the cluster itself behaves as a large particle against the external stress thus causing a high impact strength, whereas optically it behaves as if it were divided into the original individual particles.

Another interesting property of this polymer system is that only a faint stress-whitening is observable when a properly prepared sample is subjected to bending or tensile stress under a relatively low strain rate. This is unusual for a rubber-modified plastic. Figure 2 shows the electron micrographs of

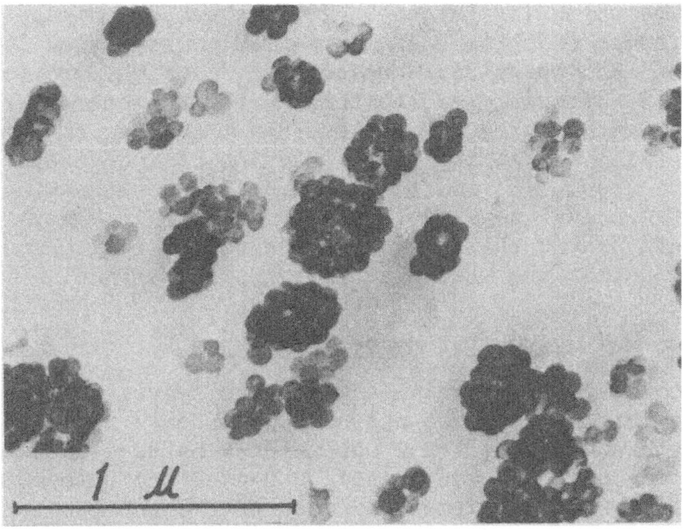

Ultrathin section of ABS-modified PVC prepared by adding salt at the grafting process.

Fig. 1

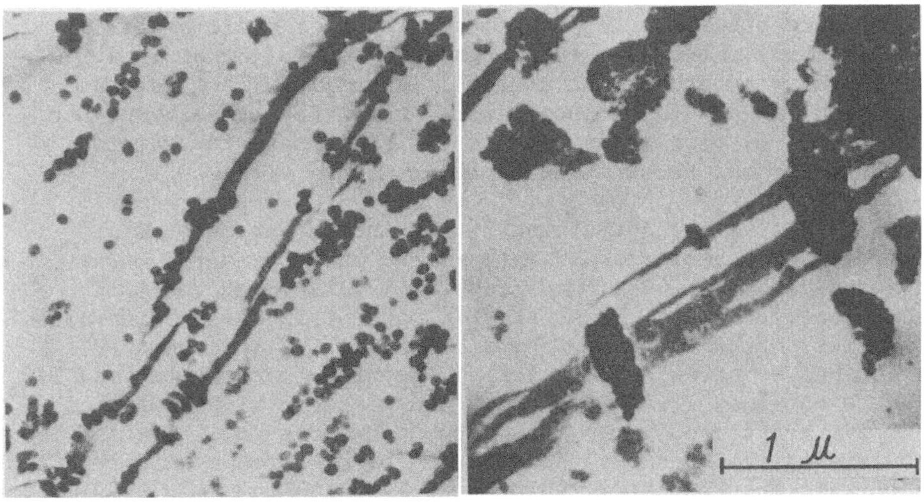

Crazing behavior of ABS-modified PVC.
Left: faint whitening; right: extensive whitening.

Fig. 2

the portion of the specimen under stress for the two extreme samples: one with a faint whitening (left) and the other with an extensive whitening (right). It is clear from these pictures that the former exhibits a single particle dispersion (ca. 700 Å in diameter) and the stress-crazes,[5,6,7] which were found to be the cause of whitening, are as thin as the diameter of the single particle. On the other hand, the craze thickness of the latter is fairly wide, reflecting the clustering of individual particles to a great extent. It is, therefore, concluded that in the ABS-modified PVC crazes develop in the same way as the straight ABS does,[7] and that when the craze thickness is small enough one can hardly detect the whitening from the appearance.

PVC-EVA GRAFT COPOLYMERS

The commercially available graft copolymers such as ABS polymers generally include unsaturated rubbers. The attempts to utilize saturated rubbers for obtaining a better weatherability have been conducted in many places. One of the interesting systems is a graft copolymer composed of PVC and ethylene-vinylacetate rubber (EVA).

PVC/EVA graft copolymers for the present study have been prepared by dissolving the EVA rubber in VCl monomer, followed by a usual suspension polymerization of the monomer. Four samples of the composition, PVC/EVA = 95/5, 80/20, 70/30 and 50/50 were prepared. For the observation of micromorphology an additional step has been introduced before the application of the Kato's method: treatment of a sample with a saturated solution of NaOH in methanol at its refluxing temperature for one hour. By this treatment the EVA phase near the sample surface becomes susceptible to the staining with OsO_4.

In the sample containing 5% EVA the impact strength is improved by a factor of 10 as compared with rigid PVC. Surprisingly however, the flow property is also improved to a considerable extent. When the EVA content is further increased a variety of properties, from a semirigid PVC to a soft PVC which has comparable mechanical properties to a plasticized PVC, can be obtained. Figure 3 shows the electron micrographs of PVC/EVA graft copolymers with varying EVA content. The sample containing 5% EVA shows that the EVA phase forms a network which separates the PVC matrix into particles with ca. 0.5μ in diameter. With the increase of EVA content, the mesh of the network decreases and the thread becomes thicker. The structure of the sample containing 50% EVA can be described as the EVA matrix containing dispersed PVC particles. It is likely that both components are contaminated with each other to some extent.

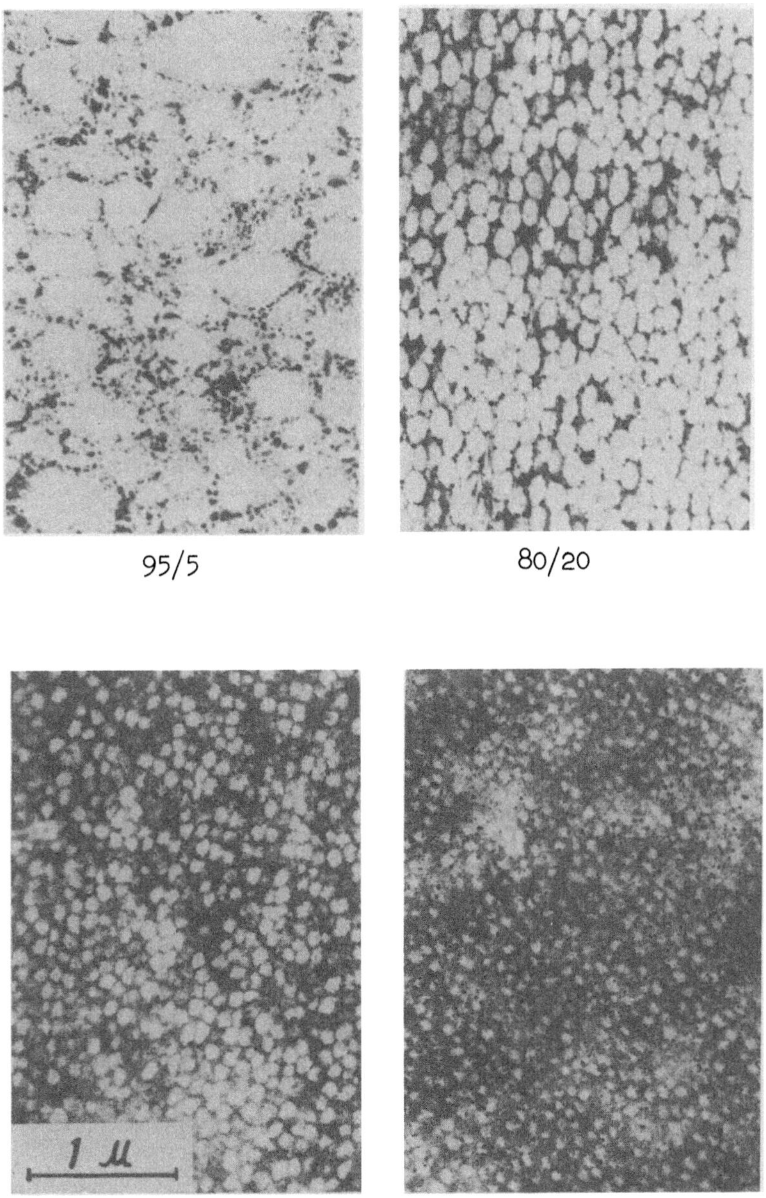

95/5 80/20

70/30 50/50

Ultrathin sections of PVC/EVA graft copolymers

Fig. 3

From these micrographs one can get some information on the properties described above. At first, it is reasonable to consider that the network formation of the rubber phase in a plastic matrix may serve to prevent an interfacial separation. Moreover, it favors the craze initiation and propagation without forming serious cracks as already discussed in a previous paper.[7] This may be a good reason for high impact strength. Secondly, the excellent flow property can be understood, if it is assumed that PVC phases separated by the thin EVA network flow as one flow unit.[8] In other words, it is likely that the EVA phase, which may be contaminated partially with PVC, serves as a lubricant at high temperature and binds the PVC phases together when the sample is cooled down to room temperature. On the other hand, the morphologies of PVC/EVA system with much higher EVA content seem to resemble EVA rubber filled with PVC particles. In this case, again both components are considered to be contaminated with each other. It has been reported that this type of structure, i.e., the elastomer filled with plastic particles, is found in styrene-butadiene block copolymers, where the plastic particles contribute to the reinforcement of the elastomer matrix.[9] It is thus easy to understand the excellent properties of this material as a soft PVC.

POLY(URETHANE-UREA)/POLYACRYLATE INTERPENETRATING POLYMER NETWORKS[10]

Interpenetrating polymer networks (IPN) have been prepared by carrying out a polymerization or crosslinking operation of one component in the presence of the other component polymer.[11,12] Although they are slightly different from graft copolymers, they are similar to them in the sense that the phase structure governs the properties. In the present study, poly(urethane-urea) latex provided by Wyandotte Chemical Corp. and self-curable polyacrylate latex supplied by B. F. Goodrich Chemical Co. were used as the starting materials. They are mixed together homogeneously and dried, followed by heating at 120°C to induce a crosslinking in both components. Dynamic mechanical measurements showed two distinct transition regions, at -20°C and 30°C, each corresponding to the glass transition temperature (Tg) of component homopolymer.

Figure 4 shows typical electron micrographs of ultrathin sections for three IPN's, A/U = 70/30, 50/50 and 30/70, where A and U designate polyacrylate and poly(urethane-urea) components, respectively. The dark phase represents the poly(urethane-urea) phase selectively stained by OsO_4. It is clear that for A/U = 70/30 and 50/50 the poly(urethane-urea) particles of 1 to 5μ in diameter are dispersed in the polyacrylate matrix. At the composition A/U = 30/70, however, the poly(urethane-urea) particles seem to be in contact with each other to form a continuous phase. In other words, phase inversion occurs from "U particles in A matrix" to "A particles in U matrix" at the composition A/U = 30/70 as the U component increases. It is interesting to note that this

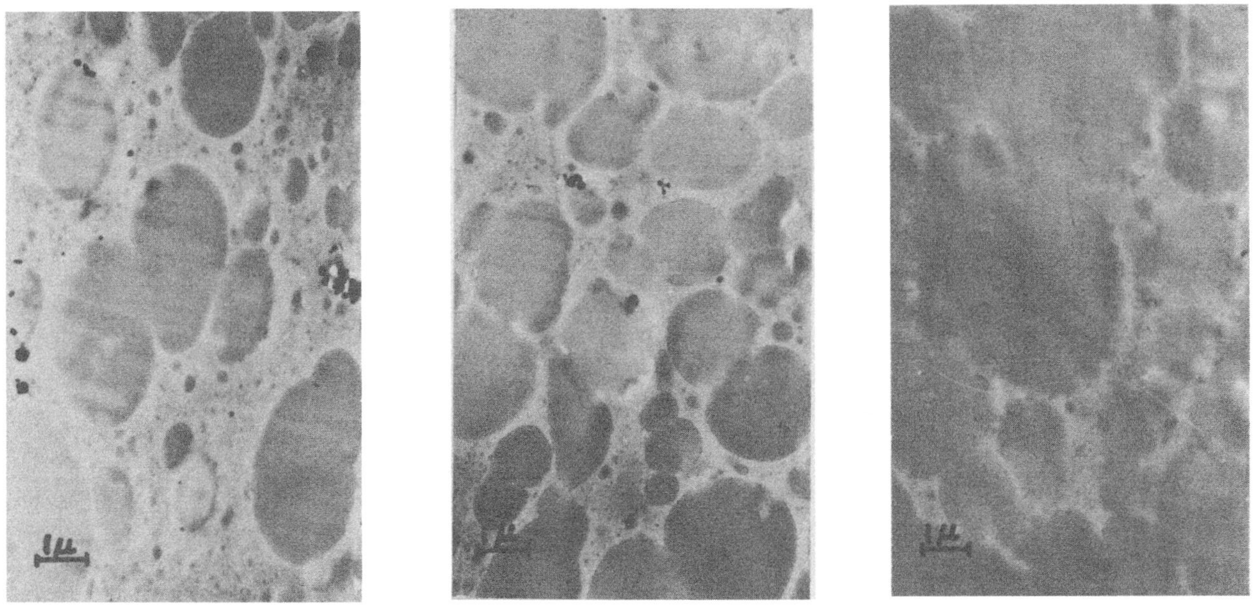

A/U = 70/30 50/50 30/70

Ultrathin section of poly(urethane-urea)/polyacrylate interpenetrating polymer networks.

Fig. 4

composition corresponds to that of the closest packing of particles.

Tensile measurements were carried out with the strain rate of 5 min.$^{-1}$ Results are shown in Fig. 5. Tensile strength does not increase until the polyacrylate (rigid) phase becomes continuous at A/U = 40/60. It increases thereafter almost linearly as the rigid phase increases. The elongation at break decreases rather rapidly until the rigid phase becomes continuous (at A/U = 40/60) and then remains almost constant. It is thus concluded that the tensile properties are considerably governed by the phase structure formation.

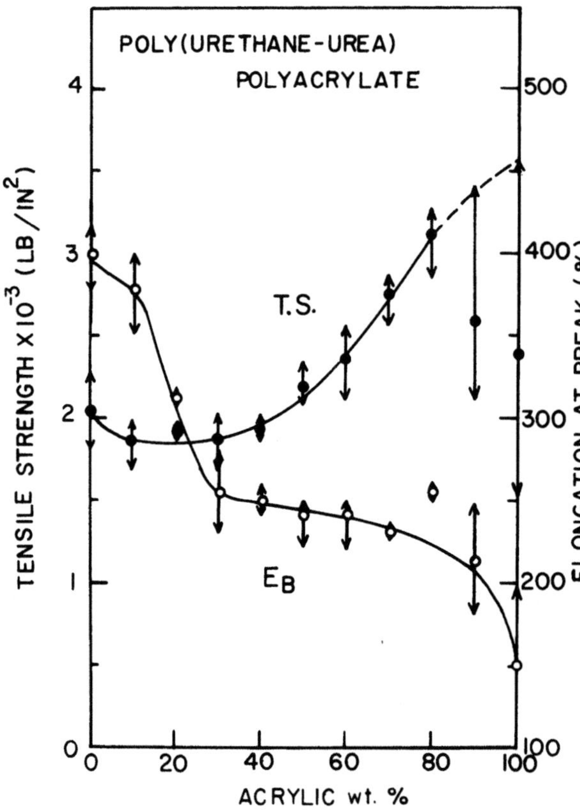

Effects of tensile strength and elongation at break of poly(urethane-urea)/polyacrylate system measured at the strain rate of 5 min.$^{-1}$ on polyacrylate content (rigid phase)

Fig. 5

BLOCK COPOLYMERS COMPOSED OF STYRENE AND BUTADIENE SEQUENCES

Block copolymers composed of styrene and butadiene sequences (designated as S and B, respectively) give interesting properties because of contrasting natures of component homopolymers. For example, a copolymer with predominant B fraction gives a cure-free thermoelastic rubber, where S particles are dispersed in B matrix.(13) A copolymer with predominant S fraction, however, exhibits a structure similar to rubber-toughened plastics, i.e., B particles are dispersed in S matrix.(14) The preparation and characterization of the block copolymers for the present study were already shown in the previous paper.(14,18)

Compression Molded Sheet

Specimens were prepared by a compression molding at 160°C of the flakes which had been obtained by dissolving the copolymers in benzene followed by precipitating in methanol.

One of the important factors which predominantly influence the mechanical properties of the compression molded sheet is a sequence arrangement. The authors already reported(14) that the block copolymers with SB- and BSB-type arrangements were as brittle as pure polystyrene, whereas that with SBS-type exhibited remarkable toughness. Here, comparison was made under the constant S/B mole ratio of 60/40. Fracture occurred in the latter sample with extensive whitening (i.e., crazing) in a manner similar to that of high-impact polystyrene.(6) Electron micrographs of these three copolymers are shown in Fig. 6. Although there is a slight difference in morphology between the SB- or BSB-type and the SBS-type, these are considered to be identical in the sense that B phase is dispersed separately in S matrix. It is thus difficult to explain the difference in tensile behavior from the difference in the phase structure. It is instead explained by the difference in chain configuration in the B phase as schematically shown in Fig. 7. The stress applied is concentrated at the B phases as in case of ABS plastics (7). The B phases of the SB- and BSB-types break easily by a chain slippage mechanism since the chains aggregating together within the B phase can transmit a load only by contacting or making an entanglement at the chain ends (Fig. 7A). On the other hand, the cluster of B phase in the SBS-type is quite strong due to a continuity of molecular chains inside the particle (Fig. 7B). The latter can sustain a large amount of crazing before serious crack occurs. This may give a possible explanation of the toughness of SBS-type block copolymer.

Next, the effects of S/B mole ratios on mechanical properties were studied using SBS-type copolymers. Five samples with S/B mole

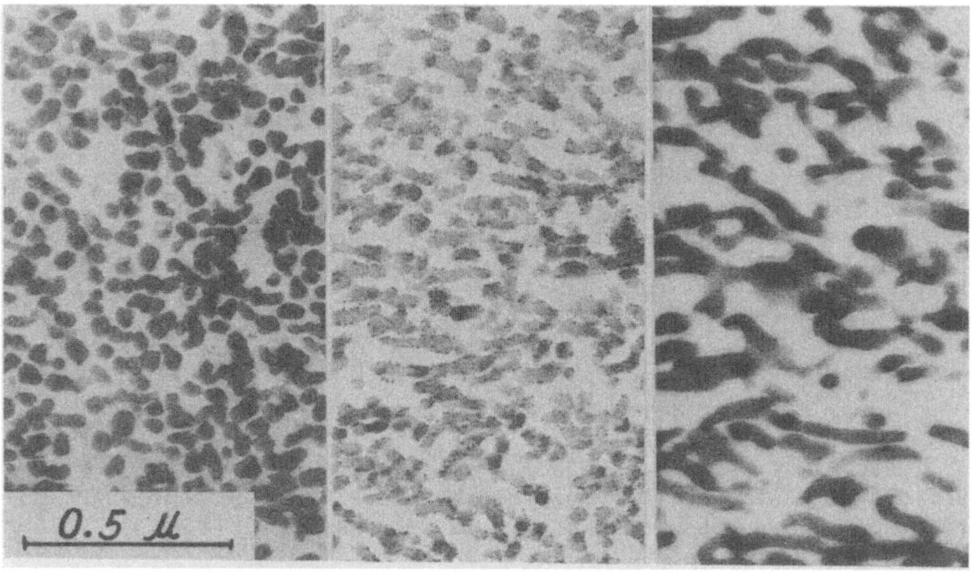

Fig. 6 Ultrathin sections of styrene-butadiene block copolymers prepared by compression molding. From the left, SBS-, BSB-, and SB-type arrangements with the constant S/B mole ratio of 60/40.

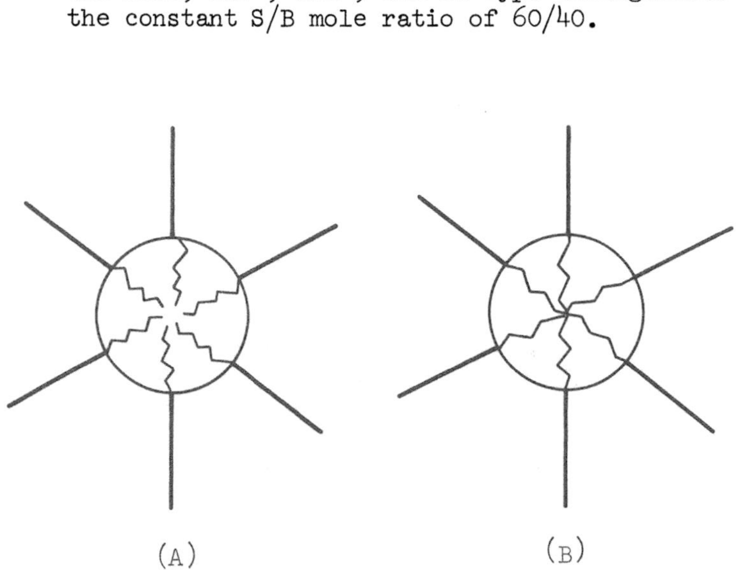

Fig. 7 Schematic representations of chain configuration within the block B phases of styrene-butadiene block copolymers. Left: SB- or BSB-type; right: SBS-type.

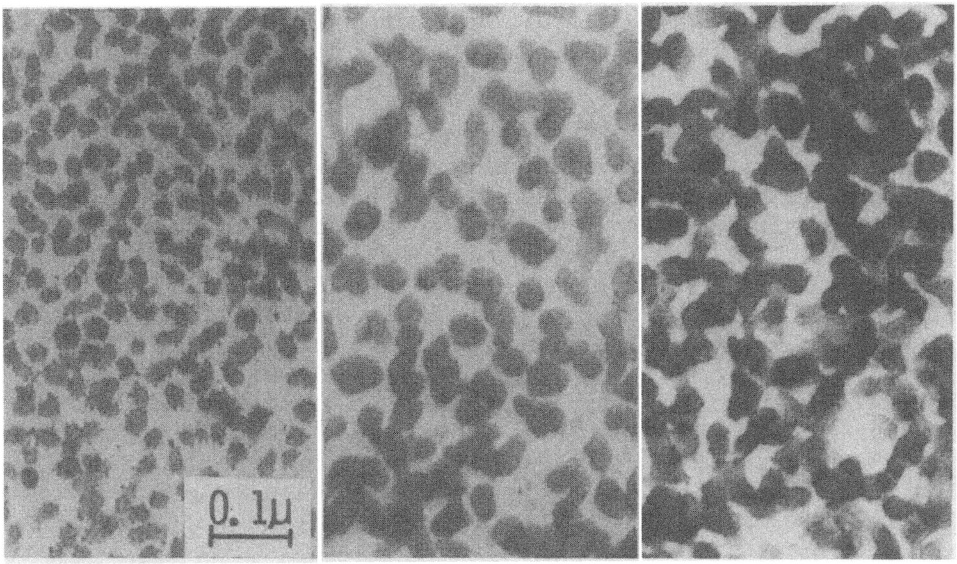

Ultrathin sections of styrene-butadiene block copolymers prepared by compression molding. From the left, SBS-1 (S/B = 80/20), SBS-3 (S/B = 60/40) and SBS-5 (S/B = 40/60).

Fig. 8

ratios of 80/20, 70/30, 60/40, 50/50 and 40/60 were prepared.[14,18] They are designated as SBS-1 to -5 according to the order of increasing B fraction. The morphologies of SBS-1, -3 and 5 are shown in Fig. 8. All of these structures are basically expressed as "rubber particles in plastic matrix" as in case of ABS plastics, although the rubber phases in SBS-5 seem to be partly in contact with each other. The yield and tensile strength decreased monotonically with increasing B fraction as shown in the part of Fig. 13. This result will be discussed again later in comparison with the properties of solvent cast film.

Solvent Cast Films

Another interesting morphological observation of styrene-butadiene block copolymers is a regular microphase separation occurring when the copolymers are cast from solution.[14-19] The authors (18) already reported that there was a progressive change in the structure of SBS-type cast from toluene solution from "B spheres in S matrix" through "B rods in S matrix" to "S and B alternate sheets" as the S/B ratio decreased from 80/20 to 40/60. In the present study another solvents such as cyclohexane and methylethylketone (MEK) were employed. The micromorphologies and tensile properties were observed and compared with those of compression molded sheets.

Figures 9, 10 and 11 demonstrate the microphase separation of SBS-1, -3 and 5, respectively, cast from cyclohexane solution. Ultrathin sections cut normal (left) and parallel (right) to the cast film surface are shown to visualize their theee dimensional structures. It is apparent that, in SBS-1, B phase forms rough networks which pile up one after another in S matrix tending to orient their surfaces parallel to the film surface (Fig. 9). This pattern is completely different from that of toluene cast film where B spheres are dispersed separately in S matrix. In SBS-3, B phase forms rods which arrange themselves regularly in S matrix parallel to the cast film surface (Fig. 10). Beautifully separated S and B alternate sheet structure is clearly observed in SBS-5 (Fig. 11). Some portions of the parallel section were either completely white or completely black, although a combination of black and white portions are shown in Fig. 11 (right). This implies that these sheets tend to arrange their surfaces parallel to the film surface. These structures of SBS-3 and -5 are similar to those observed in toluene cast films.

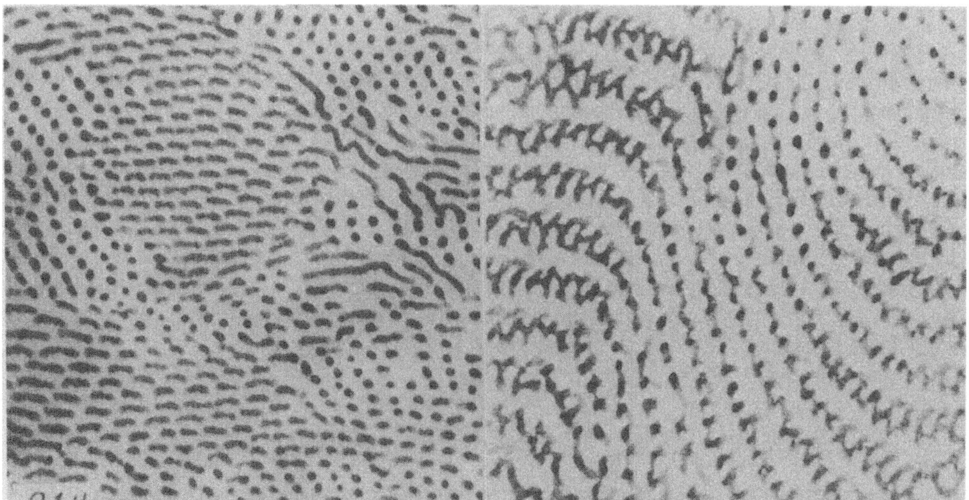

Fig. 9 Ultrathin section of SBS-1 cast from cyclohexane solution. Sections cut normal (left) and parallel (right) to the film surface (piled network structure).

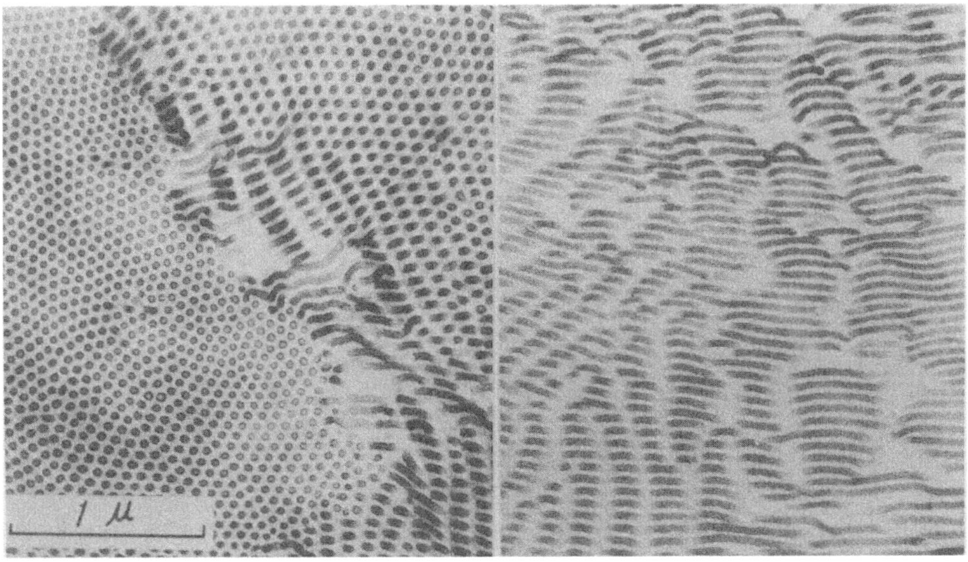

Fig. 10 Ultrathin sections of SBS-3 cast from cyclohexane solution. Sections cut normal (left) and parallel (right) to the film surface (rod structure).

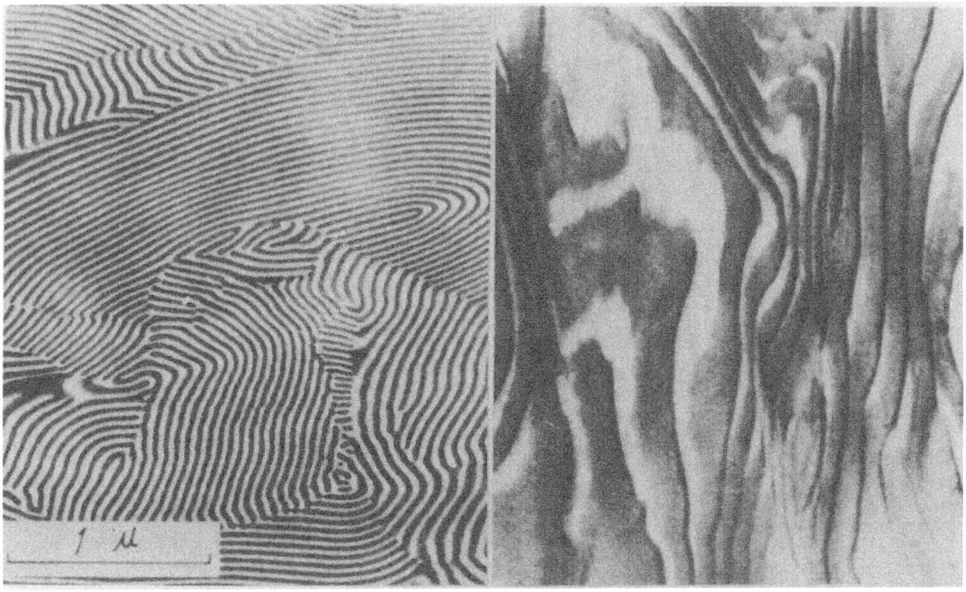

Fig. 11 Ultrathin sections of SBS-5 cast from cyclohexane solution. Sections cut normal (left) and parallel (right) to the film surface (alternate sheet structure).

The structures in MEK cast films were rather different from those of toluene and cyclohexane cast films. SBS-1 showed similar sphere structure as in the case of a toluene film. However, in SBS-3 the sphere structure was again observed. SBS-5 showed neither rod nor sheet structure but very short rod structure as demonstrated in Fig. 12. These results are attributed to the poor solubility of the B component in MEK. All the structures observed so far for SBS-1, -3 and -5 are listed in Table 1.

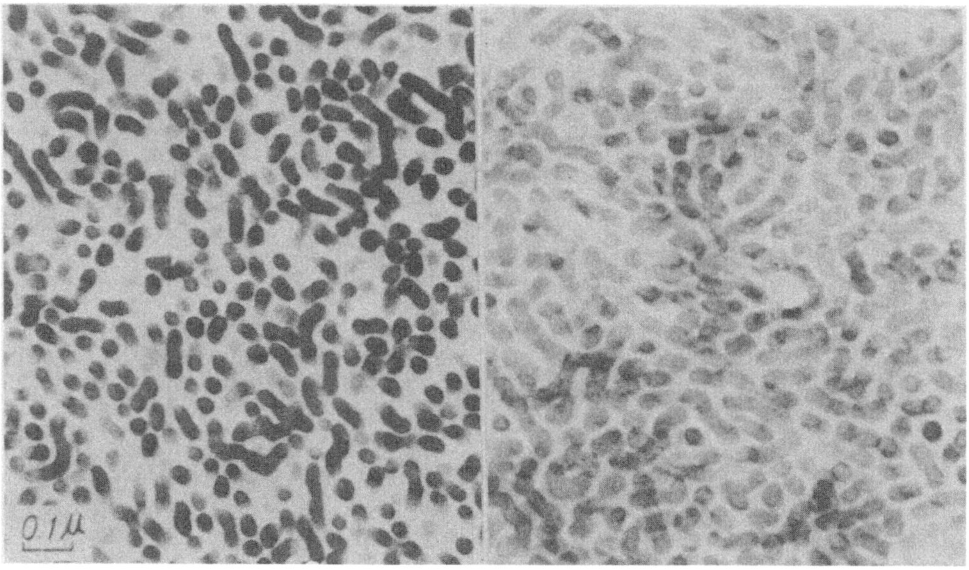

Fig. 12 Ultrathin sections of SBS-5 cast from MEK solution. Sections cut normal (left) and parallel (right) to the film surface (short rod structure).

TABLE 1

Structures of block B Phase

Samples	Casting solvents		
	toluene	cyclohexane	MEK
SBS-1	sphere	rough network	sphere
SBS-3	rod	rod	sphere
SBS-5	alternate sheets of S and B	Alternate sheets of S and B	short rod

Tensile measurements were carried out on the SBS-1, -3 and 5 cast from cyclohexane and MEK solutions. However, all the MEK cast films were so brittle that we could not even punch out the tensile specimens from the cast films. Therefore, only the results for cyclohexane cast films are shown and compared with those of compression molded sheets (Fig. 13). As already mentioned above the tensile strength of compression molded sheet decreases hand in hand with the yield stress as the B fraction increases. It forms a convex curve against the composition axis. On the other hand, the tensile strength of cyclohexane cast film first drops rapidly and then keeps constant with the increase of B fraction, whereas the yield stress continues to drop off. This striking difference is obviously attributable to the different micromorphology in these samples. For example, the B phase of SBS-1 cast film has some continuity throughout the specimen (Fig. 9). It causes lower yield and lower tensile strength than that of a compression molded specimen where the B phase disperses separately in S matrix as already shown in Fig. 8. The SBS-5 cast film, however, gave the tensile strength about three times higher than that of a compression molded specimen, although the yield stresses were almost identical. A combination of continuous thin layers of S and B components existing in SBS-5 cast film may have contributed to the higher tensile strength.

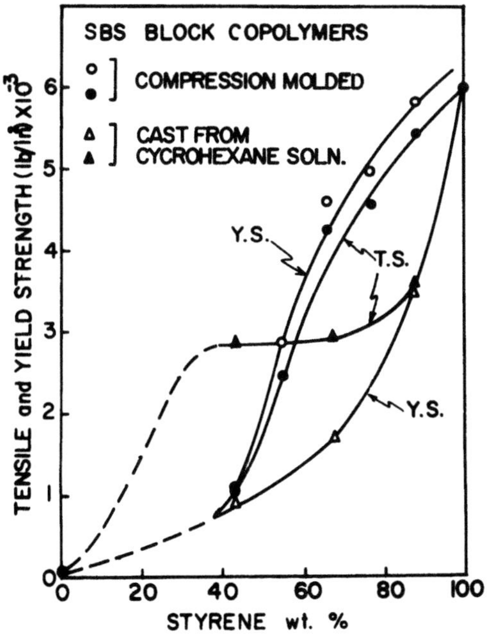

Fig. 13 Comparison of tensile and yield strengths of compression molded sheet with those of cyclohexane cast film in SBS-type styrene-butadiene block copolymers.

CONCLUSION

With the aid of Kato's OsO_4 technique diversified range of micromorphologies in plastic/rubber polymer systems has been disclosed so far. As far as we know, the micromorphologies observed so far are represented by seven simplified patterns as schematically drawn in Fig. 14, where the black and white portions correspond to rubber (R) and plastic (P) phases, respectively. Poly(urethane-urea)/polyacrylate system, SBS compression molded sheet, and ABS plastics are examples of A-series. Elastomers filled with plastic particles correspond to B-series. PVC/EVA graft system and also PVC/NBR-20* blend system which was reported previously[20] are classified into C-series. No actual system has been found for D-series. The patterns of E-, F- and G-series are known as the characteristic structures of styrene-diene block copolymers, the part of which has been discussed in this paper. The peculiar structure of piled network formation in SBS-1 cast from cyclohexane solution can be considered as a modification of the G-series. It is important to note that the dimensions of all the domain structures lie in the range of 0.01μ to 10μ; the working range of the electron microscope. It is believed that each structure has an advantage in some mechanical properties as in the case of ABS plastics where high impact strength is obtainable without sacrificing rigidity. It is thus important to accumulate the observations on micromorphology-property relationships and to find a general rule which governs them.

As far as the tensile strength is concerned, there seems to be a general trend which correlates well with the micromorphology, although further accumulation of data is needed. Fig. 15 summarizes the tensile strength vs. composition curves of three typical systems. The first system is the SBS compression molded sheet which has "R-particle in P-matrix" structure. The tensile strength forms a convex curve against the composition axis (open circles). One can find the similar convex curve in poly(urethane-urea)/polyacrylate system which has the similar "R-particle in P" structure. The second system is PVC/NBR-20 blend whose structure was found in the previous paper[20] to be "R-network in P-matrix." Its tensile strength (filled circles) decreases rather rapidly, forming a concave curve as the rubber content increases.[21] In view of these two examples, the unusual tensile behavior of SBS-type block copolymers cast from cyclohexane solution is prominant. Especially the "R-rods in P-matrix" structure of SBS-3 (S = 65% by wt.) and the "P-R alternate sheets" structure of SBS-5 (S = 43% by wt.)

---

*Acrylonitrile-butadiene copolymer rubber with containing 20% acrylonitrile by weight

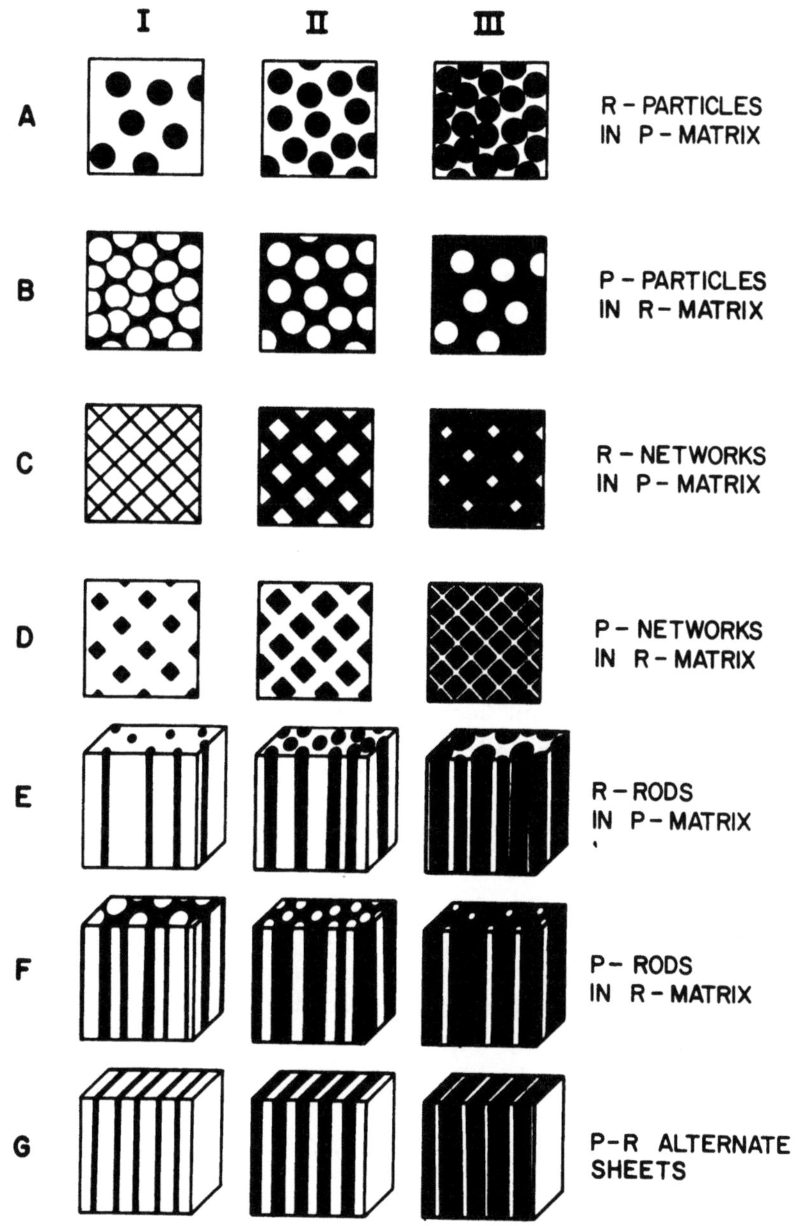

Fig. 14 Schematic illustration of various phase structures composed of plastic (P) and rubber (R) components. Rubber content increases from the left to right.

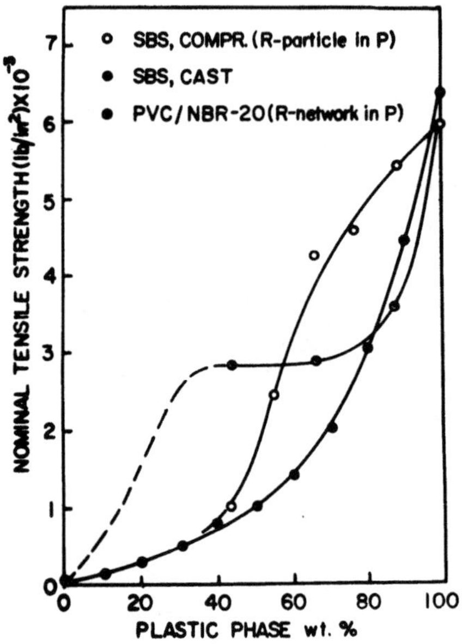

Fig. 15 Summary of tensile strength vs. plastic content curves in plastic-rubber sytems.

showed unexpectedly high tensile strength. These results may serve as a guide in designing the polymer systems with required property using plastic-rubber systems.

ACKNOWLEDGMENT

The authors wish to thank Mr. Y. Jyo of Japanese Geon Co., Ltd. and Dr. T. K. Kwei of Bell Telephone Laboratories for helpful discussions and to Miss C. Nozaki of Japanese Geon Co., Ltd. for carrying out a part of electron microscopic work.

REFERENCES

1. K. Kato, J. Electron Microscopy, 14, (3), 219 (1965); Polymer Letters, 4, 35 (1966).

2. K. Kato, Japan Plastic, 2, (2), 6 (1968).

3. M. Matsuo, Japan Plastics, 2, (3), 6 (1968).

4. JP-42-19248, 42-22541 and BE-719,425

5. C. B. Bucknall, R. R. Smith, Polymer, Lond. 6, 437 (1965).

6. M. Matsuo, Polymer, Lond. 7, 421 (1966).

7. M. Matsuo, Polymer Eng. & Sci., 9, 206 (1969).

8. A. R. Berens, V. L. Folt, Polymer Eng. & Sci., 8, 5 (1968).

9. T. L. Smith, Polymer Preprints, 10, (2), 670 (1969).

10. M. Matsuo, T. K. Kwei, D. Klempner, H. L. Frisch, Polymer Eng. & Sci., (to be published in Nov. issue, 1970).

11. H. L. Frisch, D. Klempner, K. C. Frisch, J. Polymer Sci., Part B, 1, 775 (1969).

12. L. H. Sperling, D. W. Friedman, J. Polymer Sci., Part A-2, 7, 425 (1969).

13. H. Hendus, K. H. Illers, E. Ropte, Kollidzschr. 216-217, 110 (1967).

14. M. Matsuo, T. Ueno, H. Horino, S. Chujyo, H. Asai, Polymer, Lond., 9, 425 (1968).

15. E. Vanzo, J. Polym. Sci., A-1, 4, 1727 (1966).

16. T. Inoue, T. Soen, H. Kawai, J. Polym. Sci. B, 6, 75 (1968).

17. E. B. Bradford, E. Vanzo, J. Polym. Sci., A-1, 6, 1661 (1968).

18. M. Matsuo, S. Sagaye, H. Asai, Polymer Lond., 10, 79 (1969).

19. T. Inoue, T. Soen, T. Hashimoto, H. Kawai, J. Polym. Sci., A-2, 7, 1283 (1969).

20. M. Matsuo, C. Nozaki, Y. Jyo, Polymer Eng. & Sci. 9, 197 (1969).

21. M. Matsuo, unpublished data.

THE EFFECT OF ENVIRONMENT ON THE MORPHOLOGY OF STYRENE-BUTADIENE BLOCK COPOLYMERS

E. B. Bradford

The Dow Chemical Company

Midland, Michigan 48640

INTRODUCTION

The characterization and morphology of block copolymers have been the subject of many investigations in the last few years. Many block copolymer solutions are colored and films cast from them possess ordered structures. In a previous publication (1) ordered structures from clear and colored styrene-butadiene block copolymer films were discussed. The purpose of this investigation was to show, by electron microscopy, the alteration in copolymer structure which results from various treatments, and thus, aid in the elucidation of molecular behavior.

EXPERIMENTAL

The block copolymers were prepared by anionic polymerization using butyllithium as the initiator. The polymerizations were carried out in solutions of benzene or toluene. Each reaction was carried out in separate steps; usually the styrene was polymerized in the first step to give "living chains", then butadiene was added to give the second block of the copolymer.

Two electron microscopes were employed to observe the structure of these copolymers, a RCA model EML and a JEM model 6A. Ultrathin sections of films were prepared by embedding in epoxy resin; the embedments were trimmed, faced, stained with osmium tetroxide (2), and then microtomed into sections about 750 A thick in the normal manner.

MORPHOLOGY OF A 60/40 STYRENE-BUTADIENE DIBLOCK COPOLYMER

This copolymer exhibited a blue color in solutions of 20% copolymer in toluene. The gel permeation chromatograph (GPC) showed a single peak distribution with a slight skewedness toward the lower molecular weight elutions. The peak of the curve occurred at the same elution volume as a polystyrene calibration sample with a molecular weight of 630,000. Assuming a 60/40 copolymer composition, its molecular weight is 378,000/252,000. Films dried from this solution also exhibited a blue color. Cross-sections of the film were composed of layered or striped regions 1 to 3 microns across oriented in many directions; drying the film very slowly enlarged these ordered regions (Fig. 1-A).

The addition of polystyrene of a lower molecular weight than the polystyrene block segments was expected to expand the polystyrene block domains. A film from a solution containing 7% copolymer and 7% regular polystyrene (M.W. = 250,000) possessed elongated, tangled strands of polybutadiene in a polystyrene

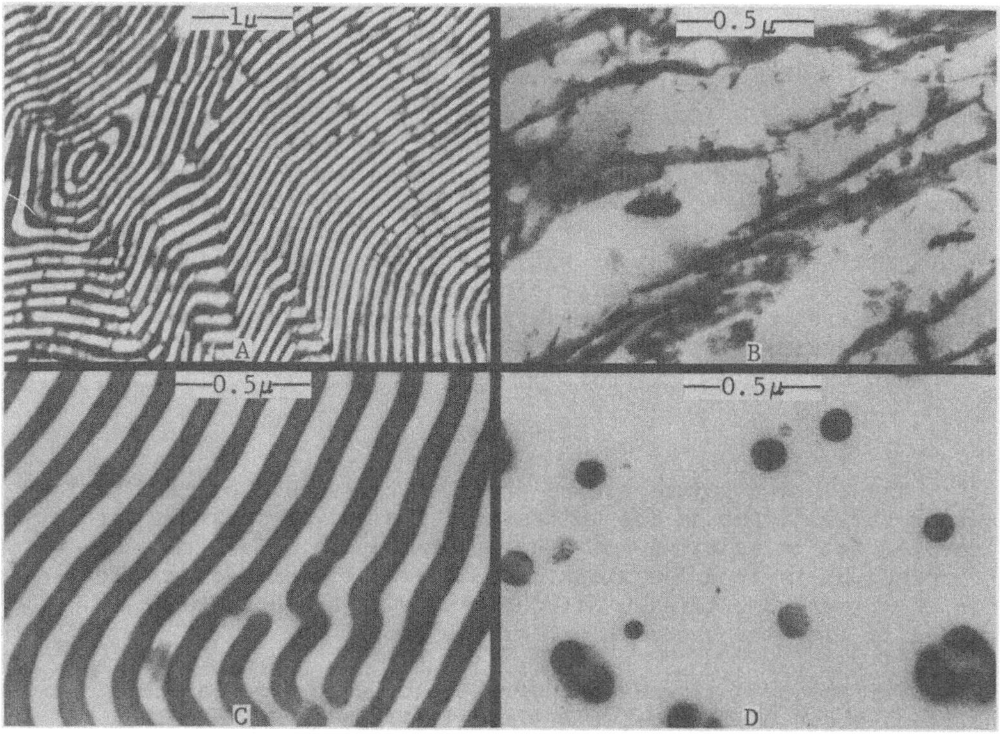

Fig. 1. CONFIGURATIONS OF A 60/40 STYRENE-BUTADIENE DIBLOCK COPOLYMER: A. Slowly dried film; B. Film containing 50% added polystyrene; C. Original film stained before embedment; D. Embedded in methacrylate.

matrix (Fig. 1-B). Since the total polystyrene content of the film was 80%, one might expect (3,4) to observe spheres of polybutadiene in a polystyrene matrix. Inoue, et al. (5) observed spheres of polyisoprene in a styrene-isoprene diblock copolymer film containing only 70% polystyrene. Perhaps extended ageing is required to obtain equilibrium structures in 14% copolymer solutions.

Since the previous films had been embedded in epoxy resin before staining, a portion of the 60/40 styrene-butadiene film was stained before embedment to determine if the epoxy resin had altered the unstained copolymer structure appreciably. The regular striped patterns were still present (Fig. 1-C). However, when a drop of the 10% copolymer in toluene solution was added to a 50/50 blend of methyl and butyl methacrylate which was then polymerized without agitation at 50°C, the striped patterns were not observed (Fig. 1-D). The copolymer was greatly diluted by the methacrylate; the block copolymer structures appear to be in the process of forming spheres of polybutadiene.

The film structure can be altered in a predictable manner by the addition of selective solvents. Cyclohexane, a selective solvent for polybutadiene, was slowly added to the original 20% copolymer in toluene until the solution contained 1% copolymer. This formed a film composed of a polybutadiene matrix containing spheroidal and chain-like aggregates of polystyrene (Fig. 2-A). Thus, this selective solvent for polybutadiene condensed the polystyrene segments into the dispersed phase even though polystyrene comprised the major component of the copolymer and the solution contained 4% of the nonselective solvent, toluene. The general structure of the film was not altered when a similar solution contained 19% methylene chloride, a low boiling nonselective solvent (Fig. 2-B). A reversal in structure was observed when similar experiments were carried out using methyl ethyl ketone, a preferential solvent for polystyrene. A film from a 1% copolymer solution was composed of spherical particles of polybutadiene in a polystyrene matrix (Fig. 2-C). Spheres remained when the solution contained 19% methylene chloride, but some larger elongated spheres were also seen (Fig. 2-D). Hence, as expected, the polymer dissolved in the selective solvent becomes the continuous phase, and approximately 20% of a lower boiling nonselective solvent does not cause gross alterations but affects the occluded rubber particles more than those of polystyrene.

When the original copolymer solution was added to n-heptane, a more selective preferential solvent for polybutadiene than cyclohexane, a precipitate with a cloudy supernate was formed. The sample dried to a rubbery, bluish-white film. It was composed of spheroidal and short, chain-like occlusions of polystyrene in a

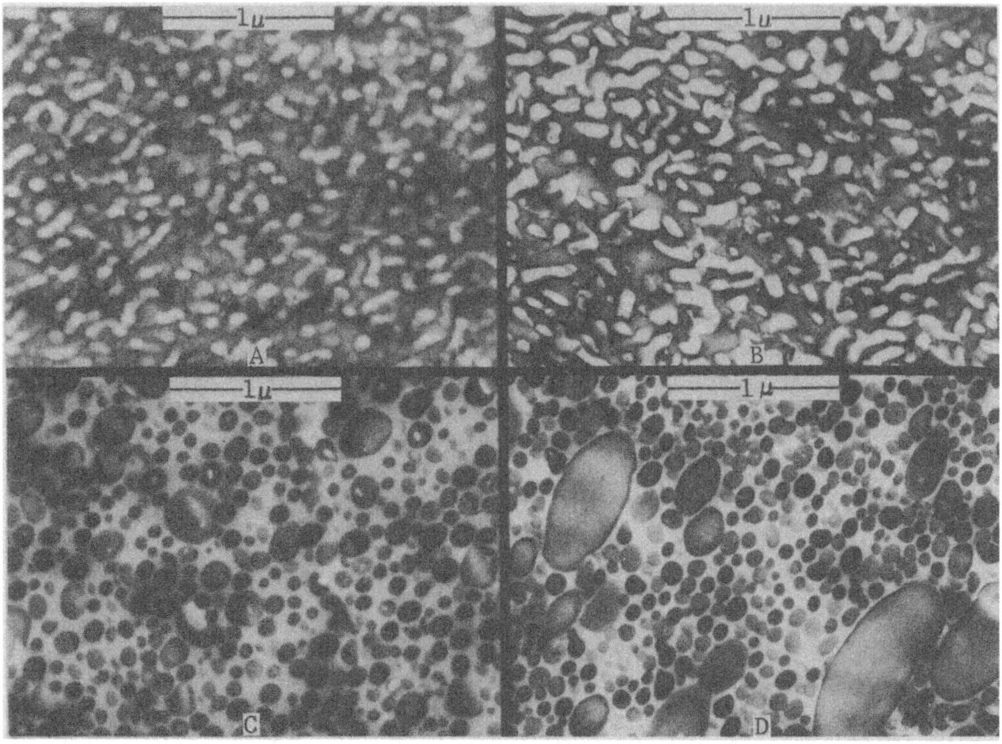

Fig 2. FILMS OF THE 60/40 COPOLYMER DRIED FROM PREFERENTIAL SOLVENTS: A. Cyclohexane with 4% toluene; B. Solution for A containing 19% methylene chloride; C. Methyl ethyl ketone with 4% toluene; D. Solution for C containing 19% methylene chloride.

polybutadiene matrix (Fig. 3-A). Since n-heptane is a preferential solvent for polybutadiene it was anticipated that polybutadiene would be the continuous phase. A portion of the supernate was diluted with additional n-heptane; a drop was dried onto a carbon substrate, stained with OsO_4 vapors, and then shadowed with chromium. This resulted in a dispersion of circular structures, each possessing a slightly darker nucleus (Fig. 3-B). Since shadows are observed on both sides of the edge (arrow A and opposite edge), the circular structure cannot be flat and is a polymeric loop, possibly containing a thin polymeric layer. All of the nuclei are attached to the interior of the loops. One explanation of these structures is that the nucleus is the polystyrene portion of the copolymer and the loop is the stained polybutadiene portion dried from a highly swollen state. Apparently the supernate contained swollen copolymer which did not contain enough polystyrene to flocculate in the n-heptane.

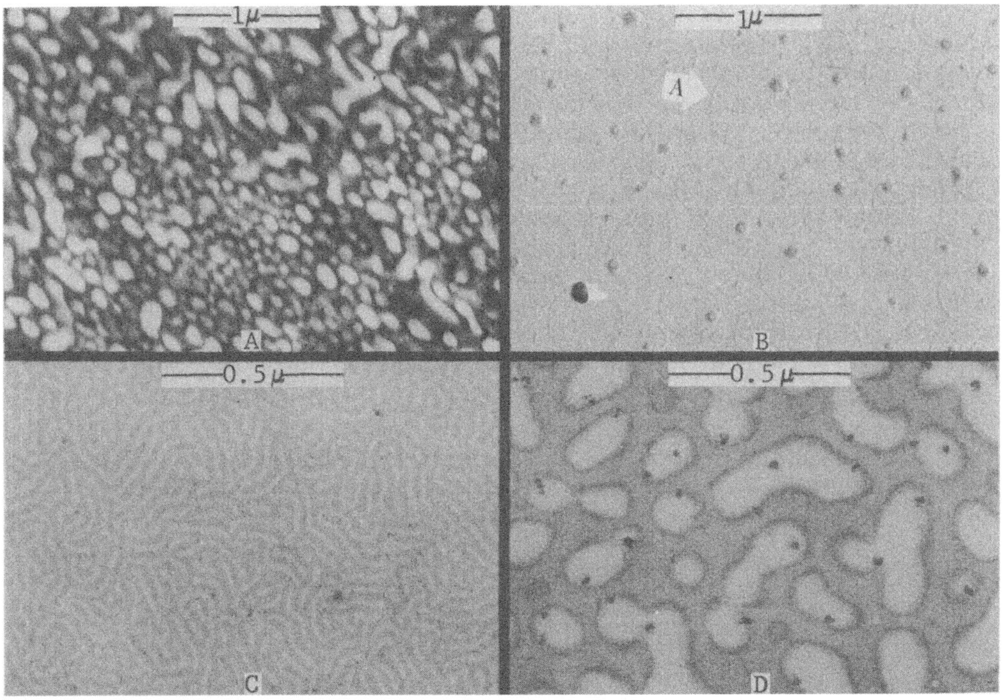

Fig. 3. MORPHOLOGY OF THE 60/40 COPOLYMER: A. Film precipitated with n-heptane; B. Dried drop of supernate above precipitate; C. Dried drop of dilute copolymer in toluene solution; D. Dried drop of dilute copolymer in methylene chloride solution.

A dried and then stained drop of the copolymer which had been diluted to approximately 10 ppm in toluene showed islands of very thin films in some regions while other regions were devoid of copolymer. The thin films did not possess striped patterns, rather random strings of polystyrene approximately 170 A in diameter in a polybutadiene matrix (Fig. 3-C). Similar structures were seen by Heinze (6) in both cast thin films and ultrathin sections of thick films of a 35/65 styrene-butadiene diblock copolymer. However, this configuration was not observed when our original copolymer was similarly diluted with a different nonselective solvent, methylene chloride (Fig. 3-D). This indicates that the interaction between the copolymer molecules and these nonselective solvents is different, and that the size of the unstained structures does not depend solely upon the molecular dimensions of the copolymer molecules. The small dense particles are probably an osmium precipitate which sometimes results when staining with osmium tetroxide vapors.

An attempt was made to preserve the copolymer structure in dilute solution by staining directly in solution. A drop from a

0.5 ppm copolymer in methyl ethyl ketone solution was dried and then stained. The aggregated particles were composed of polybutadiene cores with swollen polystyrene at the periphery (Fig. 4-A). This was expected since methyl ethyl ketone is a selective solvent for polystyrene. However, when a portion of the same solution was stained by the direct addition of staining solution, the polybutadiene block segments were hardened and did not agglomerate as the drop dried (Fig. 4-B); the micrograph clearly shows many small (300-500 A) particles of polybutadiene dispersed in islands of polystyrene. When a similar experiment was carried out using cyclohexane, selective for polybutadiene, the drop dried before staining displayed spherical and chain-like aggregates of polystyrene in a polybutadiene matrix (Fig. 4-C); but when the staining was carried out in the cyclohexane solution, the dried drop showed only dense chain-like aggregates (Fig. 4-D). Apparently the polystyrene block segments are buried in the center of the aggregates and are not observed (due to the efficient electron scattering of the stained polybutadiene shell).

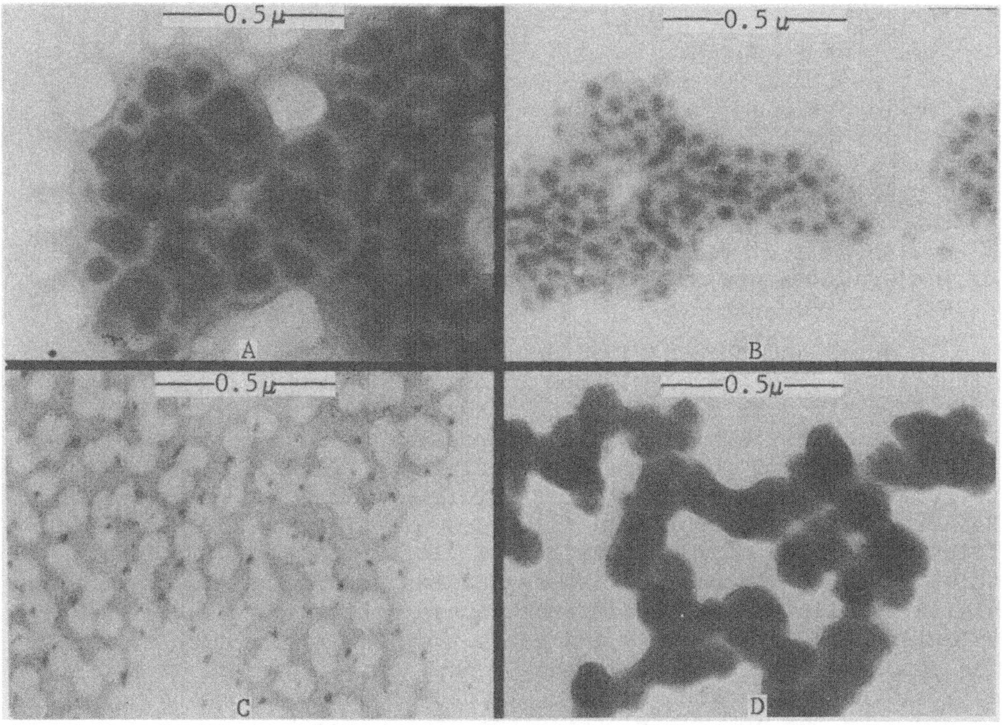

Fig. 4. EFFECT OF STAINING CONDITIONS: A. Drop dried from 0.5 ppm copolymer in methyl ethyl ketone solution then stained; B. Drop from the same solution stained; C. Drop dried from 0.5 ppm copolymer in cyclohexane solution then stained; D. Drop from solution for C stained.

To preserve the structure of the 20% copolymer in toluene solution, a drop was placed directly in the staining solution. The drop blackened immediately but was left in the staining solution 24 hrs. for additional staining. The sphere possessed a hard, black shell and a soft, darkened interior which became hard as the toluene evaporated. The interior displayed large striped regions (Fig. 5-A) which were larger than those observed in films dried from the same solution and then stained. These striped regions were not observed in the shell (Fig. 5-B); instead, short polybutadiene strings or ribbons were seen in a polystyrene matrix. The ordering may have been disrupted at the surface by viscous flow of the copolymer solution through the medicine dropper. When the solution was diluted to 5% copolymer, the drop spread out on the surface of the staining solution. Sections of this film did not show aligned structures (Fig. 5-C); the fine structure was composed of both irregular aggregates of small polybutadiene nodules in a polystyrene matrix and small polystyrene nodules in a polybutadiene matrix.

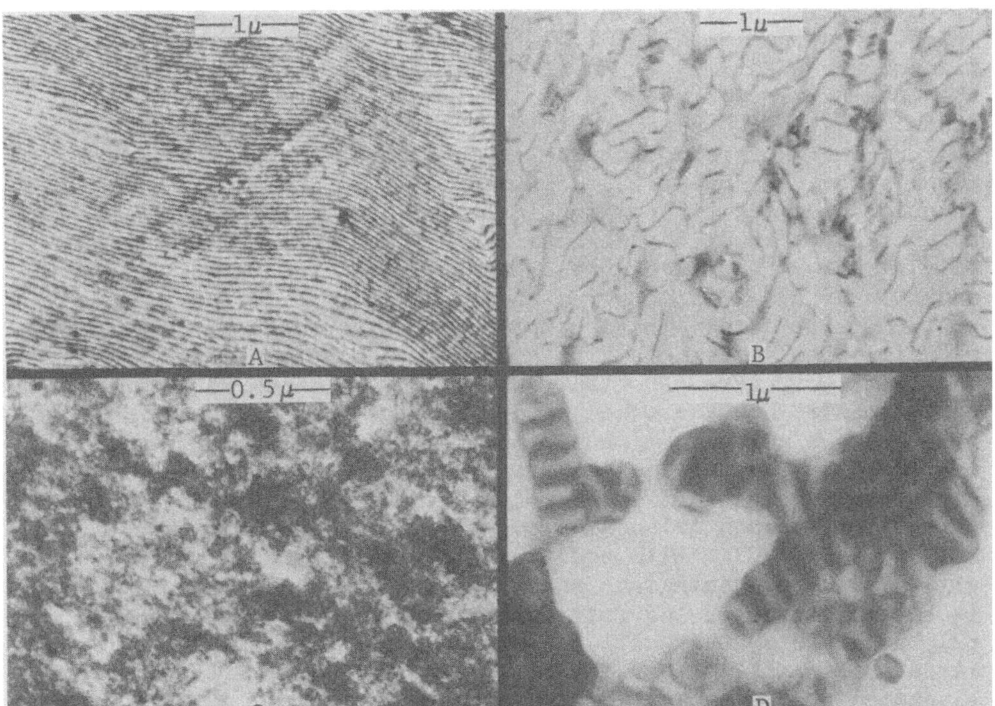

Fig. 5. EFFECT OF STAINING CONDITIONS: A. 20% copolymer in toluene dropped in staining solution; B. Shell of same drop; C. 5% copolymer in toluene dropped in staining solution; D. Dried, then stained drop of copolymer in methylene chloride containing 20% isopropanol.

The effect of adding 20% isopropanol, a higher boiling non-solvent for both block segments, to a dilute copolymer in methylene chloride solution is shown in Figure 5-D. The drop was dried before staining. Isolated particles show a polystyrene center surrounded by polybutadiene. Most of the particles are in chain-like aggregates which display alternate bands of polystyrene and polybutadiene. These bands are consistent with the layers observed in sections of cast films and might be expected if the polybutadiene exterior of the particles agglomerated.

MORPHOLOGY OF A 40/60 STYRENE-BUTADIENE DIBLOCK COPOLYMER

Since solutions of this copolymer displayed a blue-green color, it was expected that the copolymer possessed a narrow molecular weight distribution with very little, if any, homopolymer. However, GPC showed that a sizable polybutadiene homopolymer fraction existed. The styrene-butadiene ratio of the copolymer determined by GPC (40.45/59.55) agreed well with the charge ratio of 40/60. The peak of the copolymer curve indicated a molecular weight of approximately 300,000 based on the polystyrene calibration curve. Sections of an air dried film of this copolymer showed a pattern of polystyrene stripes (approximately 550 A in width) in a polybutadiene matrix (Fig. 6-A). Apparently the homopolymer was uniformly distributed throughout the polybutadiene block segments. However, the addition of 30% polybutadiene by solution blending (5% polymer in toluene) produced puddles of polybutadiene among the stripes in the cast film. The original film structure was even more extensively altered by a similar addition of 30% regular polystyrene. In this film the initial striped pattern was almost completely destroyed, leaving only an occasional polystyrene stripe among the occlusions (Fig. 6-B). The total polystyrene content was 58% but it appeared as the dispersed phase. On the other hand, when the same amount of low molecular weight polystyrene (approximately 52,000, from GPC peak) was added to the original solution in the same manner, an inversion resulted. This film showed a dispersion of spheroidal and irregularly-shaped islands of polybutadiene in a polystyrene matrix (Fig. 6-C). With additional low molecular weight polystyrene, the polybutadiene phase appeared as two configurations. At 76% polystyrene, the film consisted of a uniform dispersion of polybutadiene spheres with diameters approximately 850 A in a polystyrene matrix (Fig. 6-D); also present were large (2-3 microns) elongated polybutadiene particles containing small spheroidal occlusions of polystyrene. Thus, the morphology of the block copolymer can be altered appreciably by adding homopolymer of different molecular weight.

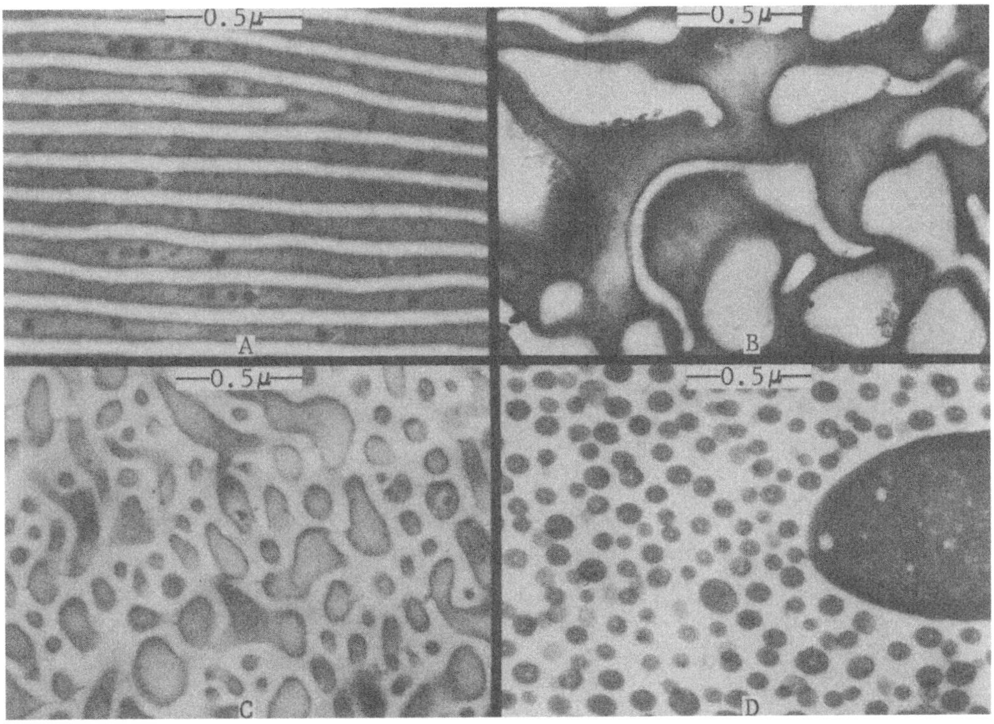

Fig. 6. FILMS OF A 40/60 DIBLOCK COPOLYMER CONTAINING HOMOPOLYMER: A. As prepared, containing polybutadiene contaminant; B. Containing 30% added polystyrene; C. Containing 30% added low molecular weight polystyrene; D. Containing 60% added low molecular weight polystyrene.

MORPHOLOGY OF A STYRENE-BUTADIENE-STYRENE TRIBLOCK COPOLYMER

The solutions of this copolymer containing 47% polybutadiene were colorless. GPC showed a polystyrene homopolymer fraction along with a narrow molecular weight distribution copolymer. Based on the polystyrene calibration curve, the peak of the copolymer curve occurred at a molecular weight of approximately 166,000 while the molecular weight of the homopolymer peak was approximately 37,000. A film from this solution showed a pattern of stripes (Fig. 7-A). Since islands of polystyrene were not seen, the homopolymer and polystyrene block segments must be quite compatible. Irregularly shaped islands of homopolymer were observed among the striped regions in the film when only 10% polybutadiene had been solution blended with the copolymer. The addition of 30% low molecular weight polystyrene did not uniformly expand the polystyrene portions of the striped regions and 60% destroyed much of the striped patterns (Fig. 7-B). Most of the

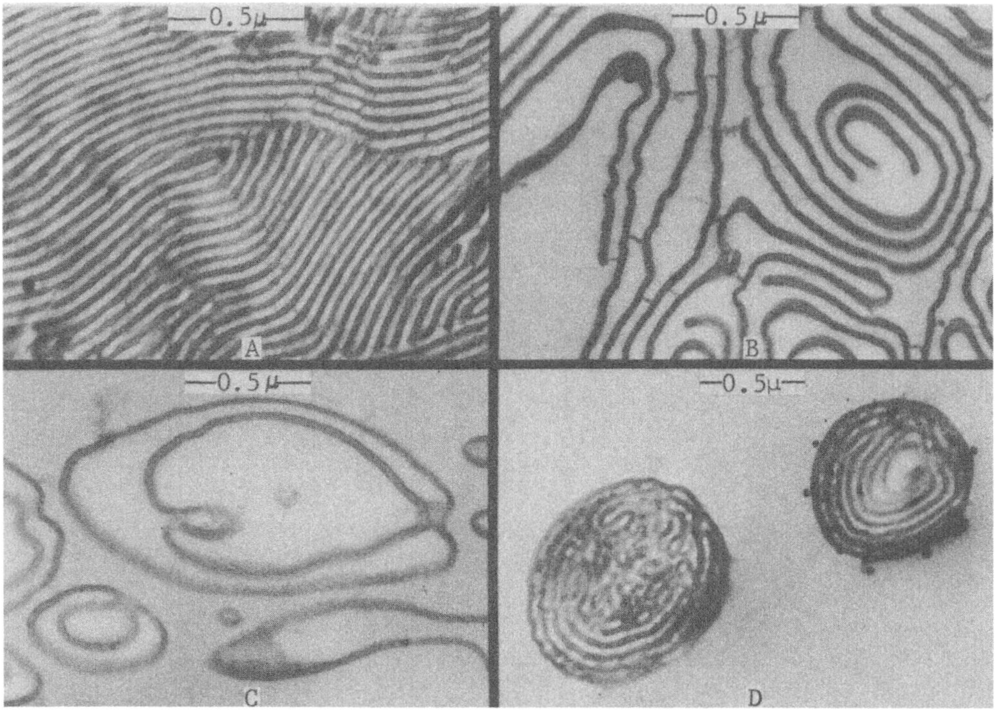

Fig. 7. FILMS OF A STYRENE-BUTADIENE-STYRENE TRIBLOCK COPOLYMER CONTAINING HOMOPOLYMER: A. As prepared, containing polystyrene contaminant; B. Containing 60% added low molecular weight polystyrene; C. Containing 99% added low molecular weight polystyrene; D. Containing 99% added polystyrene.

polybutadiene segments appeared as ribbons in various configurations. Spherical polybutadiene particles were not seen even though the film contained over 80% polystyrene. Solution blending 99% low molecular weight polystyrene with 1% of this copolymer did not produce polybutadiene spheres in cast films (Fig. 7-C), only ribbons in the form of loops in a sea of polystyrene were seen. Molau and Wittbrodt (7) obtained elliptical structures when 5-15% of a styrene-butadiene diblock copolymer was dissolved in styrene which was subsequently polymerized. Again polybutadiene spheres were not obtained when polybutadiene comprised a small portion of the sample. This indicates that copolymer content alone does not determine the formation of spherical polybutadiene domains. This was further exemplified in another experiment where regular polystyrene was substituted for the 99% low molecular weight polystyrene. The film possessed large spherical particles containing an ordered structure (Fig. 7-D), probably due to the increased incompatibility of the added higher molecular weight polystyrene.

Several different configurations were seen in films of the triblock copolymer containing 50% added polystyrene when a 1% polymer in toluene solution was subjected to ultrasonic irradiation for 1 hr. before drying. In some areas the size of the striped regions was enlarged while in other areas the patterns were confined to spherical identities suspended in a polystyrene matrix. When the film was dried in the ultrasonic field some of these identities appeared to be disintegrating into short rods and spheres of polybutadiene in a polystyrene matrix. These copolymer structures appear to disintegrate with great reluctance.

Often the morphology of block copolymer films can be predicted with confidence, but in some cases the structures observed in the electron microscope were not as anticipated. Several experiments showed the sensitivity of copolymer structure to experimental variables. Many factors contribute to the morphology observed in block copolymers.

ACKNOWLEDGEMENTS

The assistance of L. E. Morford and W. D. Keiser in preparing the micrographs is deeply appreciated, also the GPC data by J. R. Runyon and J. E. Jones.

REFERENCES

1. E. B. Bradford and E. Vanzo, J. Polym. Sci., A-1, 6, 1661 (1968).

2. K. Kato, Polym. Eng. Sci., 7, 38 (1967).

3. H. Hendus, K. L. Illers and E. Ropte, Kolloid Z. Z. Polym., 216/217, 110 (1967).

4. M. Matsuo, S. Sagae and H. Asai, Polymer, 10, 79 (1969).

5. T. Inoue, T. Soen, H. Kawai, M. Fukatsu and M. Kurata, J. Polym. Sci., B, 6, 75 (1968).

6. D. Heinze, Die Makromol. Chemie, 101, 166 (1967).

7. G. E. Molau and W. M. Wittbrodt, Macromolecules, 1, 260 (1968).

LIGHT SCATTERING FROM TWO-PHASE POLYMER SYSTEM HAVING SPHERICAL DOMAIN STRUCTURE

M. Moritani*, T. Inoue, M. Motegi**, and H. Kawai,
Department of Polymer Chemistry, Faculty of Engineering, Kyoto University, Kyoto, Japan, and
K. Kato, Development Research Laboratories, Toray Co. Ltd., Ohtsu, Shiga-ken, Japan

INTRODUCTION

The heterogeneity in the structure of polymer solids greatly influences the bulk properties, and the light scattering may be a useful technique for quantifying the heterogeneity of the colloidal dimension.

The theoretical studies of the light scattering from solid materials were first carried out by Debye and Bueche (1) for isotropic but inhomogeneous system, made more general by Goldstein and Michalik (2), and further extended by Stein and his co-workers (3-6) to the anisotropic system of crystalline polymers. In accordance with these theories, the intensity distribution of scattered light could give some statistical parameters which are related to the fluctuations of optical density and orientation within the system, i.e., the heterogeneity and local anisotropy. However, the physical meaning of the parameters in relation to the heterogeneity of the system has not been fully understood, except for, for example, some indirect investigations where the parameters are compared with the results obtained from adsorption experiments (7). This may be due to the lack of test specimens appropriate for making direct comparison of the parameters with the heterogeneity of the system.

* Present adress: Research Laboratory, Niihama Mill, Sumitomo Chemical Co. Ltd., Niihama, Ehime-ken, Japan.

** Present adress: Film Laboratory, Toray Co. Ltd., Ohtsu, Shiga-ken, Japan.

Recent development of synthesizing block and graft copolymers have made it possible to control the microphase separations of the block segments and of the grafted and backbone segments during casting the copolymers into solid specimens (8-15). In other words, it is possible to control the domain structures of the copolymers, micro-heterogeneous structures, not only in the shape but also in the size of the domains.

In this paper, the theory of light scattering from the density fluctuating materials will be first checked in terms of the correlation function $\gamma(r)$, which was defined by Debye and Bueche, by measuring the intensity distribution of scattered light from the film specimens of block copolymers of styrene and isoprene (A-B type) and graft copolymers of butadiene and styrene-acrylonitrile (ABS resin type) both having well-controlled domain structures in colloidal dimensions. Then, heterogeneity and statistical parameters, which are further defined by approximating the correlation function $\gamma(r)$ to be given in a form of summation of short and long range fluctuations, will be discussed in relation to the heterogeneity of the domain structures of the specimens which is observed from their electron micrographs.

THEORETICALS

For a definite two-phase systems of colloidal dispersion, where the difference of refractive indices between the two phases is very small, the treatment proposed by Guinier et al. (16) in terms of the Rayleigh-Gans scattering may be appliciable.

Applying Debye-Bueche theory of light scattering from isotropic but inhomogeneous system (1) to the two-phase system, an identical result with that calculated by Debye et al. for the porous materials (7) can be derived. That is, the intensity distribution of scattered light from the two-phase system may be given by

$$I(\underline{h}) = CV\rho^2\phi_A(1 - \phi_A) \int_V \gamma(\underline{r}) \exp [\, i(\underline{h}\cdot\underline{r}) \,] \, d\underline{r} \qquad (1)$$

where C is a constant, V is a volume of the system irradiated by the incident beam, ϕ_A is volume fraction of the A phase, ρ is a difference of polarizabilities between the A and B phases, $h = k\underline{s}$, $k = 2\pi/\lambda'$, and $\underline{s} = 2\sin(\theta/2)$ and where λ' is wavelength of the scattered light within the system and θ is the scattering angle.

$\gamma(r)$ is the so-called correlation function which is defined by

$$(1/V) \int_V \eta(\underline{x}) \cdot \eta(\underline{x} + \underline{r}) d\underline{x} = \overline{\rho^2} \cdot \gamma(\underline{r}) \qquad (2)$$

where $\eta(\underline{x})$ is the fluctuation from the mean value of the polarizability at a point defined by the vector $\underline{x}$ and $\overline{\rho^2}$ is mean-square fluctuation in polarizability, which is equal to $\rho^2\phi_A(1 - \phi_A)$ for the two-phase system.

As recognized from Eq. (1) the dependence of indensity of scattered light on θ is mostly attributed to the function $\gamma(\underline{r})$. However, it is difficult to draw a picture of this heterogeneous system from a knowledge of the correlation function, because the function is mearly one statistical description of the heterogeneous structure and dose not completely and unambiguously define the structure. In order to introduce some statistical parameters having simple but discrete meaning rather than $\gamma(\underline{r})$, let us consider a probability function $Z_{AA}(\underline{r})$ to fined an A-point separated by vector $\underline{r}$ from the other A-point in the system. If this function has non-directional dependence for the isotropic system; i.e., just a function of distance r, this will be unity when r=0, be ϕ_A as $r \to \infty$, unless there is a long range correlation, and may be given by

$$Z_{AA}(r) = \phi_A + (1 - \phi_A)\cdot \gamma(r) \tag{3}$$

It is easily shown that $\gamma(r)$ in Eq.(3) is an identical function defined by Eq. (2). (7)

From the general properties of the function $\gamma(r)$, the following parameters may be introduced. First, taking the derivatives of $Z_{AA}(r)$ with respect to r at r = 0;

$$[\, dZ_{AA}(r)/dr \,]_{r=0} = -S/4V \tag{4}$$

where S is the interfacial area between the two phases within V, then S/V in Eq.(4), which is denoted as S_{sp}, is a parameter which characterizes the "specific surface" of the domains and is related to $\gamma(r)$ by

$$[\, d\,\gamma(r)/dr \,]_{r=0} = \frac{-1}{4\,\phi_A(1 - \phi_A)}\; S_{sp} \tag{5}$$

Next, by integrating $Z_{AA}(r)$ over a range of r from 0 to L; i.e.,

$$\int_0^L Z_{AA}(r)dr = L\phi_A + (1 - \phi_A)\int_0^L \gamma(r)dr$$

$$= L\phi_A + (1 - \phi_A)\int_0^\infty \gamma(r)dr \tag{6}$$

when L is large enough, and defining $\int_0^\infty \gamma(r)dr \equiv \bar{l}_c/2$, then l_c is a parameter called as the "distance of heterogeneity" of the system.

Furthermore, by integrating $Z_{AA}(r)4\pi r^2$ over a range of r from O to R; i.e.,

$$\int_0^R Z_{AA}(r)4\pi r^2 dr = v\phi_A + (1-\phi_A)\int_0^R \gamma(r)4\pi r^2 dr$$
$$= v\phi_A + (1-\phi_A)\int_0^\infty \gamma(r)4\pi r^2 dr \qquad (7)$$

when R is large enough, and defining $\int_0^\infty \gamma(r)4\pi r^2 dr \equiv \bar{v}$, then $\bar{v}$ is a parameter called as the "volume of heterogeneity" of the system.

The heterogeneity parameters, S_{sp}, l_c, and v, represent the surface area of a sphere, its size and volume, respectively when the sphere is isolated in a dilute system.

$\gamma(r)$ may be determined from I(h) by means of the Fourier transformation of Eq.(1), provided that I(h) is obtained for a wide range of h. This seems, however, to be difficult from experimental point of view. From the experimental viewpoint, there have been several proporsals for the expression of $\gamma(r)$ in terms of either of an exponential (1,17) or a Gaussian function (18,19) or of summation of the two functions (7). On the other hand, we have showed theoretically that $\gamma(r)$ of the two-phase system of the spherical domain structure can be approximated by the summation of the two functions (20); i.e., exponential function for the density fluctuation for short distance, which arises from the spherical domain itself, mostly from its interface, and Gaussian function for the fluctuation for long distance, which arises from correlation between the spherical domains. Namely

$$\gamma(r) = f\exp(-r/a_1) + (1-f)\exp[-(r/a_2)^2] \qquad (8)$$

where f is the fractional contribution factor of the exponential function to $\gamma(r)$. The intensity distribution of scattered light I(h) can be given by substituting Eq.(8) into Eq.(1) as follows:

$$I(h) = cv\rho^2\phi_A(1-\phi_A)4\pi[(2fa_1)^3/\{1+(ksa_1)^2\}^2$$
$$+ (1-f)(\pi^{1/2}a_2^3/4)\exp\{-(ksa_2/2)^2\}] \qquad (9)$$

As recognized from Eq.(9), the plot of $I^{-1/2}$ against s^2 gives a linear relationship at large scattering angles, providing that the contribution of the second term in the righthand side in Eq.(9) is much smaller than the first one, from which the parameter a_1 may be determined as follows:

$$a_1 = (\lambda'/2\pi)\left(\frac{\text{slope of the linear relationship}}{\text{intercept of the linear relationship at } s^2=0}\right)^{1/2} \tag{10}$$

The plot of the logarithm of the difference between the observed intensity and the intensity calculated from the first term in the right hand side in Eq.(9) against s^2, gives also a linear relationship, from which the parameter a_2 may be determined as follows:

$$a_2 = (\lambda'/\pi)\left(\frac{\text{slope of the linear relationship}}{0.4343}\right)^{1/2} \tag{11}$$

The fractional contribution factor f may be given by

$$f = [\,1 + (8/\pi^{1/2})(a_1/a_2)^3\,A\,]^{-1} \tag{12}$$

where A is the ratio of the Gaussian term to the exponential term at $\Theta = 0^{\circ}$.

The relationship between the parameters, a_1, a_2, and f, and the parameters, S_{sp}, $\bar{l}_c$, and $\bar{v}$, may be given by

$$f/a_1 = S_{sp}/[\,4\,\phi_A(1 - \phi_A)\,] \tag{13}$$

$$\bar{l}_c/2 = fa_1 + (1 - f)a_2\,\pi^{1/2}/2 \tag{14}$$

$$\bar{v} = 4\,\pi\,[\,2fa_1^3 + (1 - f)\,\pi^{1/2}a_2^3/4\,] \tag{15}$$

TEST SPECIMENS AND EXPERIMENTAL PROCEDURES

Fig. 1 shows electron micrographs of ultrathin sections of the first series of test specimens having a domain structure of dispersed spheres of styrene component (white spheres unstained by OsO_4) within isoprene matrix (black matrix stained by OsO_4). For this series, the diameters of the dispersed spheres are kept constant but the distance between the spheres is changed systematically to become wider in the order from specimens designated 20/80 block copolymer through hi 30. These are prepared by casting from 5% toluene solutions of binary systems of different compositions of the 20/80 block copolymer (A-B type block copolymer of styrene and isoprene, M_n: 27.8 x 10^4, 18.5 wt% styrene) and a homo-polyisoprene (M_n: 39.8 x 10^4), as indicated in a triangular diagram in the figure from 0 to 30 wt% homo-polyisoprene (11).

Fig. 2 shows electron micrographs of ultrathin sections of the second series of test specimens having a domain structure of dispersed spheres of isoprene component within a styrene matrix. For this series, the volume fraction of the spherical domain is kept constant but the average diameter of the spherical domains is changed systematically to become larger in the order from Co 75 through Co 10. These are prepared by casting from 5% toluene solutions of ternary systems of different compositions of 70/30 block copolymer (A-B type block copolymer of styrene and isoprene, M_n: 104 x 10^4, 73 wt% styrene), homo-polyisoprene (M_n: 39.8 x 10^4), along an isopleth line in a triangular diagram in the figure; i.e., the composition changes in fraction of the block copolymer from 75 to 10 wt% but the ratio of total styrene sequences to total isoprene sequences is identical with that of the copolymer, 73/27 (11).

Fig. 3 shows electron micrographs of ultrathin sections of a series of ABS latices embedded in a mixture of polybutadiene and agar-agar and stained by OsO_4. The series of ABS latices was made by a particular grafting technique in emulsion system of polybutadiene latices dispersed in a medium containing styrene and acrylonitrile monomers so that the grafting occurs mostly onto the surface of the latices, not within the latices, and that the degree of grafting varies from 20 through 80 wt% for the series. As seen in the figure, polybutadiene spheres (black spheres stained by OsO_4) are surrounded by thicker layer of the grafted chains of styrene-acrylonitrile copolymer (white layer unstained by OsO_4) as the degree of grafting increase (21).

Fig. 4 shows electron micrographs of ultrathin sections of the third series of test specimens cast from 5% monochlorobenzene solutions of mixtures of a styrene-acrylonitrile copolymer, whose molecular composition is the same as that of

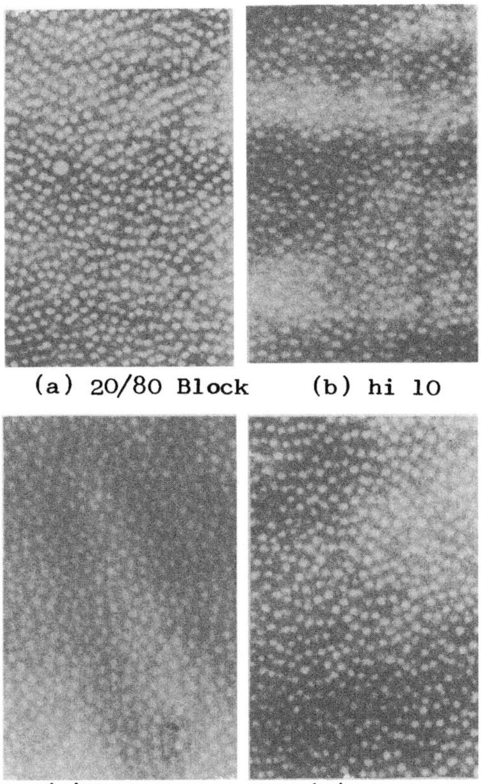

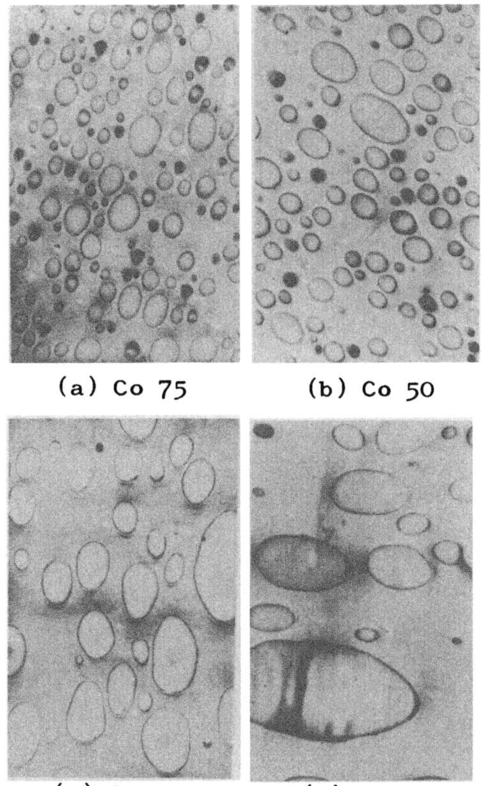

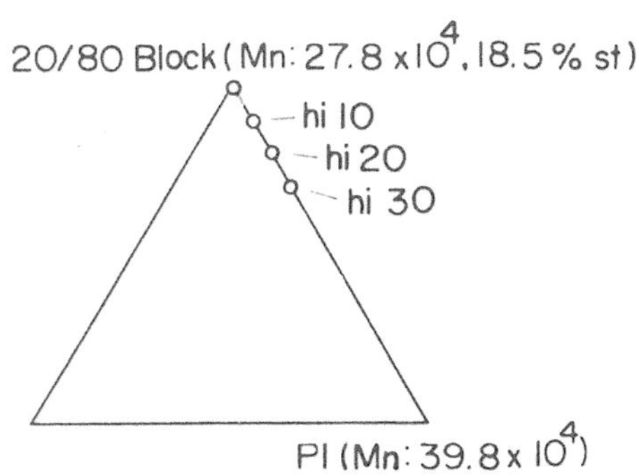

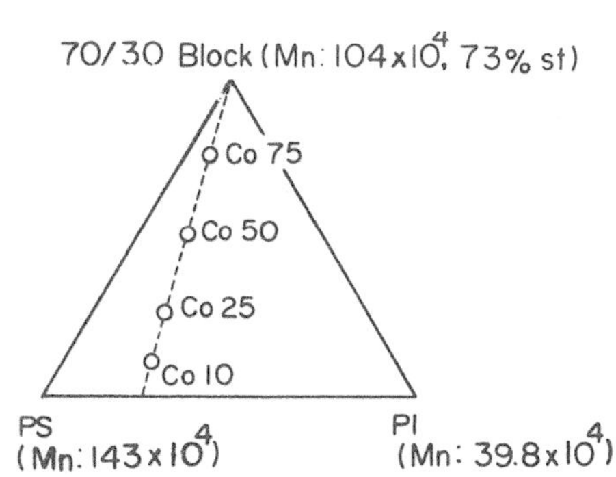

Fig. 1. The first series of test specimens.

Fig. 2. The second series of test specimens.

These are electron micrographs of ultrathin sections of about 350 Å thickness cut normal to the surfaces of test specimens pre-stained by OsO_4.

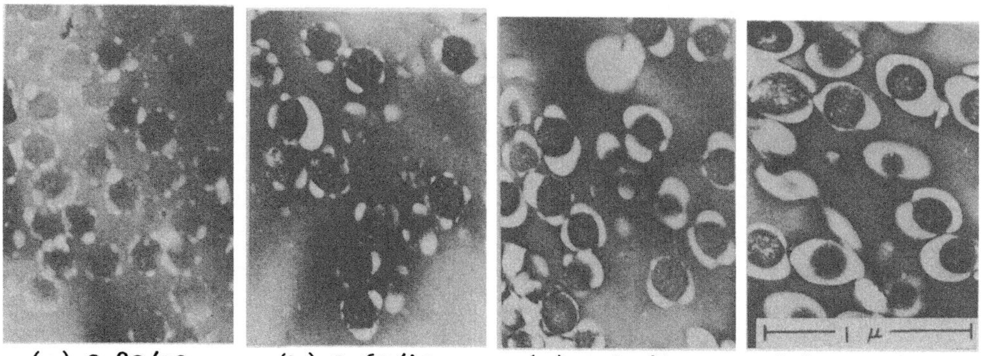

(a) G 80/20 (b) G 60/40 (c) G 40/60 (d) G 20/80

Fig. 3. Electron micrographs of ultrathin sections of a series of ABS latices embedded in a mixture of polybutadiene and agar-agar and stained by OsO_4. The series of ABS latices was made so that the degree of grafting of styrene-acrylonitrile copolymer onto the surfaces of polybutadiene latices varies from 20 through 80 wt%.

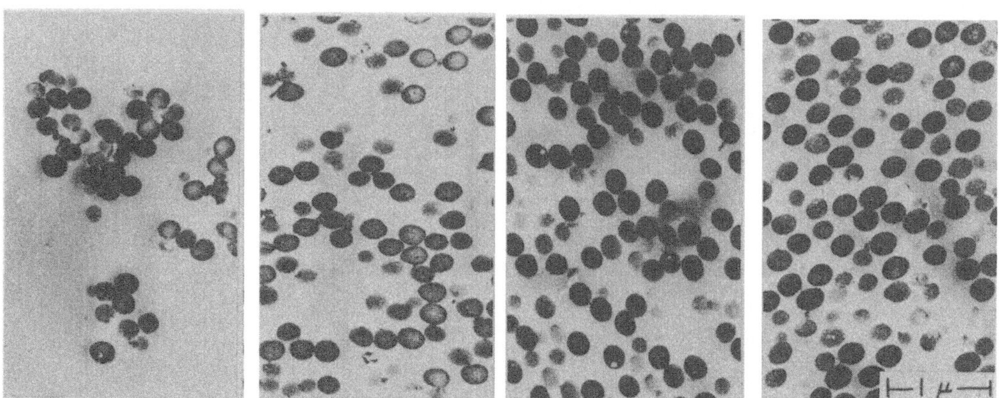

(a) G 80/20 (b) G 60/40 (c) G 40/60 (d) G 20/80

Fig. 4. Electron micrographs of ultrathin sections of about 350 Å thickness cut normal to the surfaces of the third series of test specimens pre-stained by OsO_4. The test specimens were cast from 5% monochlorobenzene solutions of mixtures of a styrene-acrylonitrile copolymer with each ABS latex in Fig. 3. For mixing the two components, the fraction of each latex was controled so that the weight fraction of polybutadiene is always kept constant, as a whole, as 20% for every specimen.

the grafted copolymer, with each ABS latex. For mixing the two components, the fraction of each ABS latex was controlled so that the weight fraction of polybutadiene is always kept constant, as a whole, as 20% for every specimen. As seen in the figure, the spherical domains of polybutadiene (black spheres stained by OsO_4) which are dispersed in a matrix of styrene-acrylonitrile copolymer (white matrix unstained by OsO_4) are very uniform in size for every specimen, but less uniformly dispersed as the degree of grafting of the ABS latex decreases. This systematic change of dispersity should be understood, as suggested in the previous paper (11), in terms of the coalescence barrier which prevents demixing of the polymeric oil-in-oil emulsion of ABS latices in the monochlorobenzene solution during casting the system. That is, the greater the degree of grafting on the ABS latices, the stronger the entropic repulsion between approaching droplets of the ABS latices, becomes resulting in the more uniform dispersion of the spherical domains of polybutadiene in the matrix of styrene-acrylonitrile copolymer. This repulsion arises, as suggested by Clayfield (22) and Meier (23), from decrease in configurational entropy of the grafted chain due to the loss of possible configurations as the space available to the grafted chains is reduced between the approaching droplets.

The test specimens which are cast into films of about 0.1 mm thickness, were placed normal to the incident beam, provided by a He-Ne gas laser apparatus, Model NAL-705Z, Nippon Kagaku Kogyo Inc.. The incident beam was well-collimated, polarized to the vertical direction by using a polarization rotator, Model 310, Spectra Physics Inc. and monochlomatized at wavelength of 6328 Å. The intensity of scattered light from the specimen was detected as function of scattering angle θ and azimuthal angle Ω by scanning with a photo-tube in a horizontal plane and rotating the specimen around its film normal, respectively. The analyser was placed just before the photo-tube so that its polarization direction can be rotatable within a plane perpendicular to the scattered ray at a given θ and that any combination of polarization condition, such as V_v and H_v polarization condition was mainly used because of very weak H_v scattering being probably due to smallness of the local anisotropy of the system.

RESULTS AND DISCUSSION

For every specimens, the intensity distribution of scattered light was measured under V_v polarization condition. There was hardly any variation of the intensity with azimuthal angle, confirming the isotropy of the systems to have non-directional dependence of the heterogeneity.

In order to discuss the parameter a_2, the long range correlation distance, the intensity distribution of scattered light from the first series of test specimens in Fig. 1 was measured under the V_v polarization condition. Fig. 5 shows the intensity variation with the scattering angle in terms of the plot of log(I) vs. s^2.* The values of a_2 determined from Eq.(11) by using the slope of this linear plot, are listed in Table I. As seen in the table, a_2 becomes larger as the concentration of the dispersed spheres decreases; i.e., as the distance between the spheres becomes larger.

In order to discuss the other parameters, the intensity distribution of scattered light from the second series of the

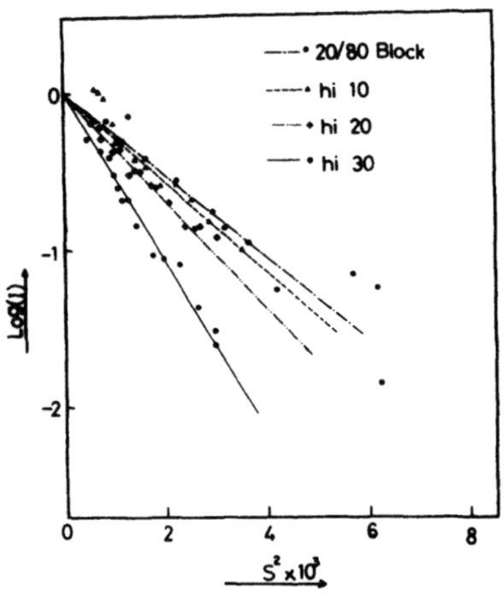

Fig. 5. Plot of log I vs. s^2 for the specimens in Fig. 1.

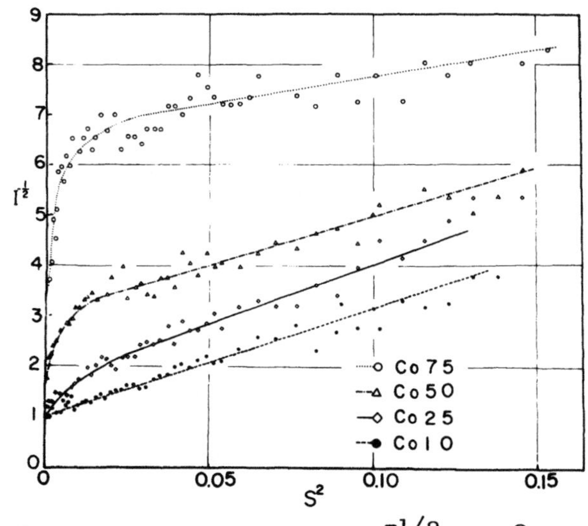

Fig. 6. Plot of $(I)^{-1/2}$ vs. s^2 for the specimens in Fig. 2.

---

* Here, log I, not log I_{diff} as used for deducing Eq.(10), is plotted against s^2. However, the contribution of the first term in the righthand side of Eq.(9) at small angles is usually very small so that the value of a_2 is hardly affected by this omission.

Table I. The heterogeneity and statistical parameters

Specimen	a_1(Å)	a_2(u)	f	$S_{sp}(\mu^{-1})$	$\bar{l}$ (Å)	$v(10^9 Å^3)$
1-st series						
20/80 Block		1.01				
hi 10		1.06				
hi 20		1.17				
hi 30		1.43				
2-nd series						
Co 75	885	1.2	0.999	9.48	1980	30
Co 50	1870	1.6	0.983	4.42	4000	370
Co 25	2490	1.5	0.985	3.32	5400	660
Co 10	3020	——	——	2.78	6040	690
3-rd series						
G 20/80	1350	1.38	0.994		2830	149
G 40/60	1500	1.31	0.983		3340	296
G 60/40	1900	1.16	0.944		5140	651
G 80/20	2000	1.13	0.918		5320	845

test specimens in Fig. 2 was measured also under the V_v polarization condition. Fig. 6 shows the intensity variation with the scattering angle in terms of the plot of $(I)^{-1/2}$ against s^2. As seen in the figure, the linear relationship at large scattering angles, as expected for deducing Eq.(10) from Eq.(9), is followed. The separation of the scattering angle dependence of intensity into two parts, at small and large scattering angles, becomes obvious as the size of the dispersed spheres becomes small. In other words, the rapid decrease at small scattering angles diminishes to make the dependence just a linear relationship as the size of the dispersed spheres becomes large.

The heterogeneity parameters determined from the results by using Eqs.(10) through (15) are listed also in Table I for the second series of test specimens. As seen in the table, a_2 is much larger than a_1 for every specimen, ranging around a micron. The value of f is very close to unity for every specimen, which suggests that the fluctuation of optical density is mostly ascribed to the fluctuation in the short range correlation. The heterogeneity parameters, $\bar{l}_c$ and $\bar{v}$, become large, and, in contrast, S_{sp} becomes small both as the size of the dispersed spheres becomes large. R_{scatt} is a mean diameter of the dispersed spheres determined from S_{sp}. The comparison of the R_{scatt} with $R_{E.M.}$, which is determined directly from the electron micrograph in Fig. 2, is shown in Fig. 7, giving a

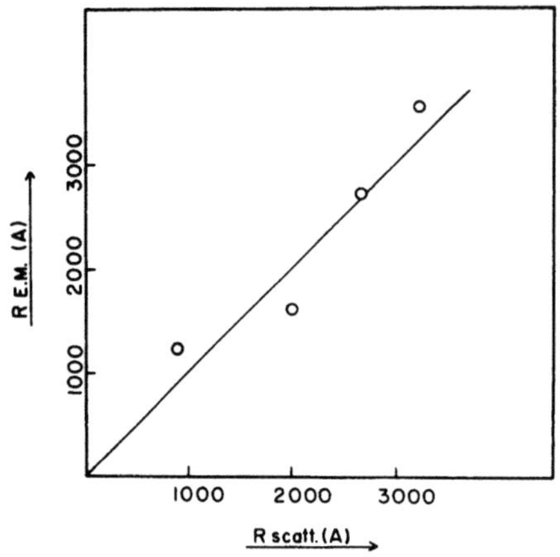

Fig. 7. A plot of R_{scatt}, determined from S_{sp}, against $R_{E.M.}$, determined from electron micrograph, for the second series of test specimens.

fairly good linear relationship passing through the origin with a slope of 45°.

Furthermore, in order to discuss the parameters in more detail, the intensity distribution of scattered light from the third series of test specimens in Fig. 4, for which the spherical domains are dispersed in different regularity but keeping the size of the domain uniform and the volume fraction of the domains constant, was also measured. The values of the parameters determined from the same procedures as the above, are listed also in Table I.

As recognized from Table I, the less uniform the dispersion of the spherical domains, the greater the value of a_1, the short range correlation distance, which increases in the order from G 20/80, G 40/60, G 60/40, and G 80/20. This must be understood, as expected from Eq.(13), in terms of the local coalescence of the dispersed spheres, which decreases the "specific surface" S_{sp} of the system when the dispersion becomes less uniform. In contrast, the more uniform the dispersity, the greater become the value of a_2 and f, the long range correlation distance and the fractional contribution factor of the short range fluctuation to the total fluctuation $\gamma(r)$. In addition, the "distance of heterogeneity" $\bar{l}_c$ and the "volume of heterogeneity" $\bar{v}$ increase as the dispersion becomes less uniform. Any quantitative discussion on the "specific surface" S_{sp} of the third series of the specimens, as for the second series of the specimens, may be meaningless, because of some possibility of polymerization of styrene-acrylonitrile copolymer within the spherical domains of polybutadiene, as seen in the electron micrographs in Fig. 4, which must affect the value of S_{sp} in an unexpected direction.

CONCLUSIONS

Characterizing the heterogeneity of the two-phase polymer systems, three kinds of heterogeneity parameters, "specific surface" S_{sp}, "distance of heterogeneity" l_c, and "volume of heterogenity" v, as well as some statistical parameters, such as "short and long range correlation distances" a_1 and a_2 and "fractional contribution factors of short and long range fluctuations" f and (1 - f), are introduced theoretically. These parameters were determined from the intensity distribution of scattered light from the specimens of the various two-phase structures, and were discussed in relation to the morphology of the specimens observed from their electron micrographs. Some systematic effects of the morphological factors on those parameters became apparent as follows: the larger the size of the spherical domain, a_1 and a_2, as well as l_c and $\bar{v}$ increase, and, in contrast, S_{sp} decreases; the larger the distance between the dispersed spherical domains, a_2 becomes large; the less uniform the dispersion of the domains, a_1 as well as $\bar{l}_c$ and $\bar{v}$ increase, while a_2 and (1 - f) decrease.

On the other hand, the turbidity is obtained by integrating a Rayleigh ratio reduced from I(h) over a whole solid angle. So, the transparency of the two-phase system may be discussed quantitatively in relation to the morphology by means of the above relationship between the morphological factors and the heterogeneous parameters.

ACKNOWLEDGEMENTS

The authors are deeply indebted to Mr. M. Nishimura, Development Research Laboratories, Toray Co. Ltd., for kindly arranging for preparation of the electron micrographs.
A part of this work was supported by the grant from the Japan Synthetic Rubber Co. Ltd., Tokyo, Japan.

REFERENCES

1) P. Debye and A.M. Bueche, J. Appl. Phys., 20, 518 (1949)
2) M. Goldstein and E.R. Michalik, J. Appl. Phys., 26, 1450 (1955)
3) R.S. Stein and M.B. Rhodes, J. Appl. Phys., 31, 1873 (1960)
4) R.S. Stein and P.R. Wilson, J. Appl. Phys., 33, 1914 (1962)
5) R.S. Stein, P.R. Wilson, and S.N. Stidham, J. Appl. Phys., 34, 46 (1963)

6) R.S. Stein, P. Erhart, J.J. van Aartsen, S. Clough, and M.B. Rhodes, J. Polymer Sci., C-13, 1 (1966)
7) P. Debye, H.R. Anderson, and H. Brumberger, J. Appl. Phys., 28, 679 (1957)
8) H. Hendus, K.H. Illers, and E. Ropte, Kolloid-Z. Z. für Polymere, 216-217, 110 (1967)
9) M. Matsuo, S. Sagae, and H. Asai, Polymer, 10, 79 (1969)
10) T. Inoue, T. Soen, T. Hashimoto, and H. Kawai, J. Polymer Sci., A-2, 1283 (1969)
11) T. Inoue, T. Soen, T. Hashimoto, and H. Kawai, Macromolecules, 3, 87 (1970)
12) K. Kato, J. Electron Microscopy, Japan, 14, 219 (1965)
13) G. Molau and H. Keskkula, J. Polymer Sci., A-1, 4, 1595 (1966)
14) R.J. Williams and R.W.A. Hudson, Polymer, 8, 643 (1967)
15) T. Ono, M.S. thesis presented to the Department of Polymer Chemistry, Faculty of Engineering, Kyoto University, March 11th, 1969
16) A. Guinier and G. Fournet, Small-angle Scattering of x-rays, John Wiley and Sons, Inc., New York, 1955, Chapter 2.
17) J.J. Keane and R.S. Stein, J. Polymer Sci., 20, 327 (1956)
18) A.E.M. Keijzers, J.J. van Aartsen, and W. Prins, J. Appl. Phys., 36, 2874 (1965)
19) E.V. Beebe, R.L. Coalson, and R.H. Marchessault, J. Polymer Sci., C-13, 103 (1966)
20) M. Moritani, T. Inoue, M. Motegi, and H. Kawai, Macromolecules, 4, July-August issure (1970)
21) K. Kato, Japan Plastics, 19, No.10, 89 (1968)
22) E.J. Clayfield and E.C. Lumb, J. Colloid Interface Sci., 22, 269 (1966)
23) D.J. Meier, J. Phys. Chem., 71, 1861 (1967)

DEFORMATION MECHANISM OF ELASTOMERIC BLOCK COPOLYMERS

T. Inoue, M. Moritani*, T. Hashimoto**, and H. Kawai

Department of Polymer Chemistry, Faculty of Engineering, Kyoto University, Kyoto, Japan

INTRODUCTION

In recent years, a class of linear amorphous block copolymers, such as styrene-butadiene block copolymers and polyester-MDI block copolymers, has attracted attension. These are named as "Thermoplastics" or "Segmented Elastomers", because of their characteristic mechanical properties similar with those of the conventional reinforced rubber vulcanizates at service temperature without both crosslinking and filler-modifying processes (1,2). The elastomeric properties of these materials have been investigated by means of tensile stress--strain relation and speculated in terms of their molecular architectures without taking account of the deformation mechanism of their microheterogeneous structures, which are resulted from micro-phase separation of the block segments in close relations to the molecular architectures and the preparation conditions of the bulk materials (3).

In order to understand the elastomeric properties in detail, some rheo-optical studies using birefringence (4-6) and infrared dichroism (7) have been carried out. However, it should be noted that such optical quantities as the above are based on the optical anisotropies of chemical bonds and are just average values of orientation of segments of chain molecules, which are

---

* Present address: Research Laboratory, Niihama Mill, Sumitomo Chemical Co. Ltd., Niihama, Ehime-ken, Japan.

** Present address: Polymer Research Institute, University of Massachusetts, Amherst, Mass. 01002.

too small to get definite information on the deformation mechanism of the microheterogeneous structures; i.e., the domain structures, of these materials. In contrast, the light scattering is very sensitive phenomenon to the fluctuations of optical-anisotropy and density of the system having correlation distances of colloidal dimension. The fluctuations having shorter correlation distances than the above, i.e., several hundreds angstroms, may be investigated by a small-angle x-ray scattering study. These scattering techniques must be powerful for investigating the deformation mechanism of the two-phase polymer systems in close relation to the change of the domain structure.

In this paper, some results of rheo-optical study using light any x-ray scattering techniques, will be demonstrated for uniaxial stretching of film specimens of styrene-isoprene block copolymers and of mixtures of the copolymers with the corresponding homopolymers both having spherical domains of one component dispersed within a matrix of the other component. Some fundamental deformation mechanisms of the domain structures will be suggested.

EXPERIMENTAL PROCEDURES AND RESULTS

Test Specimens

A-B type block copolymer of styrene and isoprene ($\bar{M}n$; 27.8×10^4, 18.5 wt% styrene) designated as "SI", A-B-A type block copolymer of styrene-isoprene-styrene ($\bar{M}n$; 59.5×10^4, 23.0 wt% styrene) designated as "SIS", homo-polystyrene ($\bar{M}n$; 3.34×10^4) designated as "PS", and homo-polyisoprene (Mn; 39.8×10^4) designated as "PI", were synthesized by living polymerization technique initiated by n-BuLi in THF at -78 °C. Film specimens were prepared by casting from 5% toluene solutions of mono or bi components of these polymers. The domain structure of the cast films was examined under electron microscope by means of the OsO_4 fixation technique.

Fig. 1 shows electron micrographs of ultrathin sections of about 350 Å thickness cut normal to the film surfaces. The domain structure of SI(Fig. 1-a) may be characterized as spheres of styrene component dispersed in a matrix of isoprene component. Radius of the sphere is about 200 Å. (8) Specimen hi-20 is a binary mixture of 80 wt% SI and 20 wt% PI. As seen in Fig. 1-b, PI is solubilized into the matrix of isoprene block of the copolymer SI, which increases the distance between the spherical domains in quantitative way when compared with Fig. 1-a. (9)

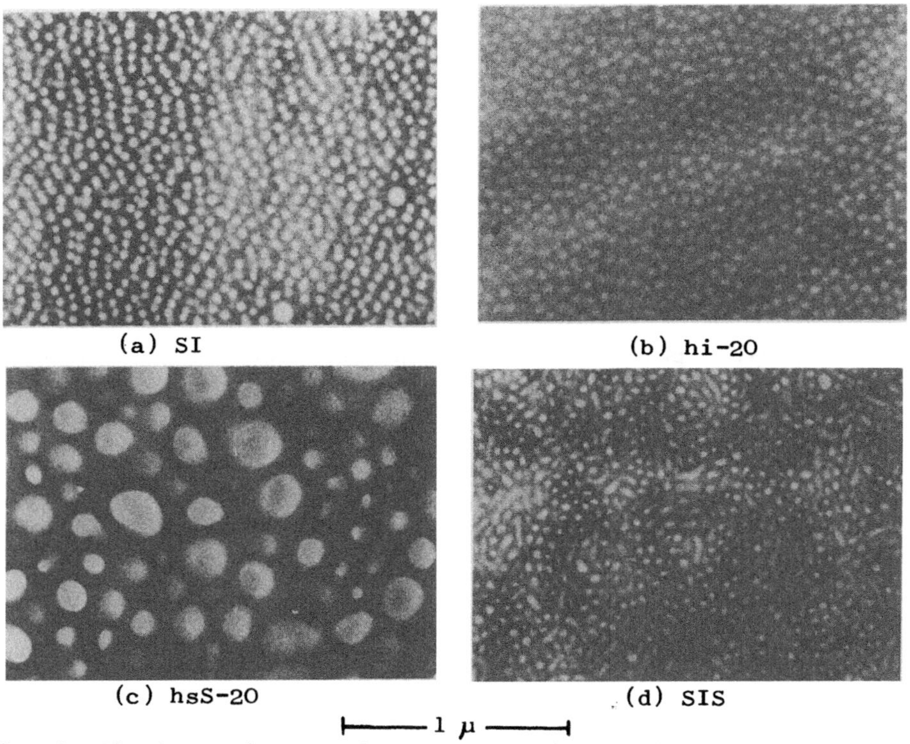

Fig. 1. Electron micrographs of ultrathin sections of film specimens

Specimen hsS-20 is a binary mixture of 80 wt% SI and 20 wt% PS. As seen in Fig. 1-c, PS is solubilized, in contrast, into the spherical domains of styrene block of the copolymer SI, which enlarges the size of the spherical domains when compared with Fig. 1-a. (9) The domain structure of SIS (Fig. 1-d) may be characterized as spheres of styrene component dispersed in a matrix of isoprene component, and some of these spheres are interconnected to form short rodlike domains.

In any case, every specimen has a domain structure of hard plastic spheres dispersed in soft rubber matrix, just like a filler-reinforced rubber.

Stress-strain Behavior

The tensile stress-strain relation of each film specimen of about 0.1 mm thickness, was measured under a constant rate of elongation, 16%/min., at 20°C. Every specimen behaves, as illustrated in Fig. 2, like rubber vulcanizates. The specimen,

SIS, is the stiffest and most elastomeric, while the specimen, hi-20, is the softest and less elastic showing rather low stress of leveling off and lacking the rapid re-increase of the stress at higher elongations. It is noted that the SIS, whose styrene content is 23 wt%, has higher initial Young's modulus and larger tensile strength than the hsS-20, whose styrene content is 36 wt% in total. This is inconsistent to the reinforcing effect of the styrene component. This apparent contradiction will be explained later.

V_v Light Scattering Patterns

The photographic light scattering technique (10, 11) was used for this small-angle light scattering experiment. The incident beam was provided by a He-Ne gas laser apparatus, Model NAL-705Z, Nippon Kagaku Kogyo Inc. The beam, which is well-collimated, polarized parallel to the stretching direction by using a polarization rotator, Model 310, Spectra Physics Inc., and monochromatized at a wavelength of 6328 Å, was radiated normal to the film surface.

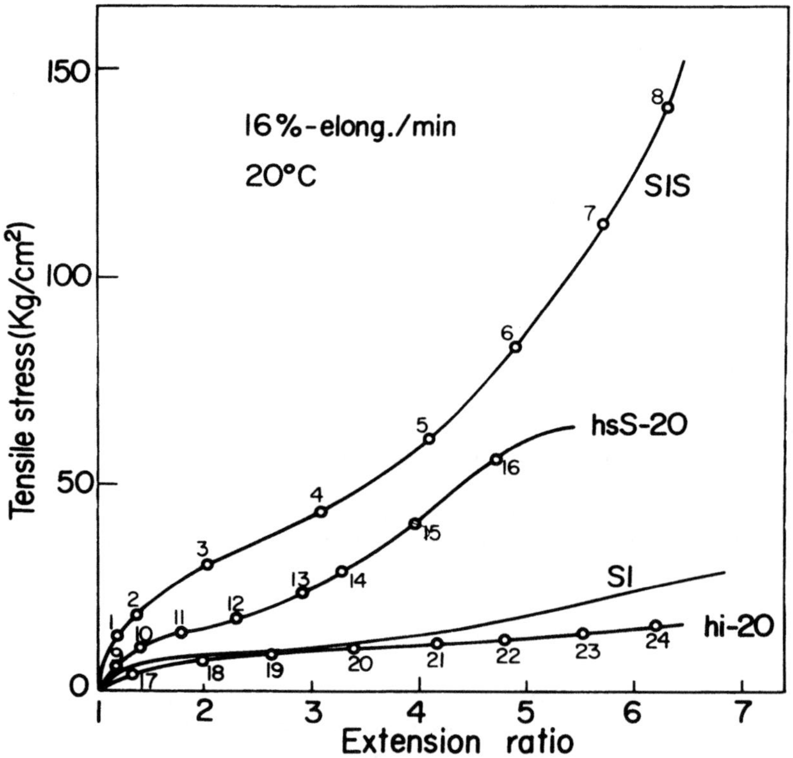

Fig. 2. Tensile stress-strain relations of the film specimens.

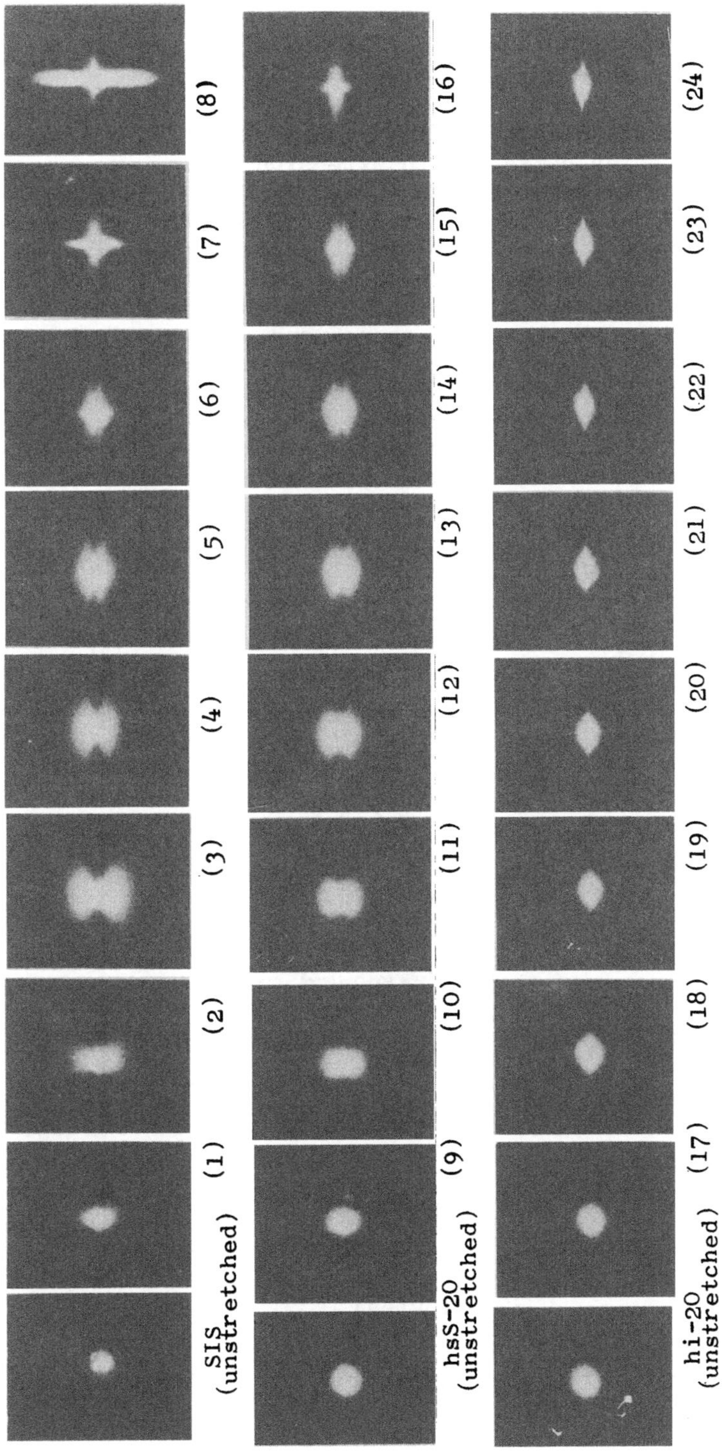

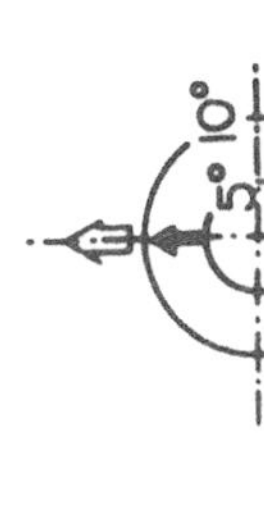

Fig. 3. V_v light scattering patterns from the film specimens at various stretched states indicated by the same numbers as those in Fig. 2. The light source is a He-Ne gas laser having a wavelength of 6328 Å.

The H_v (cross polarized) light scattering pattern could not be detected from any film specimen at any stretched state along the stress-strain relation in Fig. 2.

On the other hand, characteristic change of the V_v (parallel polarization) scattering pattern with extension ratio was observed for every specimen, as illustrated in Fig. 3 together with the scale of scattering angles and the indications of stretching direction of the film specimen (white arrows) and of polarization directions of optical system (black arrows) both with respect to the scattering patterns. The numbers from 1 through 24 attached to the patterns correspond to those in Fig. 2 along the stress-strain relations, giving respective states of stretching. In Fig. 3, where the scattering patterns from the specimen SI are omitted because of almost the same feature as the SIS and hsS-20, it may be noted that such drastic change of the pattern with extension as observed for SIS and hsS-20 is degenerated for the hi-20.

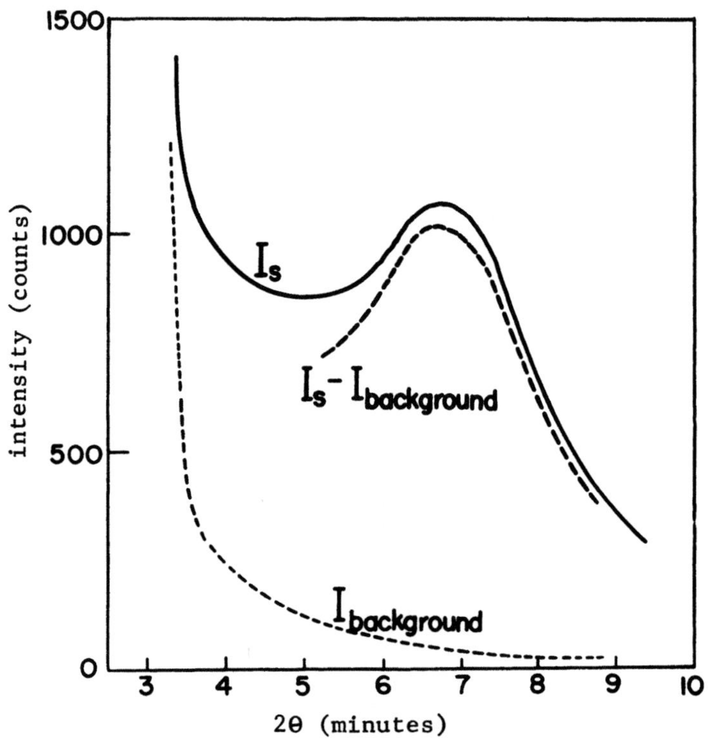

Fig. 4. Small-angle x-ray scattering from the unstretched SI film.

Small-Angle X-Ray Scattering

Small-angle x-ray scattering studies were carried out by using a rotating anode x-ray generator, Model 2C, Rigaku Denki Inc. and a goniometer equipped by a slit collimator. Fig. 4 shows an intensity distribution of the scattered x-ray from the unstretched SI film of ca. 0.5 mm thickness, curve I_s, and that of the blank testing, i.e., the scattered x-ray from the collimation apparatus and air, curve $I_{background}$. A peak found in the scattering curve I_s at the scattering angle $2\theta = 6.6'$ must arise from a regularity in the para-crystalline arrangement of the spherical domains. Assuming this scattering behavior as a sort of diffraction phenomenon, the position of this peak can be used to calculate a characteristic distance from the Bragg's relation; a Bragg distance. (12-14) The Bragg distance obtained from the reduced curve (I_s- $I_{background}$), broken line in Fig. 4, is found as 800 Å, and it may be an interdomain distance being consistent with the result of the electron microscopic observation in Fig. 1-a.

For the stretched film specimen, scanning the diffraction intensity within a plane including the stretching direction of the film and the incident direction of the x-ray, the interdomain distance along the stretching direction of the film may be obtained by the same procedures as the above. The interdomain distance along the stretching direction increases with stretching. The extension ratio of the interdomain distance $\alpha_{interdomain}$ is plotted against the extension ratio of the bulk specimen α_{bulk} in Fig. 5. As seen in Fig. 5, the change of the

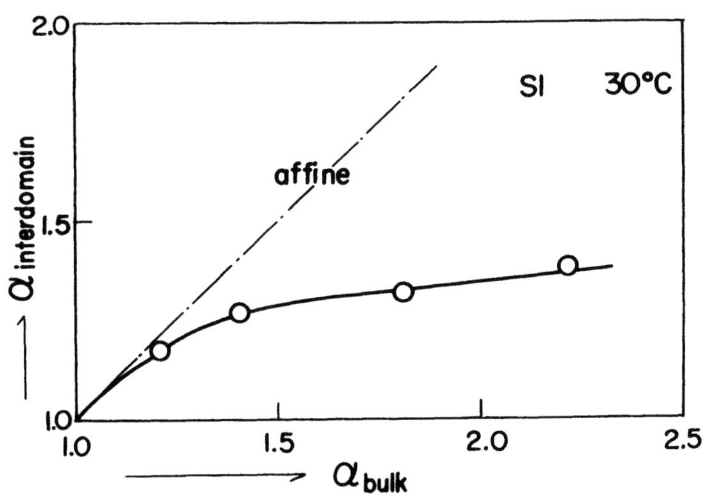

Fig. 5. Change of interdomain distance of the SI film with stretching.

inter-domain distance is followed by "affine" deformation only at very small elongations, and then deviates gradually from the affine deformation (chain line in Fig. 6).

DISCUSSION

The fact that any H_v light scattering was not detected from the specimens suggests that any correlated fluctuation of local anisotropy is not present (11) in the specimens at elongated as well as unelongated states. Therefore, the problem to be discussed must be based mostly on the density fluctuation of the system due to the extension of the specimen.

Despite of much difference of molecular architecture and domain structure among the film specimens, the characteristic change of V_V light scattering pattern with extension ratio may be summarized in Fig. 6 in terms of a common series of schematic diagrams with the exception of the hi-20, for which the change is less drastic than the others. The deformation process may be divided, as illustrated in Fig. 6, into the following three regions; the region I where the tensile stress increases rapidly with strain and C-type pattern grows up via. A- and B-type patterns, the region II where the rapid increase of stress is leveled off and the C-type pattern becomes flattened to give D-type pattern, and the region III where the stress increases rapidly again and a streak spreading in the stretching direction grows and becomes stronger to result in E-type pattern, while the D-type pattern degenerates.

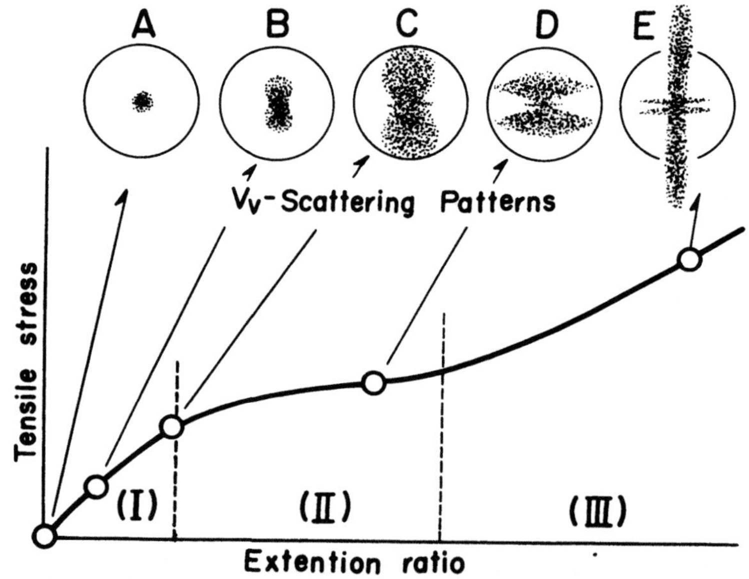

Fig. 6. Schematic diagrams showing the typical change of light scattering patterns from the film specimens with extension ratio.

The circularly symmetric pattern at the unelongated state of each specimen (A-type pattern in Fig. 6) is ascribed to the density fluctuation due to the two-phase structure itself. More strictly speaking, it must be arised from the long-range correlation of the spherical domains dispersed in the matrix (15). Anisotropy in the long-range correlation, which may be generated by the stretching, would give a change of the circular A-type pattern to an elliptical pattern, whose major axis is perpendicular to the stretching direction, in a range of almost the same scattering angles as those of the A-type pattern. (16) This may be realized by the case of hi-20, the third series in Fig. 3. In addition, the deformation of the plastic domains of styrene component, if any, might not affect the scattering pattern in such a range of small angles as seen in B- through E-type pattern, because the dimension of the spherical domains of the specimens ranges from about 200 through 500 Å. In other words, the characteristic change of the V_v patterns from B- through E-type cannot be interpreted in terms of any changes of the heterogeneity parameters of the two-phase structure (15), and one cannot help but speculates that a new fluctuation in optical density within the system be generated by the stretching of the specimen. The problem to be discussed is the entity of the new fluctuation and its geometrical relation.

The two-phase structure of all of the specimens is composed of the spherical domains of styrene component dispersed in the rubbery matrix of isoprene chains, which are sequrely anchored in the domains and diffuse from their interfaces to compose the matrix of themselves. The structural part to be deformed most easily by the external stress must be the rubbery matrix between the plastic domains. As suggested by the small-angle x-ray diffraction studies, the interdomain distance increases with the stretching. The deformation may be accompanied with decrease of the segment density of the matrix between the domains along the stress direction, because the domains are relatively closely packed so that the anchored isoprene chains are prevented to flow freely into the lower density region and compensate the lower density. The lower density region in the matrix might not be distributed unifomly at everywhere between the domains. An inhomogeneous deformation composed of the lower density region (V-region) and the region undisturbed so much by the external stress (H-region) may be assumed to represent a new model for the structural element resulting in the V_v scattering patterns in response to the external stress; i.e., a spherical H-region of radius R surrounded by a spheroidal V-region of major axis of αR, as illustrated in the upperhalf in Fig. 5, may be assumed irrespective of the size of the domain structures of the specimens.

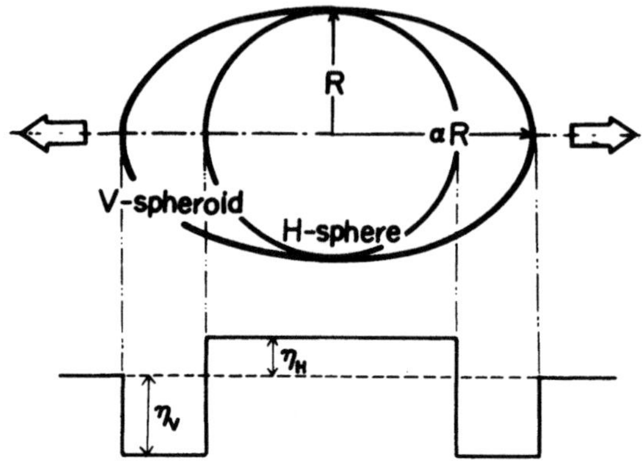

Fig. 7. A new model for the structural element responding to external mechanical excitation and its polarizability profile.

A polarizability profile of the model is also illustrated in the lowerhalf in Fig. 5, where η_H and η_V are the mean fluctuations in the polarizability of the H- and V-regions, respectively. The intensity of scattered light from the model may be given by

$$I = (3CV)^2 \left\{ (\eta_H + \eta_V)(\sin U - U\cos U)/U^3 - \alpha\,\eta_V(\sin U' - U'\cos U')/U'^3 \right\}^2 \qquad (1)$$

where $U = hR$, $U' = U[1 + (\alpha^2 - 1)\cos^2(\theta/2)\cos^2\mu]^{1/2}$, $h = (4\pi/\lambda')\sin(\theta/2)$, θ is the scattering angle, μ is the azimuthal angle, λ' is wavelength of the scattered light within the system, C is a proportional constant, and V is volume of the spherical H-region.

Eq.(1) gives various types of the scattering patterns, when one varies the values of R/λ', α, and (η_V/η_H). Among them, a series of V_V patterns with $R/\lambda' = 6.0$ gives the most close variation of the patterns to the experimental results in Fig. 3. Some of the results, varying the values of α from 1.0 through 4.0 and of (η_V/η_H) from 0.1 through 1.5, are selected and arranged in Fig. 8 to follow the change of the pattern in Fig. 6.

The formation process of the C-type pattern in Fig. 6 may be realized by (a) → (b) → (c) in Fig. 8, and the flattening

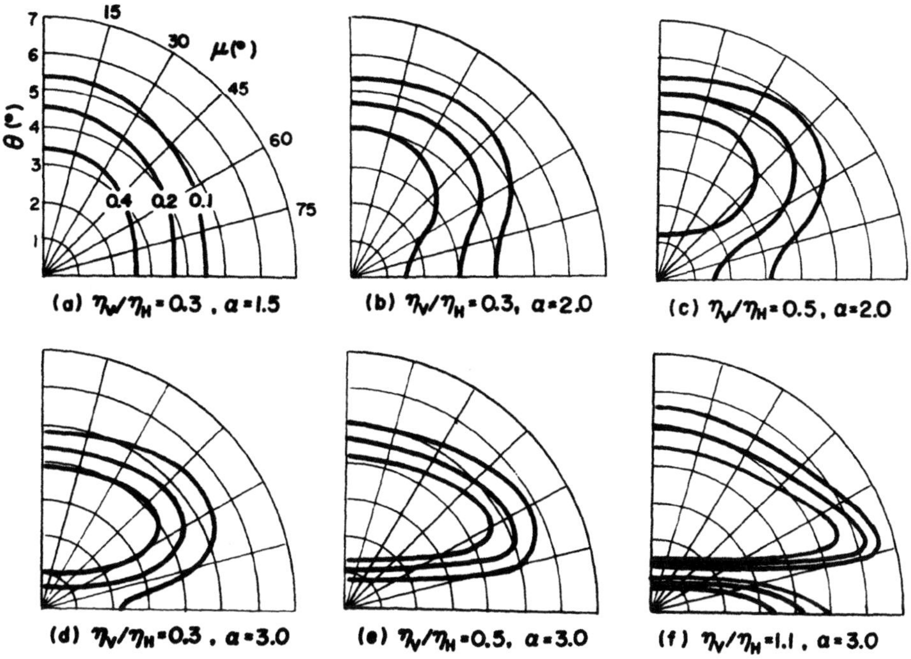

Fig. 8. Contour plot of V_V light scattering patterns obtained from Eq.(1) by assuming $R/\lambda' = 6.0$ and varying (η_V/η_H) from 0.3 to 1.1 and α from 1.5 to 3.0.

process of the C-type pattern into D-type pattern in Fig. 6 may be realized by the change from (d)→ (e)→ (f) in Fig. 8. Namely, according to this model, the deformation process in the region I in Fig. 6 may correspond to the formation of the V--region in the specimen, while the deformation process in the region II may correspond to the process where the density of the V-region decreases still more with maintaining the size of the region almost constant (increase of η_V/η_H with constant R and α).

On the other hand, the E-type pattern which has a strong streak spreading in the stretching direction, may correspond to a growth of a sort of micro-void at the ultimate stage of the V-region. The degenerate of the D-type pattern and, in contrast, the generate of the streak in the region III suggest a process of the transformation of some of the V-regions into the micro-voids. Judging from the spreading of the streak to the stretching direction, the void may propagate in the direction perpendicular to the stretching direction. (16)

The deviation of the change of interdomain distance with the stretching from the affine deformation, as shown in Fig. 5, may suggest that the deformation energy is mostly consumed by the formation of V-region and only a part of the energy is transformed into the separation of the interdomain distance. The relatively sharp diffraction curve of the stretched film may suggest the regularity in the domain arrangement at the H-region is not distributed so much.

The concept of the formation of the V-region in these two-phase materials has been supported by the investigations of the change of morphology with stretching of solvent cast films of Kraton 101 under electron microscope by using the OsO_4 fixation technique by Aggarwal et al. (16). They have suggested that the deformation of spherical domains of polystyrene does occur concurrently to different groups of domains in an area of about 1μ square. This dimension is in the same order of our result; i.e., $R = 6.0\lambda' = 2.5\ \mu$.

According to the concept of the heterogeneous deformation, it may be explained that any drastic change of patterns, such as shown in Fig. 6, is not observed for the specimen hi-20, a binary mixture of SI and PI, in which the PI is solubilized into the matrix of the isoprene block of the SI. As the solubilized PI chains enlarge the distance between the spherical domains of styrene component and are not unchored to the domains, the PI chains easily flow into the lower density region to average out the density uniform, preventing the formation of the V-regions and the drastic appearance of the characteristic change of the V_V patterns.

On the other hand, in the case of SIS, the parmanent entanglements of the middle blocks of A-B-A type block copolymer are formed, as has been demonstrated by the swelling experiments (18, 19). The entanglements make somewhat difficult to deform the matrix and decrease the segment density of the matrix between the domains along the stress direction. As the result, much energy is necessary for the deformation. This effect in the case of SIS would overcome the reinforcing effect in the case of hsS-20, resulting in the higher elastomeric behavior of the A-B-A type block copolymer than the others.

CONCLUSION

Deformation mechanisms of the elastomeric block copolymers, heterogeneous system of the spherical domains of plastic component dispersed in a matrix of rubber component, was investigated by means of simultaneous measurements of tensile stress-strain relation and light scattering and the small-angle x-ray

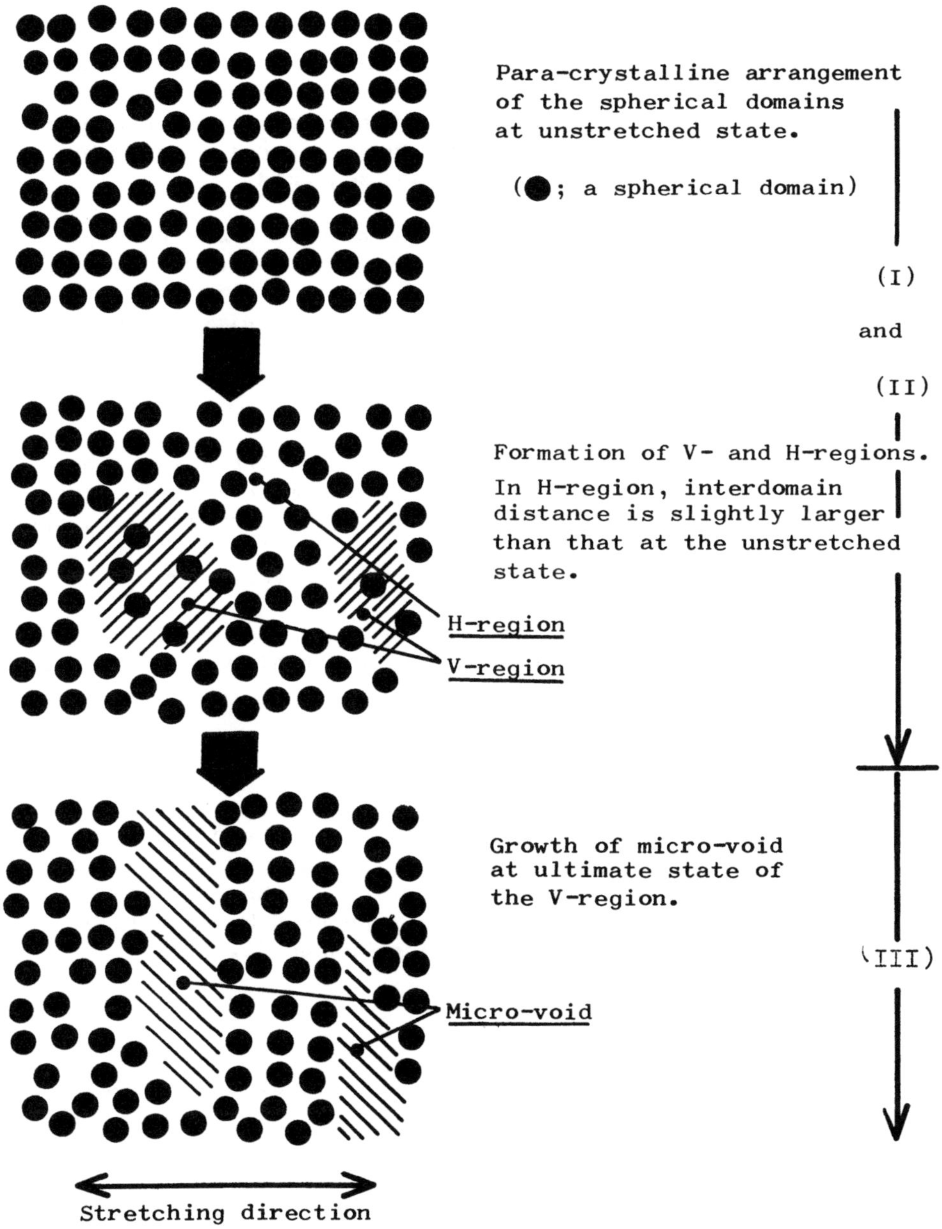

Fig. 9. Fundamental deformation mechanism of the domain structure.

diffraction studies. The results can be summerized as follows:

(1) The deformation of these materials is accompanied with the formation of the heterogeneity in density over considerably wide area to be called by "mechanical domain" ranging a few microns, which is much larger in size than the microheterogeneous structures of the materials, the domain structures, formed by the micro-phase seqpration of the block segments.

(2) The deformation process of the materials may be classified into three regions; region I with the formation of lower density region in the mechanical domain, region II with further decrease of the density in the lower density region in the mechanical domain while keeping the size of the region unchanged, and region III with transformation of the mechanical domain into the micro-void.

(3) The elastomeric properties of these materials may be ascribed to whether each rubber chain in the matrix is unchored to the plastic domain with one chain-end or both chain-ends, and whether each rubber chain have enough mobility to flow into the lower density region between the domains and average out the density uniform or not.

(4) Morphological changes with the stretching are summarized schematically in Fig. 9.

ACKNOWLEDGEMENTS

The authors are deeply indebted to Dr. K. Kato and Mr. M. Nishimura, Development Research Laboratories, Toray Co. Ltd., for kindly arranging for prepatation of the electron micrographs. A part of this work was supported by scientific research grant from the Japan Synthetic Rubber Co. Ltd..

REFERENCES

1) M. Morton, J.E. McCrath, and P.C. Juliano, J. Polymer Sci. C, 26, 99 (1969)
2) S.L. Cooper and A.V. Tobolsky, Polymer Preprints, ACS, 8, 52 (1967)
3) R. Zelinski and C.W. Childers, Rubber Chem. Technol., 41, 161 (1968)
4) E. Fischer and J.F. Henderson, J. Polymer Sci. C, 26, 149 (1969)
5) R.S. Stein and G.L. Wilkes, J. Polymer Sci. A-2, 7, 1525 (1969)

6) D.G. LeGrand, Polymer Letters, 7, 579 (1969)
7) R.S. Stein, private communication
8) T. Inoue, T. Soen, T. Hashimoto, and H. Kawai, J. Polymer Sci. A-2, 7, 1283 (1969)
9) T. Inoue, T. Soen, T. Hashimoto, and H. Kawai, Macromolecules, 3, 87 (1970)
10) A. Plaza and R.S. Stein, J. Polymer Sci., 40, 267 (1959)
11) R.S. Stein and N.B. Rhodes, J. Appl. Phys., 31, 1837 (1960)
12) H. Hendus, K.H. Illers, and E. Ropte, Kolloid-Z. Z. Polymere, 216-217, 110 (1967)
13) G.A. Harpell and C.E. Wilkes, Polymer Preprints, ACS, 10, No.2, 678 (1969)
14) D. McIntyre and E. Campos-Lopez, "Block Polymers" edited by S.L. Aggarwal, Plenum Press, N.Y., 19 (1970)
15) M. Moritani, T. Inoue, M. Motegi, and H. Kawai, Macromolecules, 3, July-August issue (1970)
16) A. Guinier et al., "Small-Angle Scattering of X-rays", John Wiley and Sons., N.Y., (1955), Chapter 2.
17) J.F. Beecher, L. Marker, R.D. Bradford, and S.L. Aggarwal, J. Polymer Sci. C, 26, 117 (1969)
18) E.T. Bishop and S. Davison, J. Polymer Sci. C, 26, 39 (1969)
19) T. Uchida, M.S. thesis presented to the Department of Polymer Chemistry, Faculty of Engineering, Kyoto Univ., March 11 th, 1970.

THE SYNTHESIS AND DILUTE SOLUTION PROPERTIES OF STYRENE-SILOXANE ABA BLOCK COPOLYMERS

W.G. Davies and D.P. Jones

Midland Silicones Limited,

Barry, Glamorgan, U.K. CF6 7YL

ABSTRACT

Narrow molecular weight distribution ($\bar{M}_w/\bar{M}_n \sim 1.3$) polystyrene (A)-polydimethylsiloxane (B) ABA block copolymers have been prepared using anionic polymerisation techniques. These copolymers are structurally similar to polystyrene-polybutadiene block copolymers, as they possess well-defined block segments. The narrowness of their molecular weight distribution has been confirmed by direct measurement of $\bar{M}_w$ and $\bar{M}_n$ and by gel permeation chromatography. Light scattering, using the Benoit and Krause approach to copolymer solutions, osmometry and viscometry have been used to study the molecular parameters of these block copolymers. The evidence from both light scattering and viscometry suggests that the preferred configuration of the copolymers in solution is that of randomly interpenetrating coils.

INTRODUCTION

The preparation of narrow molecular weight distribution polystyrene-polybutadiene ABA block copolymers and the demonstration of their useful and unique thermoplastic elastomeric properties[1] prompted us to investigate similar block copolymers in which the rubbery segment (B) was polydimethylsiloxane. As part of this study of these novel AB and ABA copolymers, an investigation of their dilute solution properties was undertaken. The solution properties of wholly organic block copolymers have been discussed principally by Benoit[2], Urwin[3], Inagaki[4] and Utracki[5]. These authors have demonstrated the influence of segment A-segment B interactions on the solution properties of the copolymers, but so far have been unable to present a unified model for the interpretation of the complex situation that exists within these solutions.

Polydimethylsiloxane-polystyrene (PDMS-PS) copolymers represent a new series of copolymers in which the glassy and rubbery segments are much more incompatible than their organic counterparts. The difference in compatibility may be exemplified by solubility parameter[6] (δ) data, $\delta_{PS} = 9.1$ $\delta_{PDMS} = 7.3$ (cf. $\delta_{polybutadiene} = 8.6$). This large difference in compatibility should be reflected in increased segment A-segment B interactions and thus give rise to expanded coil configurations of the copolymers in solution.

The discussion which we present here has been restricted to ABA block copolymers in which the segments are A≡ polystyrene and B≡ polydimethylsiloxane, and both segments possess narrow molecular weight distributions.

RESULTS AND DISCUSSION

ABA block copolymers can be unambiguously prepared by the controlled sequential anionic copolymerisation of styrene and hexamethylcyclotrisiloxane (D_3). The copolymerisation may be carried out in an aprotic solvent or in an aprotic solvent mixture, such as tetrahydrofuran-benzene, at room temperature. D_3 was used because it has significantly higher strain energy than any higher cyclic siloxanes (octamethylcyclotetrasiloxane, etc.), and can be rapidly polymerised in aprotic solvents by n-butyl lithium[7], without the formation of oligomeric products by detrimental chain cleavage reactions. In this manner it is possible to prepare polydimethylsiloxane segments of narrow molecular weight distribution by the initiation of D_3 by polystyryl lithium[8,9]. Unfortunately the reactivity of the siloxanolate anion is much less than that of the styryl anion and so $Si\bar{O}$ cannot be used to re-initiate styrene polymerisation. As a consequence of this, a coupling reaction[10], using a dichlorosiloxane, has to be finally employed to form the ABA copolymer. The reaction is illustrated in the scheme below.

The effectiveness of the coupling stage was determined by Gel Permeation Chromatography (GPC) and examples of AB, ABA and mixed homopolymer chromatograms are shown in figure 1.

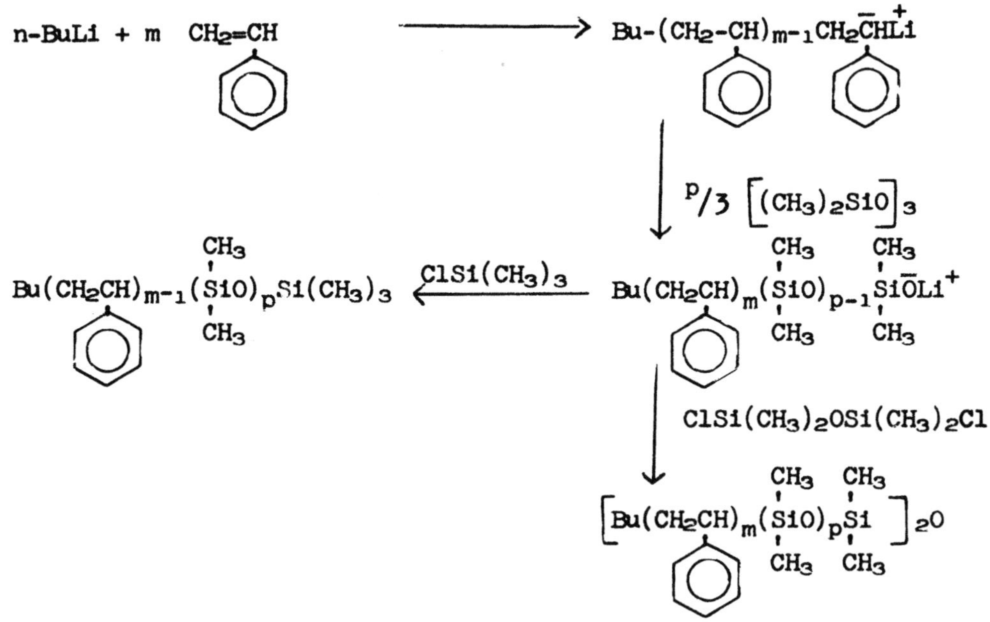

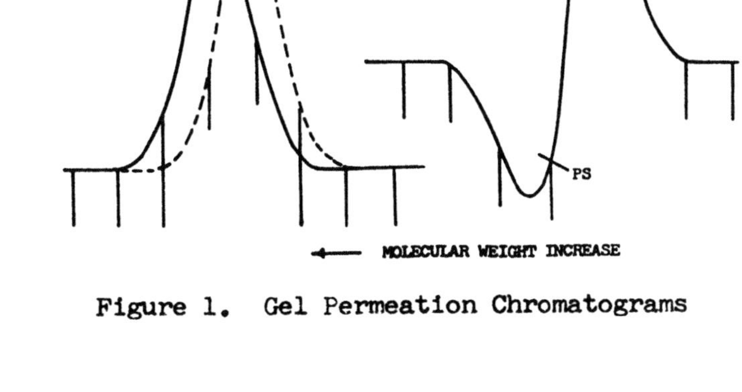

Figure 1. Gel Permeation Chromatograms

Molecular weight data for the ABA copolymers prepared by the described route are listed in table 1.

Table 1

Copolymer	mole % styrene	M_k x 10^{-4}	$\bar{M}_n$ x 10^{-4}	$\bar{M}_w$ x 10^{-4}	$\bar{M}_w/\bar{M}_n$	M_{GPC} x 10^{-4}	$\bar{M}_w/\bar{M}_n$ GPC
16A	14.8	1.6	2.05	--	--	2.0	1.06
25A	30.8	9.5	8.2	10.6	1.29	8.5	1.35
24A	11.9	10.0	10.5	14.6	1.39	11.0	1.35
27A	2.2	12.6	13.1	--	--	12.5	1.52

M_k is the calculated molecular weight assuming each initiator molecule initiates a single polymer chain. The discrepancies between M_k and $\bar{M}_n$ are attributed to the loss of catalyst by reaction with impurities present in the system.

Figure 2 shows the osmotic pressure/concentration results for these copolymers from which it is noted that the normal π/c relationship is obeyed. The second virial coefficient, A_2, for the copolymers was higher than the corresponding values for PDMS or PS, and were apparently independent of composition. GPC data, although calibrated for PDMS, showed good agreement with the directly measured $\bar{M}_n$ and $\bar{M}_w/\bar{M}_n$ values.

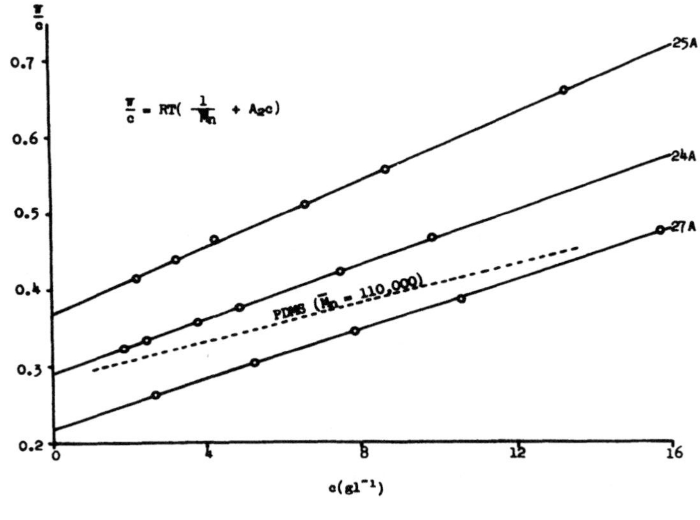

Figure 2. Osmometry Data in Toluene at 37°C

The light scattering approaches proposed by Krause[11], and Bushuk and Benoit[12] were used for the direct determination of $\bar{M}_w$. Four solvents, benzene, toluene, tetrahydrofuran and ethyl acetate were used to obtain values of the apparent molecular weight (M_{app}) in order to solve equations 1 and 2 for $\bar{M}_w$.

Bushuk and Benoit

$$M_{app} = i_A i_B \bar{M}_w + \frac{i_A(i_A - i_B)}{i^2_C} \cdot x_A M_A + \frac{i_B(i_B - i_A)}{i^2_C} (1-x_A) M_B. \quad (1)$$

i_A, i_B and i_C are the refractive index increments for homopolymer A and B and copolymer respectively; x_A is the weight fraction of component A present; M_A and M_B are representative of the weight average molecular weight of segments A and B.

Krause

$$M_{app} = \bar{M}_w + \frac{2b}{i_C} < M\Delta x > + \frac{b^2}{i^2_C} < M(\Delta x)^2 > \quad (2)$$

where $b = i_A - i_B$ and $<M\Delta x>$ and $<M(\Delta x)^2>$ are representative of the compositional and molecular weight distribution of the copolymer.

Both approaches gave good agreement to within ± 3%.

Comparison of $\bar{M}_w$ and $\bar{M}_n$ shows that the copolymers have narrow molecular weight distributions; this is also confirmed by calculation from GPC data. Analysis of the distribution within each segment using equation 1 revealed that the majority of polydispersity arises in the polystyrene segments, for example copolymer 24A has values of M_A and M_B of 37,500 and 121,800 respectively, whereas the values based on composition are 25,000 and 121,000. The broadening of the PS segments arose from the rapid polymerisation of styrene due to the use of high concentrations of tetrahydrofuran, which were required for the subsequent polymerisation of D_3.

Whilst studying the light scattering behaviour of the copolymers, large excluded volume effects were observed at concentrations greater than 0.5 wt.%. These effects resulted in marked curvature and distortion of the Zimm plots, see figure 3, and were most pronounced in solvents which were good solvents for both A and B segments, such as toluene. Structural ordering effects in solution, which could give rise to micelle formations, were initially considered to be the main contributing factors. However, a study of the change of refractive index with

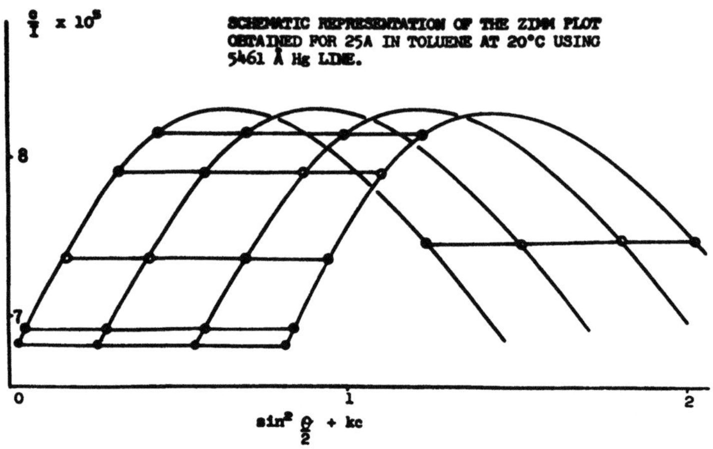

Figure 3.

concentration through this concentration region showed no deviation or inflection. This observation suggested the lack of micelle formation. Similarly, the Zimm plots obtained showed no dissymmetry effects before or after the curvature, again demonstrating the lack of association.

In order to further elucidate the conformation of the tri block copolymers in dilute solution we investigated their viscosity behaviour, using the Flory-Huggins equation.

$$\frac{\eta_{sp}}{c} = [\eta] + k'[\eta]^2 c \qquad (3)$$

Details of the viscosity behaviour are illustrated in figure 4 for the three solvents used. The solvents were, toluene at 25.0°C, which is a good solvent for both segments, and M.E.K. at 20.0°C and cyclohexane at 34.0°C, which are theta solvents for PDMS and PS respectively.

In good solvents the intrinsic viscosity, $[\eta]$, of a homopolymer can be related to the short (K) and long (B') range interactions by the Stockmayer-Fixman[13] equation (4).

$$[\eta] = KM^{1/2} + B'M \qquad (4)$$

The equation has very limited use for block copolymers as

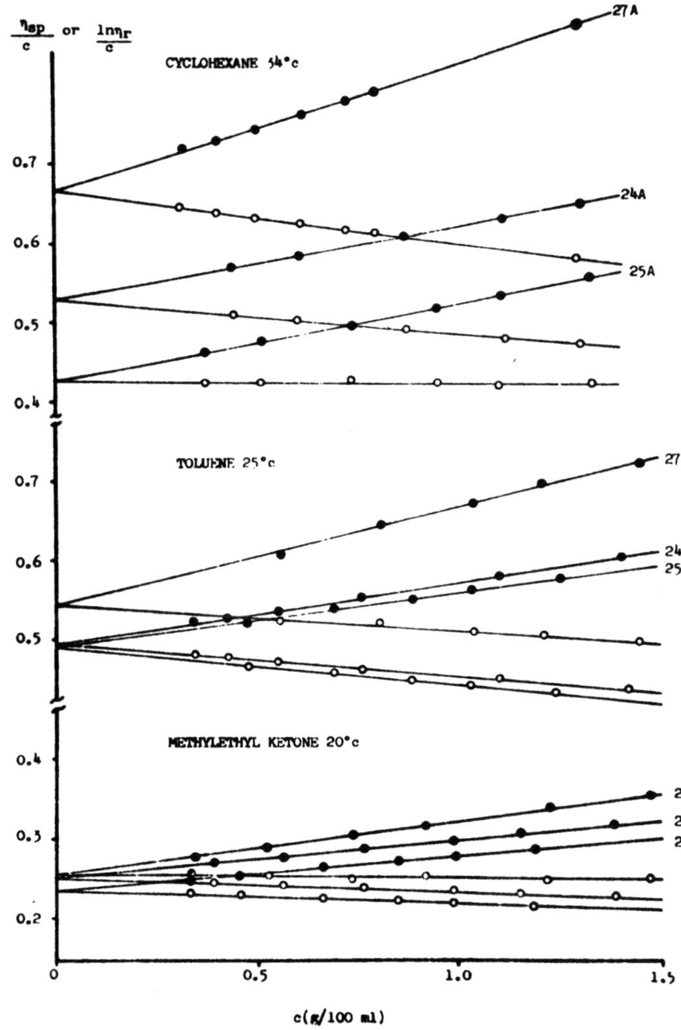

Figure 4. Solution Viscosity Plots

interactions between chain segments of blocks A and B must be considered. In a solution of an ABA block copolymer there are five types of interaction possible. As [η] is measured at infinite dilution it will only be influenced by intramolecular interactions and as such will only depend on interactions involving near neighbours within the co-ordination sphere of the considered chain segment. In the ABA case there will only be two pure A-B links per molecule and unless an A chain segment enters the co-ordination sphere of a B chain segment, the A-B interaction will contribute little to the total interaction energy. If this assumption is correct, it should be possible to calculate $[\eta]$ from existing Mark-Houwink data of the homopolymers. By considering the copolymer as a binary mixture of homopolymers[14] the empirical expression:-

$$[\eta]^* = n_A[\eta]_A + n_B[\eta]_B \tag{5}$$

may be written.

This equation can be further expanded with the Mark-Houwink relationship to give:

$$[\eta]^* = n_A K_A M^{a_A} + n_B K_B M^{a_B} \tag{6}$$

where $[\eta]^*$ is the calculated intrinsic viscosity, n_A and n_B are the mole fractions of A and B present, and M is the overall number average molecular weight for the copolymer. Using equation (6) a qualitative estimation of the A-B interaction and hence chain configuration of the copolymer was obtained. The results listed in table 2 give a comparison of $[\eta]^*$ with that obtained experimentally ($[\eta]$), and those calculated for PDMS of an equivalent molecular weight, [PDMS].

Table 2

		n_A	$[\eta]$	$[\eta]^*$	Δ	[PDMS]
Toluene 25.0°C	25A	0.308	0.491	0.455	7.9	0.350
	24A	0.119	0.493	0.458	7.6	0.411
	27A	0.022	0.545	0.487	11.9	0.478
MEK 20.0°C	25A	0.308	0.351	0.227	54.6	0.232
	24A	0.119	0.334	0.261	27.9	0.262
	27A	0.022	0.354	0.293	20.8	0.293
CYCLOHEXANE 34.0°C	25A	0.308	0.427	0.362	17.9	0.418
	24A	0.119	0.529	0.467	13.3	0.494
	27A	0.022	0.666	0.630	5.7	0.579

where $$\Delta = \frac{[\eta]-[\eta]^*}{[\eta]^*} \times 100 \tag{7}$$

Table 3 lists the relevant Mark-Houwink data[6].

Table 3

Polymer	Toluene at 25.0°C		M.E.K. at 20.0°C		Cyclohexane at 34.0°C	
	K	a	K	a	K	a
PDMS	2.0×10^{-4}	0.66	8.1×10^{-4}	0.50	1.59×10^{-4}	0.70
PS	4.4×10^{-4}	0.65	7.03×10^{-4}	0.71	8.2×10^{-4}	0.50

For all three solvents used, higher $[\eta]$ values were observed than calculated using equation 6, also the observed values were considerably higher than calculated for pure PDMS. These results suggest expansion of the polymer coils due to A-B interactions. In toluene the differences, as expressed by Δ, are small and represent only 6-12% of the calculated value; they are also independent of molecular weight and composition. In M.E.K., however, the theta solvent for PDMS, values of Δ are up to 54% greater than those calculated and are proportional to the mole fraction of polystyrene. Cyclohexane, a theta solvent for the PS segment, also caused expansion of the coil configuration, but to a smaller extent than M.E.K. All these data show expansion of the predicted coil configuration caused by large contributions of A-B interactions. These findings cannot be reconciled with the proposals of Benoit[2], who, from a study of polystyrene-polymethylmethacrylate block copolymers, concludes that in dilute solution, in any solvent, the blocks are segregated and thence have distinct locations. The interpretation which most readily lends itself to our results is that the copolymers form randomly interpenetrating coils. In good solvents the coils are expanded by polymer-solvent interactions and A-B interactions are a minimum, whereas in a theta solvent for one of the blocks repulsive A-B interactions contribute appreciably towards coil expansion. These large repulsive interactions are to be expected if the incompatibility of segment A with segment B is considered.

EXPERIMENTAL

Materials

Hexamethylcyclotrisiloxane was obtained from Midland Silicones pilot plant and was distilled at 134°C under an inert atmosphere before use. The styrene was supplied by British Drug Houses and was purified by reduced pressure distillation. n-Butyl lithium was supplied by Alpha Inorganics Ltd., as a 2.3M solution in n-hexane, the concentration was estimated by the Gilman double titration method[15]. Tetrahydrofuran was purified by distillation from sodium into a flask containing sodium and naphthalene; it was then re-distilled from this complex immediately before use. Other solvents were re-distilled and dried over Linde molecular Sieve 4A, where necessary.

Copolymerisations

Solutions of monomers in THF, etc. were thoroughly dried by standing in contact with Linde molecular Sieve type 4A and calcium hydride for at least 48 hours before polymerising. All polymerisations were conducted in an inert atmosphere in preflamed apparatus. Solution transfers were made via syringes through silicone rubber serum caps. ABA block copolymers were prepared by the sequential addition of D_3, in THF, to polystyryl lithium, followed by the addition of the stoichiometric amount of coupling agent (based on the initiator concentration). Block copolymer composition was determined by n.m.r. analysis.

Techniques

A Waters Gel Permeation Chromatograph, operated at 20°C, with toluene as the eluent at a flow rate of 1 ml. min^{-1}, was used to study molecular weight and molecular weight distribution changes. Four columns packed with polystyrene gel of porosities 2×10^3, 3×10^4, 2×10^5 and 8×10^5 Å were used in series. Sample solutions contained 0.25 wt.% polymer. The instrument was calibrated with narrow molecular weight distribution PDMS and PS fractions.

Number average molecular weights ($\overline{M}_n$) were measured using a Mechrolab High Speed Membrane Osmometer (Model 501) with toluene at 37°C.

Light Scattering studies were conducted using a Sofica instrument, at 20°C, employing the Hg green line (5461 Å). $\frac{dn}{dc}$ Measurements of the copolymers in the various solvents were done using a P.C.L. differential refractometer.

All viscosity data were obtained by means of calibrated Ubbelohde suspended level viscometers; kinetic energy corrections being made. Intrinsic viscosities were determined graphically using the Huggins equation, values being expressed in terms of dl g^{-1}.

REFERENCES

1. J. Polymer Sci., Part C, Polymer Symposia, 1969, 26.

2. A. Dondos, P. Rempp and H. Benoit, Makromol. Chem., 1969, 130, 233.

3. J.R. Urwin, Aust. J. Chem., 1969, 22, 1649.

4. H. Inagaki, Makromol. Chem., 1965, 86, 289.

5. L.A. Utracki, R. Simha, L.J. Fetters, J. Polymer Sci., 1968, A-2, 6, 2051.

6. 'Polymer Handbook', Ed. J. Brandrup and E.H. Immergut, Interscience, 1966.

7. W.G. Davies, B. Elliott and T.C. Kendrick, U.S. patent 3,481,898.

8. J.C. Saam, D.J. Gordon and S. Lindsey, Macromolecules, 1970, 3, 1.

9. E.E. Bostick, U.S. patent 3,483,270.

10. G. Greber and A. Balciunas, Makromol. Chem., 1963, 69, 193.
" " 1964, 79, 149.

11. S. Krause, J. Phys. Chem., 1961, 65, 1618.

12. W. Bushuk and H. Benoit, Can. J. Chem., 1958, 36, 1616.

13. W.H. Stockmayer and M. Fixman, J. Polymer Sci., 1963, part C, 1, 137.

14. B. Bohmer, D. Berek and S. Florian, Eur. Polymer J., 1970, 6, 471.

15. H. Gilman, J. Organometal. Chem., 1964, 2, 447.

PROPERTIES OF POLYSTYRENE-POLYDIMETHYLSILOXANE BLOCK COPOLYMERS

John C. Saam and F. W. Gordon Fearon

Polymer Research Laboratories

Dow Corning Corporation, Midland, Michigan 48640

ABSTRACT

Block copolymers of polystyrene and polydimethylsiloxane formed when hexamethylcyclotrisiloxane polymerized from the "living" ends of polystyryllithium. Little of the complicating intra- or intermolecular chain scission usually encountered in anionic siloxane polymerization occurred and essentially pure block copolymers were obtained. These were of the AB or BAB variety (where B signifies polydimethylsiloxane and A polystyrene) depending on the functionality of the alkyllithium initiator used in formation of the polystyryllithium. The block copolymers, owing to the disparity in solubility between the A and B blocks, showed a tendency to form microphases in the bulk. The BAB block copolymers upon conversion to $(BAB)_x$ multi-block systems gave strong thermoplastic elastomers which could be deposited as films from solution or directly molded without any added re-enforcing fillers or cross-linking. Microphase formation was also observed in the multi-block copolymers.

INTRODUCTION

The recent synthesis of well defined AB di-block copolymers of polystyrene (A) and polydimethylsiloxane (B) provides an opportunity to examine the morphological properties of this unusual copolymer system.[1] The present investigation extends the synthesis and broadens the previous observation[1] that these copolymers form microphases resembling those seen in the polystyrene-polydiene block copolymers.[2,3] Included herein are $(BAB)_x$ multi-block copolymers which should show properties characteristic of thermoplastic elastomers if microphases do indeed form in the bulk.[4] Thermoplastic elastomers containing crystalline silphenylene-siloxane blocks[5] or glassy polybisphenol-A-carbonate blocks[6,7] alternating with blocks of polydimethylsiloxane have been reported but, because of the short lengths of the blocks, these systems may often resemble random rather than true block copolymers. For example, in the polysiloxane-polybisphenol-A-carbonate system it is necessary to postulate extensive distribution of the polycarbonate blocks in the rubbery polysiloxane phase in order to explain certain mechanical and optical properties.[8] In the present study well defined $(BAB)_x$ polystyrene-polydimethylsiloxane block copolymers whose block lengths are a magnitude higher than previous systems are prepared and mechanical as well as morphological properties are examined.

MATERIALS

The AB copolymers were synthesized by the previously described polymerization of hexamethylcyclotrisiloxane, D_3, with polystyryllithium in the presence of promoting solvents such as THF or glyme.[1] This general method was extended to the polymerization of D_3 with living α,ω-dilithiopolystyrene prepared from difunctional initiators. This gives BAB block copolymers now terminated at each end with the lithium siloxanolate moiety. These or the corresponding silanol-ended polymers are condensed by known non-bond rearranging methods of siloxane chain extension[9] to give the $(BAB)_x$ block copolymers.

The BAB prepolymers as well as the derived $(BAB)_x$ species were judged to be essentially free of homopolymer based on the failure of either polystyrene or polydimethylsiloxane to separate from their solutions in selective solvents such as bromobenzene or methylcyclohexane.[1] Gel phase chromatography also shows the BAB prepolymers to be of relatively narrow molecular weight distribution and free of other polymeric species. The same analysis shows increased molecular weight as well as a broadened molecular weight distribution in the condensed $(BAB)_x$ copolymers as might be expected of a polymer obtained from a step-growth process. In some cases, a small fraction of the pre-polymer failed to condense. These observations coupled with the high yields obtained and the close agreement of the silicon analysis with calculated values further substantiates that the block copolymers were free of homopolymer. Examples of the BAB prepolymers, the derived $(BAB)_x$ block copolymers and their molecular weights as determined by membrane osmometry are shown in Table I.

DISCUSSION

Electron photomicrographs of thin films of the AB block copolymers cast from 5% toluene solutions[1] all show formation of microphases.

Special staining techniques to define the different phases are unnecessary because of the differences in electron absorption and scattering of the polystyrene and polydimethylsiloxane blocks. The darker regions in all cases represent phases containing polydimethylsiloxane. The three basic morphological structures previously reported for polystyrene-polydiene block copolymer systems-i.e., spheres, rods, and lamellae[3,10]- are also evident in the present system. Examples showing these structures may be seen in Figure 1. which shows a series of AB block copolymers all containing about 70 wt. % polystyrene but at three different molecular weights. The polymer in Figure 1C appears to be close to the lower limit of molecular weight for mocrophase formation.

A recent theoretical approach by Meier[11] allows an estimate of domain dimensions provided the molecular weight of the blocks and the interfacial tension or energy per unit area of micelle surface are known. Comparison of the observed dimensions for some of the present AB block copolymers with dimensions calculated by this method gives good agreement when a value of 15 dyne/cm. is chosen for the interfacial tension, γ. These are summarized in Table II. By comparison, the value of γ used in the polystyrene-butadiene system was slightly less than one dyne/cm.[11]

Changes in morphological structure with changes in molecular weight as noted above are unexpected when considered in terms of current theory[3] which predicts the observed change in morphology in the polystyrene-polydiene block copolymer systems when composition is changed but not when molecular weight is changed.[3,10] An important factor in the present system is the high interfacial contact energy observed above which significantly influences the molecular weight dependence of the end-to-end distance of the chains within the micelles. This in turn might be expected to affect the entropy in placing a copolymer chain within a particular type of micellular structure. This factor can be ignored in the polystyrene-diene systems presumably because the interfacial contact energies and subsequent "mirror effects" can be neglected.

Figures 1A, 2A and 2B show the effect of changing composition of the polystyrene-polydimethylsiloxane AB block copolymers on morphology. The effect is an apparent inversion in phases when one or the other components predominates as might be predicted from previous observations of the analogous diene block copolymer systems.[10] Electron photomicrographs of the multi-block, $(BAB)_x$, copolymers, on the other hand, show that their morphology does not change over a broad range of compositions. An example is illustrated in Figure 2C. The absence of any significant changes in micelle structure is attributed to the importance of the number of blocks in determining the entropy of microphase formation.[12] This would be particularly true for the present $(BAB)_x$ block copolymers where the total number of blocks per molecule is > 5.

The use of solvents which selectively solvate either the polydimethylsiloxane blocks or the polystyrene blocks in the $(BAB)_x$ block copolymers gave films which showed better definition of the microphases but gross changes in morphology were not observed. This contrasts with previous observations of the corresponding AB copolymers where profound changes in morphology were observed when selective solvents were used.[1] Improvement in resolution between the microphases in the present $(BAB)_x$ system through the use of selective solvents suggests that some of the polystyrene blocks were originally distributed in the polydimethylsiloxane microphases and that some of the polysiloxane blocks were originally distributed within the polystyrene microphases. This is again attributed to the dominating positive entropy requirements for microphase formation arising from the large number of blocks in the $(BAB)_x$ system.[12]

The $(BAB)_x$ multi-block copolymers are thermoplastic elastomers which give strong, tough compression molded specimens without any re-enforcing fillers or cure. Molding is carried out at 150°C and may be repeated several times without altering properties of the sample. Polymers which contain 50% polystyrene are tough ductile plastics and show a "knee" in the stress-strain curve. As the amount of polystyrene is decreased the block copolymers become increasingly elastomeric and the high initial moduli as well as yield points become less significant. Stress-strain curves of two polymers of differing polystyrene content are shown in Figure 3. Good mechanical properties are also evident in films of the polymers cast from solution. Selective solvents tend to favor high tensile strengths probably because of the more defined microphase formation noted above. Tables III and IV summarize some ultimate properties of the $(BAB)_x$ copolymers.

At temperatures over 50°C the mechanical properties of either the molded or solution-cast films decrease precipitously. This is illustrated in Figure 4 which is a plot of relative modulus as determined by torsional braid analysis at various temperatures. The sudden decline in mechanical properties at elevated temperatures is due to softening and flow in the plastic phases. This is typical behavior for thermoplastic elastomers in which glassy blocks are polystyrene.[13]

REFERENCES

1. J. C. Saam, D. J. Gordon and S. Lindsey, Macromolecules, 3, 1 (1970).

2. E. B. Bradford and E. Vanzo, J. Poly. Sci. Part A-1, 6, 1661 (1968).

3. T. Inoue, T. Soen, T. Hashimoto and H. Kawai, J. Poly. Sci. A-2, 1283 (1969).

4. G. Holden, E. T. Bishop and N. R. Legge, J. Poly. Sci., Part C, 26, 37 (1969).

5. R. L. Merker and M. J. Scott, J. Poly. Sci., Part A-2, 15 (1964), R. L. Merker, M. J. Scott and G. G. Haberland, Ibid, 31 (1964).

6. H. A. Vaughn, J. Poly Sci. Part B, 7, 569 (1969).

7. R. P. Kambour, Polymer Preprints, 10, (2), 885 (1969).

8. D. G. LeGrand, J. Poly. Sci., Part B, 7, 579 (1969).

9. W. Noll, Chemistry and Technology of Silicones, Academic Press, N. Y., 1968, Page 212.

10. M. Matsuo, S. Sagae and H. Asai, Polymer, 10 (2), 79 (1969).

11. D. J. Meier, J. Poly. Sci., Part C, 26, 81 (1969).

12. S. Krause, Macromolecules 3, 84 (1970).

13. T. L. Smith and Roy A. Dickie, J. Poly. Sci., Part C, 26, 163 (1969).

TABLE I

Polystyrene - Polydimethylsiloxane $(BAB)_x$ Multi-Block Copolymers

Wt % Me_2SiO	Yield (%)[a)]	Mn x 10^{-3} [b)]	x[c)]	% Si Found	% Si Calc'd
80	93	85.5	2.3	30.1	30.2
70	87	137	3.2	26.0	26.6
50	92	110	2.3	18.6	19.0
50	97	303	6.4	—	—

a) Based on both steps of the synthesis. b) Over-all Mn as measured by membrane osmometry. c) Based on the initial Mn of prepolymer as measured by membrane osmometry.

TABLE II

Micelle Dimensions of the Discontinuous Phase in Films of AB Block Copolymers

Wt % Me_2SiO	Mn x 10^{-3} a)	Type of Micelle	Diameter (Å) b)		Calculated Diameter (Å)
28.4	7.1	Spheres(?) c)	76	13	89.6
29.1	28.6	Lamellae c)	178	14	182.0
32.0	47.6	Spheres-Rods c)	231	37	232.0
61.6	48.8	Rods d)	282	15	303.0

a) As measured by membrane osmometry. b) See Reference 11. Interfacial tension is assumed to be 15 dyne/cm. c) Blocks of polydimethylsiloxane in the discontinuous phase. d) Blocks of polystyrene in the discontinuous phase.

TABLE III

Mechanical Properties of Molded Polydimethyl-Siloxane-Polystyrene $(BAB)_x$ Multi-Block Copolymers

Wt % $(Me_2SiO)_n$	Mn x 10^{-3} a)	x a)	Stress (psi) Yield	Stress (psi) Break b)	Strain (%) Yield	Strain (%) Break b)	Tear Strength c)
80	85.5	2.3	—	320	—	230	42
70	137	3.0	280	1020	5	350	170
50	110	2.3	1550	1810	8	350	460
50	303	6.4	2000	2300	4.5	13.0	—

a) As measured in Table I. b) Rate of clamp separation is 10"/min. c) "Die B" Tear, pli.

TABLE IV

Mechanical Properties of Solution-Cast Films of a $(BAB)_x$ Multi-Block Copolymer a)

Solvent	Stress (psi) Yield	Stress (psi) Break b)	Strain (%) Yield	Strain (%) Break b)
Toluene	990	1827	4.0	350
Methylcyclohexane	1260	2360	4.0	530
Bromobenzene	1760	2575	3.0	480

a) 50 wt. % polydimethylsiloxane, $\overline{Mn}$ = 110,000 and x = 2.3
b) Rate of clamp separation is 10"/min.

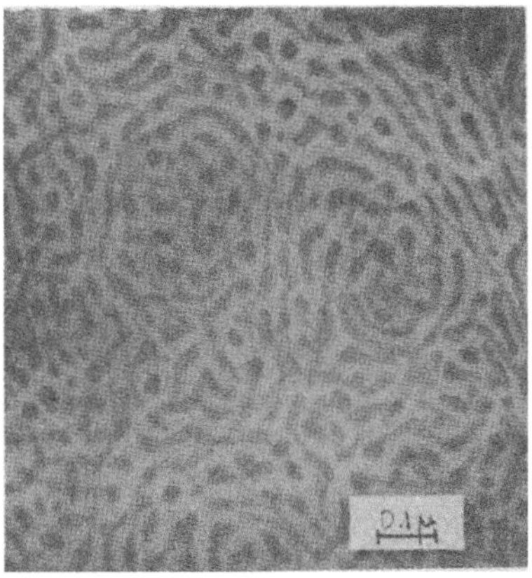

A

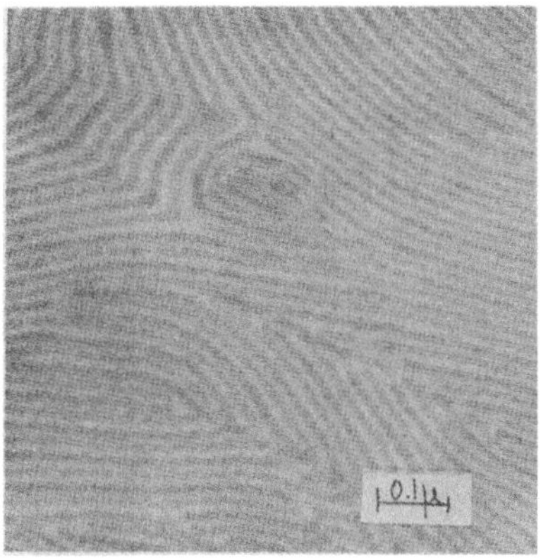

B

C

Figure 1. Electron phtomicrographs of films of AB block copolymers of varying molecular weights cast from 10% toluene solution: A, $\bar{M}n$ = 47,600, 32 wt. % $(Me_2SiO)_n$; B, $\bar{M}n$ = 28,600, 29 wt. % $(Me_2SiO)_n$; C, $\bar{M}n$ = 7060, 28 wt. % $(Me_2SiO)_n$.

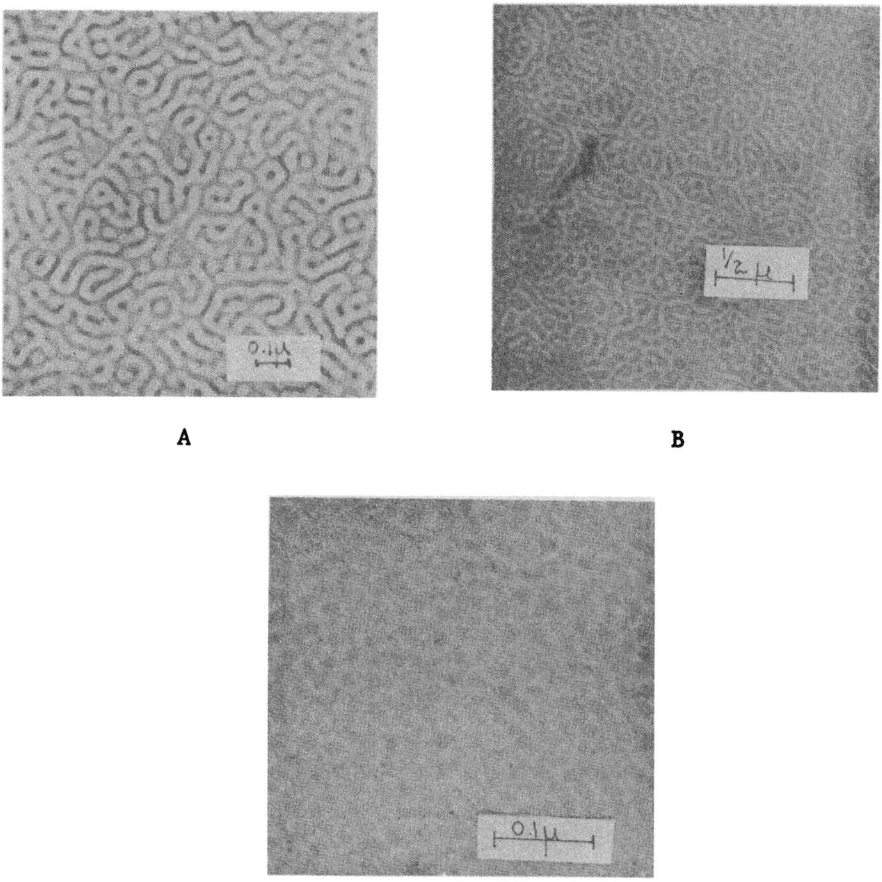

A B

C

Figure 2. Electron photomicrographs of films of block copolymers of differing composition cast from toluene solution: A, AB block copolymer, 62 wt. % Me_2SiO, $\overline{Mn}$ = 48,800; B, AB block copolymer, 82 wt. % Me_2SiO, $\overline{Mn}$ = 53,000; C, $(BAB)_x$ block copolymer cast from 1% toluene solution, 50 wt. % $(Me_2SiO)_n$, $\overline{Mn}$ = 303,000, x = 6.7.

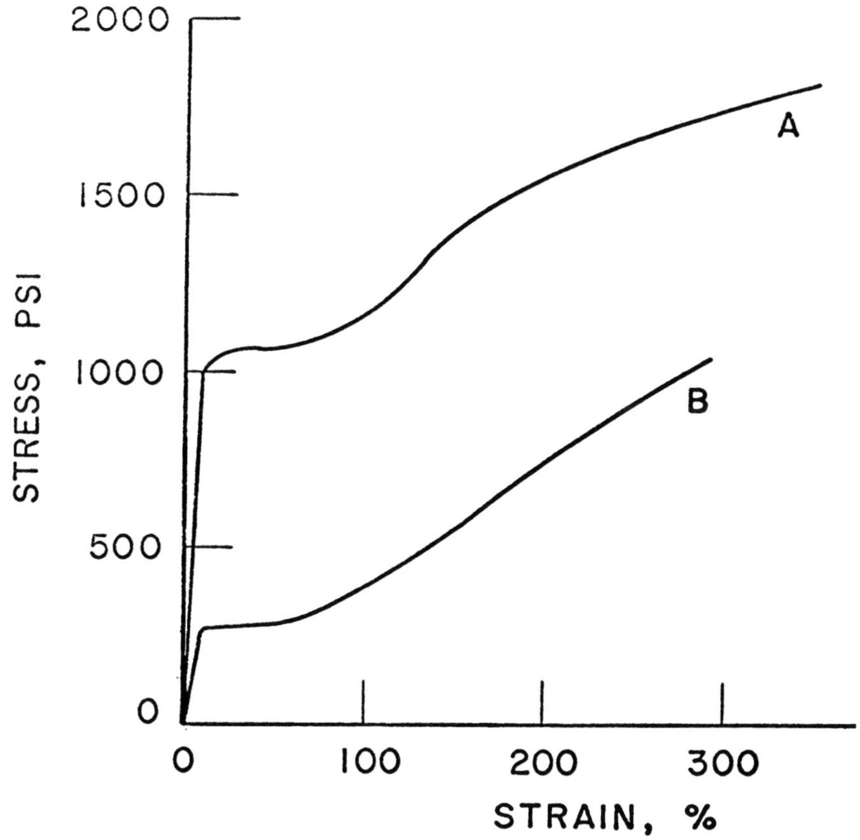

Figure 3. Stress-strain curves for compression molded samples of $(BAB)_x$ block copolymers containing (A) 50 wt. % $(Me_2SiO)_n$ ($\bar{M}n$ = 110,000, x = 2.3) and (B) 70 wt. % $(Me_2SiO)_n$ ($\bar{M}n$ = 137,000, x = 3.3).

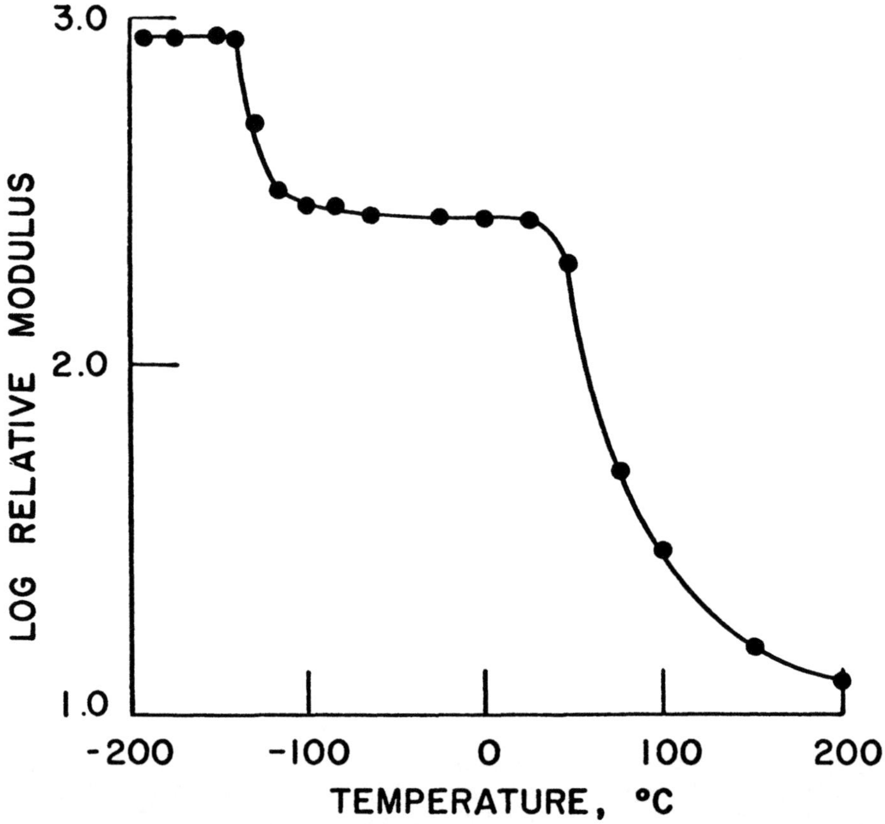

Figure 4. Relative modulus of the $(BAB)_x$ multi-block copolymer in Figure 3B measured at various temperatures by torsional braid analysis.

SOLVENT INTERACTIONS OF POLY(DIPHENYLSILOXANE-BLOCK-DIMETHYLSILOX-ANE-BLOCK-DIPHENYLSILOXANE) COPOLYMERS

A. Keith Fritzsche* and Fraser P. Price*

Physical Chemistry Branch, General Electric Research and Development Center, Schenectady, New York 12301

INTRODUCTION

Block copolymers have gained increasing scientific attention since their successful synthesis by homogeneous anionic polymerization in 1956(1). Interest was heightened by the possibility that block copolymers might be materials possessing desirable properties, a possibility originally considered by Mark(2). One important class of block copolymers are the polystyrene-polybutadiene-polystyrene block copolymers (S-B-S) introduced by the Shell Chemical Company (U.S.A.) in 1965(3). Electron microscope studies by Meier(4) on styrene-isoprene two block model and studies by Hendus, Illers, and Ropte(5) on polystyrene-polybutadiene block copolymers showed that under certain conditions of preparation the styrene exists in glassy domains interconnected by amorphous polybutadiene regions. Furthermore, Hendus, et al., showed that the size of the polystyrene domains and the morphology of the films is dependent on composition and molecular weight of the copolymer. Also, the domains of the polymers were found to be altered by the solvent from which the copolymer films were cast(6).

The synthesis of poly(diphenylsiloxane-block-dimethylsiloxane block-diphenylsiloxane) copolymers by E. E. Bostick and co-workers(7) yielded a polymer consisting of crystalline diphenylsiloxane regions connected by amorphous dimethylsiloxane blocks. In styrene-buta-

* Present Address, Polymer Science and Engineering Program, University of Massachusetts, Amherst, Massachusetts 01002.

diene or styrene-isoprene block copolymers the morphology visible in the electron microscope is revealed only by staining. In the diphenylsiloxane-dimethylsiloxane block copolymers the crystalline diphenylsiloxane blocks were visible in the electron microscope without staining, thus facilitating their study. The morphology of the system also was found to be altered by changes in the composition of the copolymer, in the solvent from which the films were cast, and in annealing(8).

The presence of these two blocks, differing as they do in solubility parameters, contributes unique properties to the polymer molecules. These unusual properties are reflected in the behavior of both the bulk polymer and its solutions. It is of non-trivial interest from both a theoretical and a practical viewpoint to describe and rationalize this behavior in quantitative terms. The work described herein was undertaken to try, at least partially, to satisfy these aims.

Below the temperature of dissolution, the diphenylsiloxane blocks, which form crystallites of varying stabilities, are tie-points, multifunctional crosslinks, which permit swelling of the dimethylsiloxane blocks but forbid the complete dissolution of the polymer molecule. The extent of swelling of crosslinked networks was shown by Gee to be dependent upon the difference between cohesive energy densities of the solvent and the polymer(9). Also the swelling can be correlated with the structure of the polymer as was shown by Flory, who was able to determine the frequency of crosslinked units in butyl rubber from swelling experiments in a variety of solvents(10).

Above the temperature of dissolution, the crystallites are dissolved (melted) by appropriate solvents. The end-to-end distance of a given block is dependent upon the difference between the solubility parameters of the block and the solvent. Thus, a given solvent will usually expand the two blocks to different degrees. Standinger, Heuer, and Husemann observed that the viscosity and swelling are related(11). They observed that those solvents which yield the largest values of specific viscosity for polystyrene solutions also produce the greatest swelling of styrene-divinylbenzene gels(11). Whitby, McNally, and Gallay had previously noted a correlation between viscosity and swelling of rubber within a homologous series of solvents(12).

Therefore, it appeared that intrinsic viscosity studies and network swelling experiments in a variety of solvents of differing solubility parameters would be a fruitful field of endeavor. The results of the swelling experiments should give an insight into the structure of the copolymer in bulk, i.e., the stabilities of the crystallites and the tiepoint distances, as well as the effect of solvent on the copolymer. The measurement of viscosities in appropriate solvents at appropriate temperatures should yield information about the interactions of the various solvents with two different blocks when the isolated molecule is under consideration. It is worth noting that with the siloxane block copolymers under study here, a distinct advantage of network swelling over intrinsic viscosity measurements is that experiments can be run at lower temperatures than the 120°C necessary for viscosity measurements. The results of the experiments are presented and interpreted, and correlation between the two types of experiments is attempted.

INTRINSIC VISCOSITY EXPERIMENTS

Theory. The capacity of high polymers to greatly increase the viscosity of the liquid in which they are dissolved is measured by the intrinsic viscosity of the polymer. The magnitude of the intrinsic viscosity varies with the solvent depending on the extent to which the polymer molecule is expanded or contracted by the solvent in question. Expansion of the polymer by the solvent is related in a semi-quantitative manner to the difference in the solubility parameters of the solute and solvent. The solubility parameter of the dimethylsiloxane block is approximately 7.5 and that of the diphenylsiloxane block is approximately 9.5 Therefore, a solvent which will effectively extend dimethylsiloxane block will contract diphenylsiloxane blocks resulting in an anomalous intrinsic viscosity.

Suitable solvents were chosen according to their solubility parameters. The solubility parameter is defined as the square root of the cohesive energy density. At constant temperature,

$$\delta = \left(\frac{-E}{v_1}\right) \tag{1}$$

where -E is the cohesive energy of the solvents, essentially the molal energy of vaporization to gas at zero pressure, v_1 is the molal volume of the liquid. -E is approximated by ΔE_v, the energy of vaporization, which can be replaced by $\Delta H_v - P\Delta V = \Delta H_v - RT$, where ΔH_v is the heat of vaporization. Therefore

$$\delta = \left(\frac{\Delta H_v - RT}{v_1}\right)^{1/2} \tag{2}$$

The heat of vaporization can be calculated from the dependence of the vapor pressure on temperature, by means of the Clausius-Clapeyron equation

$$\frac{d \ln P}{dT} = \frac{\Delta H}{RT^2} \qquad (3)$$

Usually the data necessary to calculate δ was obtained from the literature[13,14]. When solubility parameters or the necessary data could not be found in the literature for the desired solvents[15,16], the values of δ were calculated in the above manner.

It should be noted that the solubility parameters are not constant with temperature. Changes in temperature, however, change the magnitudes of the solubility parameters in the same direction. It is assumed here that among the solvents used, the differences in solubility parameter remain constant over the temperature range of interest.

Results and Discussion

The solvents used in this investigation and their solubility parameters are listed in Table I. Those which are underlined were found to dissolve some or all of the copolymers in question at 120°C and were suitable for the intrinsic viscosity studies. The others were suitable only in the network swelling experiments.

TABLE I

Solvents and Their Respective Solubility Parameters

Solvent	δ	Solvent	δ
n-octane	7.55	<u>o-dichlorobenzene</u>	9.6
cyclohexane	8.22	<u>monochlorobenzene</u>	9.4
<u>n-propylbenzene</u>	8.65	<u>o-xylene</u>	9.0
ethyliodide	9.4	<u>anisole</u>	9.5
toluene	8.9	nitrobenzene	10.0
<u>o-chlorotoluene</u>	9.1	chloroform	9.3
methylcyclohexane	7.85	n-hexane	7.3
mesitylene	8.8	pyridine	10.7
dioxane	9.73	acetone	9.7
methylenechloride	9.7	n-pentane	7.05
hexane	7.3	m-xylene	8.8
<u>p-chlorobenzene</u>	9.55		

Copolymers were chosen which had a nearly constant degree of polymerization but varying diphenylsiloxane block lengths. A dimethylsiloxane homopolymer at approximately the same degree of polymerization was also studied for comparison. The compositions of all the polymers studied in viscosity and swelling experiments are listed in Table II.

TABLE II

Compositions of ABA Block Copolymers

$A = (\phi_2SiO)_x$, $B = (Me_2SiO)_y$

Sample	$\bar{M}_n$(Me)	$\bar{y}$	$\bar{M}_n$(Ph)	$\bar{x}$	$DP_{\bar{y}+2\bar{x}}$
BC-1	159,000	2149	18,950	96	2341
BC-2	163,650	2198	30,253	153	2504
BC-3	153,000	2068	38,100	192	2452
BC-4*	296,000	≈4000	99,000	≈500	5000
H-1	179,080	2420	---	---	2420

* Values are approximate.

The results of the intrinsic viscosity determinations are compiled in Table III. Plots of solubility parameter δ versus intrinsice viscosity [η] are given in Figure 1 for the copolymers and in Figure 2 for the dimethylsiloxane homopolymer.

TABLE III

Intrinsic Viscosities of Dimethylsiloxane-Diphenylsiloxane Block Copolymers in a Variety of Solvents
N.R. Not Run; ___ Insoluble

		[η]			
Solvent	δ	BC-1	BC-2	BC-3	H-1
o-dichlorobenzene	9.6	.62	.59	.31	.49
anisole	9.5	.65	.60	.29	.42
p-dichlorobenzene	9.46	.70	N.R.	.31	.74
monochlorobenzene	9.4	.78	.67	.33	.73
o-chlorotoluene	9.1	.89	.71	.40	.77
o-xylene	9.0	.89	--	--	.89
m-xylene	8.8	.88	--	--	.74
n-propylbenzene	8.65	.96	--	--	N.R.

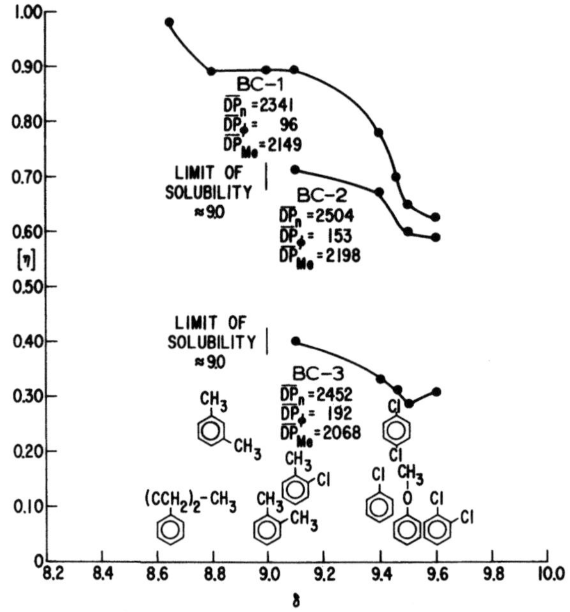

Fig. 1. Intrinsic viscosity versus solubility parameter for copolymer series. Temperature 120°C.

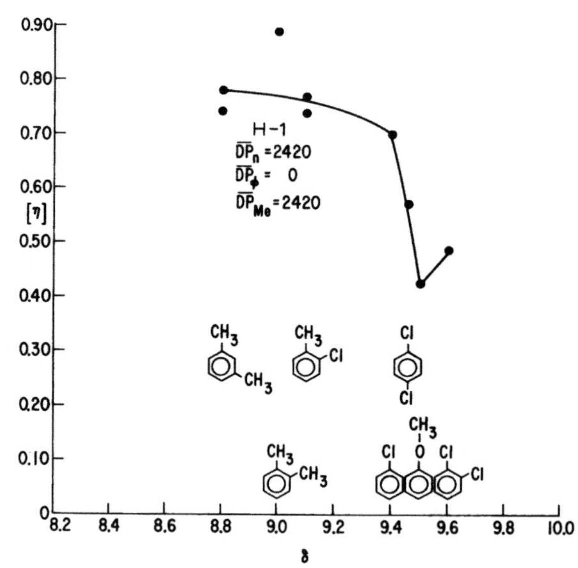

Fig. 2. Intrinsic viscosity versus solubility parameter for dimethylsiloxane homopolymer. Temperature 120°C.

From Figure 1, it is evident that [η] varies greatly with the solubility parameter of the solvent. Apparently, the dimethylsiloxane blocks become more extended as the solubility parameter of the solvent decreases toward that of the dimethylsiloxane homopolymer, resulting in a higher measured value for [η]. The magnitude of this increase is related directly to the length and relative proportion of the dimethylsiloxane blocks. At 120°C a sharp minimum is observed for [η] at δ = 9.6 for both the copolymers and the homopolymer indicating the collapse of the dimethylsiloxane blocks. Increasing the length of the dimethylsiloxane block extends the range of solubility to solvents of lower solubility parameters. At 120°C BC-3 (87 mole percent dimethylsiloxane) is completely insoluble and BC-2 (88 mole percent dimethylsiloxane) only partially soluble at δ = 9.0. At this temperature, BC-1 (92% dimethylsiloxane) and the dimethylsiloxane homopolymer, H-1, exhibit solubility throughout the range δ = 9.6 to δ = 8.65.

Even though the limit of solubility of the diphenylsiloxane blocks is exceeded at these low solubility parameters and the diphenylsiloxane blocks collapse, the high extension of the dimethylsiloxane blocks is apparently sufficient to maintain the whole molecule in solution. In solvents of low solubility parameter, the copolymers can be conceptualized as randomly coiled dimethylsiloxane chains with small, globules of contracted diphenylsiloxane units at either end.

The sharp decrease in the solubility of the dimethylsiloxane homopolymer at δ = 9.6 indicates that here the dimethylsiloxane chain collapses. The addition of relatively small numbers of diphenylsiloxane units is insufficient to raise the limit of solubility for these copolymers.

NETWORK SWELLING EXPERIMENTS

Theory

The diphenylsiloxane crystallites function as tiepoints (multifunctional crosslinks) in the poly(diphenylsiloxane-block-dimethylsiloxane-block-diphenylsiloxane) copolymers. These crystallites will have a spectrum of stabilities which will be reflected in a distribution of melting points. Therefore, the network swelling of the copolymer will not only result from the absorption of a quantity of a suitable solvent, but will also be enhanced by the dissolution of the crystallites of lower stabilities. As the temperature of the solvent is raised, more of the crystallites will be dissolved, thus increasing the swelling.

When the network is swollen by absorption, the chains between network junctions are required to assume elongated configurations, and a force related to the elastic retractive force in a rubber consequently develops in opposition to the swelling process. As swelling proceeds, this force increases and the diluting force, related to osmotic pressure, decreases. Ultimately, a state of equilibrium swelling is reached in which these two forces balance. When the swollen gel is in equilibrium with the pure solvent, the following equation is applicable[17]

$$\frac{v_e}{V_o} = \frac{[\ln (1 - v_{2m}) + v_{2m} + \chi_1 v_{2m}^2]}{V_1 (v_{2m}^{1/3} - v_{2m})} \qquad (4)$$

in which

$\frac{v_e}{V_o}$ = moles of crosslinks/cc

v_{2m} = inverse of the swelling ratio, q_m

$q_m = \frac{V \text{ (swollen)}}{V_o \text{ (dried)}} = \frac{\text{volume of swollen polymer}}{\text{volume of unswollen polymer}}$

χ_1 = nearest neighbor polymer-solvent interaction parameter

V_1 = molar volume of solvent

If χ_1 can be found, the distance between crosslinks can be estimated from v_e/V_o using the relationship

$$d = \sqrt[3]{\frac{v_e}{V_o} N}^{\,-1} \qquad (5)$$

where $N = 6.023 \times 10^{23}$/mole.

There are other assumptions that are made in the derivation of the swelling theory for rubbers which severely limits its applicability to poly(diphenylsiloxane-block-dimethylsiloxane) copolymers. The theory assumes that the crosslink between any two chains is located at random. In these polymers, the dimethylsiloxane blocks, regardless of how randomly distributed and entangled, are involved in crystallites only at their ends. There is thus a somewhat regular pattern to the crosslinking which is counter to the idea of completely random crosslinks. More importantly, the theory assumes that the bonds joining two chains together are essentially volumeless. This clearly is not the case here. In the light of these remarks, the calculated results for v_e/V_o and d at various temperatures are to be regarded as fairly educated guesses rather than very good approximations. Thus the picture that emerges will not suffice as a detailed description of the system but instead merely serves to give a better insight into the structure than previously available.

EXPERIMENT AND RESULTS

Films of BC-1, BC-2, BC-3, and BC-4 were cast from o-dichlorobenzene. BC-2 was also cast from o-chlorotoluene. BC-4, which has a different degree of polymerization, was used primarily because of the good mechanical properties of its films and its availability. Solvent traces were removed by heating under nitrogen atmosphere at 260°C for 0.5 hours. A sample of BC-4 after casting from o-dichlorobenzene was submitted for chlorine analysis and showed not detectable residual chlorine. Thermal measurements using DSC of BC-2 and BC-4 cast from o-dichlorobenzene and BC-2 cast from o-chlorotoluene gave essentially the same results as untreated BC-2 and BC-4. To determine if chemical crosslinks were formed in the process, a film of BC-2 was dissolved and the viscosity run. The viscosities of the dissolved film and the untreated film were identical within experimental error.

In the swelling experiment, the film was immersed in solvent until no noticeable change in weight occurred during a twenty-four hour period. Equilibrium was then assumed to be established. After equilibrium was attained, excess solvent was wiped away and the film was weighed, dried, and reweighed. The swelling ratio, q_m, was calculated from Z, the ratio of the weight swollen to the weight dried, by

$$q_m = 1 + (Z - 1)\frac{\rho_p}{\rho_s} = \frac{1}{v_{2m}} \qquad (6)$$

where ρ_p = density of the polymer

ρ_s = density of the solvent.

The experimentally determined value of the density of the polymer BC-4 is 1.038 g/cc.

Values of q_m for BC-4 in the solvents used are displayed in Fig. 3. This figure illustrates for the various solvents the relationship between q_m and the solubility parameter of the solvent. Two maxima are evident, one at $\delta \approx 7.8$ and the other at $\delta \approx 9.5$. The latter becomes increasingly pronounced as the temperature is raised. The first maximum, $\delta \approx 7.8$, occurs when the solubility parameter of the solvent is near that of the dimethylsiloxane blocks and represents swelling of these units. The second maximum occurs near the solubility parameter of the diphenylsiloxane blocks and is best interpreted as attack of the solvent on the crystallites resulting in their dissolution. The crystallites have varying degrees of stability. As the temperature is raised, the less stable crystallites are dissolved while the more stable remain. The removal of these less stable crystallites accounts for the increased swelling. At the temperature at which the most stable crystallites dissolve, the copolymer goes into solution.

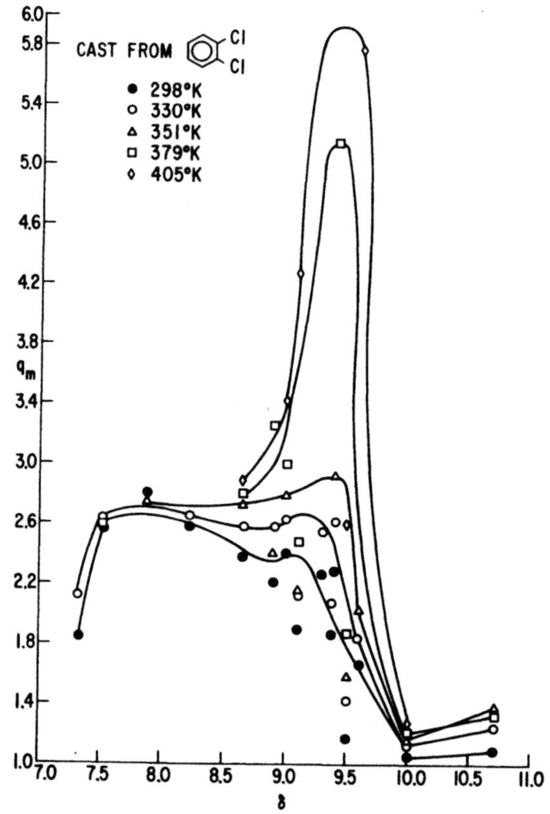

Fig.3. Swelling of BC-4 versus solubility parameter of various temperatures.

In order to obtain approximations of the moles of tiepoints/cc (v_e/V_o) and the average distance between tiepoints (d), a value of zero was assumed for χ_1 in Equation (4). This assumption is open to considerable question. However in the absence of measured values of this parameter and in the interest of simplicity it seemed the most reasonable one to make. The value of d calculated by use of Equations 4 and 5 and displayed in Figure 4 should accordingly be regarded as estimates good to within a factor of between 1.5 and 2.0.

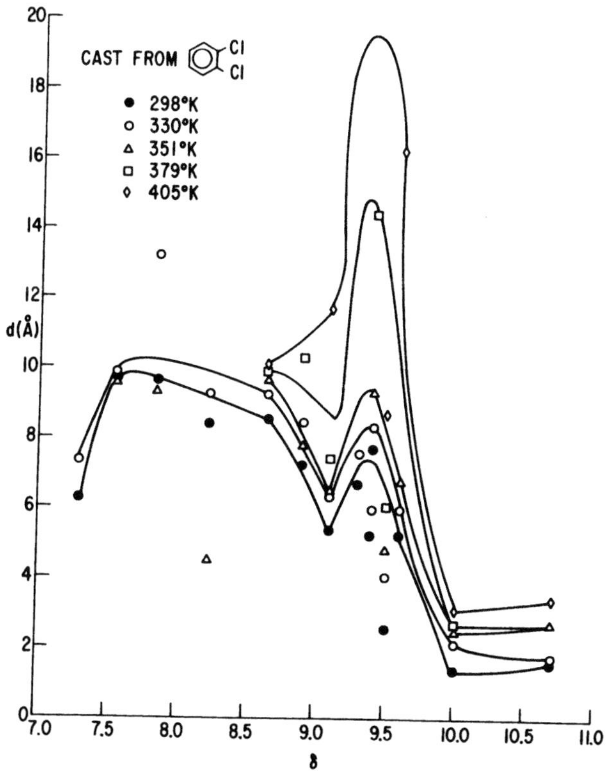

Fig. 4. Solubility parameter versus approximate tiepoint distance for BC-4.

As expected, Figure 4 shows two maxima. The maxima occur at the same solubility parameters as in Figure 3. The slight increase in swelling and tiepoint distances for the maximum at 7.8 probably result from the increase in the solvating power of the solvent at increased temperature.

The polymer BC-4 is 80 mole per cent dimethylsiloxane block (DP [Me_2 S_iO] ≈ 4000) and 20 mole per cent diphenylsiloxane block

(DP [ϕ_2 S_iO] ≈ 500]. The root-mean-square end-end distance is given(18) by

$$\sqrt{\overline{r^2}} = 730 \times 10^{-11} \sqrt{M}$$

This leads to a value of $\sqrt{\overline{r^2}}$ of 397Å for the dimethylsiloxane block in BC-4. The average end-end distance is much larger than the calculated tiepoint distances by more than an order of magnitude, showing that entanglements as well as crystallites can function as tiepoints. Also, with this long end-end distance, it seems unlikely that a copolymer can have its diphenylsiloxane blocks embedded in adjacent crystallites. Instead the ends more probably will be embedded in crystallites that are several tiepoint distances away. Since numerous diphenylsiloxane blocks are embedded in each crystallite and since there are numerous entanglements among the dimethylsiloxane blocks of different chains, the polymer is a highly interconnected network. There seems to be a high probability that every crystallite is connected to the majority of the other crystallites in its vicinity.

Attempts to investigate the effect of increased diphenylsiloxane block length on copolymers of a constant degree of polymerization were only partially successful due to crumbling of swollen films of BC-3 and loss of mechanical strength upon swelling as in BC-1. The loss of mechanical strength was so great for BC-1 that swelling measurements were impossible around $\delta = 9.5$. Therefore, although this polymer was observed to be highly swollen at $\delta = 9.5$, the temperature sensitive peak does not appear on the graph. Qualitative results are depicted in Figure 5. It can be seen that swelling is highly sensitive to the diphenylsiloxane block length. The diphenylsiloxane blocks occupy about twice as much volume in BC-3 (15.7 mole percent) than as in BC-1 (8.2 mole percent) decreasing the volume available for solvent molecules between crystallites. The crystallites in BC-1 have a spectrum of lower stabilities than those in BC-3, having a smaller size and higher surface area to volume ratio.

In order to determine whether the solvents from which the films are cast also influence the swelling characteristics, films of BC-2 were cast from o-dichlorobenzene ($\delta = 9.6$) and from o-chlorotoluene ($\delta = 9.1$). These were swollen at 25°C until equilibrium was attained, and q_m calculated. The results are shown in Figure 6. Here it is clear that the solvent from which the copolymer is cast has a definite influence on its physical properties. The maxima are altered in their location and in their relative height. The maximum for swelling of the dimethylsiloxane block changes from $\delta = 7.6$ for BC-2 cast from o-dichlorobenzene to around 8.0 for the same polymer from o-chlorotoluene. The maximum for the diphenylsiloxane block occurs at 9.6 for BC-2 from o-dichlorobenzene but at 9.3 for BC-2 from o-chlorotoluene. Evidently, the molecules maintain the relative degree of extension and contraction of the respective dimethylsiloxane and diphenylsiloxane blocks when the solvent is removed. The o-chlorotoluene (δ = 9.1) is a better solvent for the dimethylsiloxane blocks than is the o-dichlorobenzene (δ = 9.6). So, the dimethylsiloxane blocks are more extended when the copolymer is cast from it. Conversely, it is a poorer solvent for the diphenylsiloxane blocks than the o-dichlorobenzene so that the diphenylsiloxane blocks are not so highly extended as those from o-dichlorobenzene and the crystallites formed have a higher stability evidenced by their lower degree of dissolution. The decrease in the extension of the diphenylsiloxane blocks upon casting from o-chlorotoluene may result from the tighter collapse of the block and the formation of crystallites of higher stability than those cast from o-dichlorobenzene. Thus, since the crystallites may occupy less volume, more volume is available for solvents with lower solubility parameters to swell the dimethylsiloxane blocks. A plausible explanation for the shift of the peaks upon variation of solvent from which the films are cast cannot at this time be advanced.

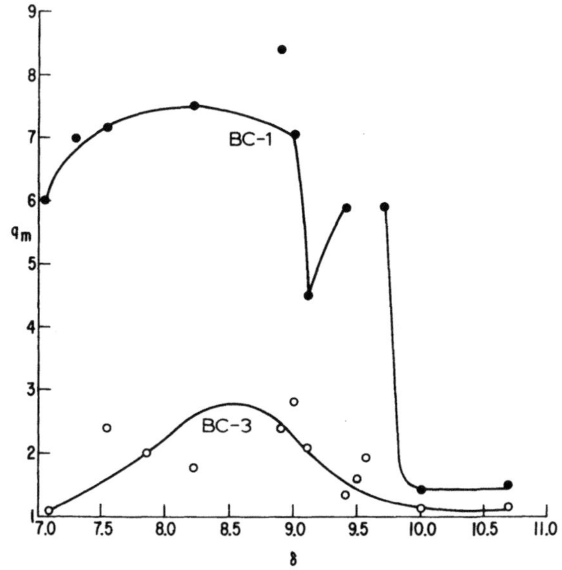

Fig. 5. Swelling of copolymers of varying composition with constant $\overline{DP}$.

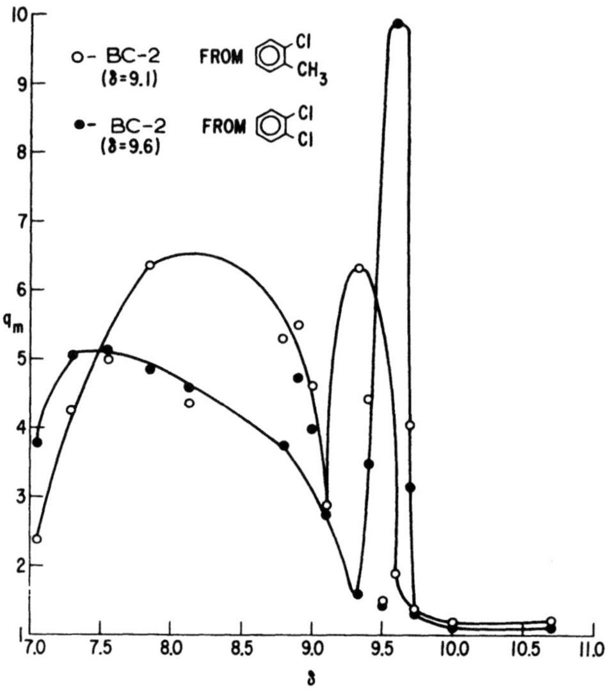

Fig. 6. Swelling at 25°C of BC-2 cast from two different solvents at 25°C.

CONCLUSIONS

The experiments show that the intrinsic viscosity varies with solvents of differing solubility parameter. There is an increase in intrinsic viscosity with an increase in the length of the dimethylsiloxane blocks in polymers of constant degree of polymerization. There is also an increase in viscosity resulting from the expansion of the dimethylsiloxane blocks by solvents of appropriate solubility parameter.

At 120°C BC-3 (84 mole percent dimethylsiloxane) is completely insoluble at δ = 9.0; BC-2 (88 mole percent dimethylsiloxane) is partially soluble at δ = 9.0; BC-1 (92 mole percent dimethylsiloxane) exhibits solubility at δ = 8.65. This increased solubility is due to the capacity of the dimethylsiloxane blocks to drag the collapsed diphenylsiloxane blocks into solution.

Swelling experiments with sample BC-4 (80 mole percent dimethylsiloxane) exhibit two maxima, one at $\delta = 7.8$ and another at $\delta = 9.5$. The peak at $\delta = 9.5$ is very temperature sensitive because of the dissolution of the crystallites. The peak at $\delta = 7.8$ is relatively insensitive to temperature because the dimethylsiloxane blocks are amorphous and dilated essentially to their maximum at all temperatures.

Swelling is highly sensitive to the diphenylsiloxane block length in polymers of constant degree of polymerization. The lower the mole percent of diphenylsiloxane, the greater is the swelling. This probably results from the smaller size of the crystallites which in turn results in both more room available for the solvent and lower stability for crystallites.

The solvent from which the copolymer is cast has an influence on the properties of the resulting film. The heights and location of the maxima in curves of swelling versus δ are different when BC-2 is cast from o-dichlorobenzene than when it is cast from o-chlorotoluene. The maximum for swelling of the dimethylsiloxane block of BC-2 cast from o-dichlorobenzene shifts from 7.6 to 8.0 when it is cast from o-chlorotoluene. The maximum also increases in height at constant temperature. The maximum of 9.6 of BC-2 cast from o-dichlorobenzene shifts to 9.3 for o-chlorotoluene and decreases in height. The increase in height of the maximum for the dimethylsiloxane blocks when cast from o-chlorotoluene results from the increased swelling of the dimethylsiloxane blocks of a solvent of lower solubility parameter. The drop in height of the diphenylsiloxane peak in a solvent of lower solubility parameter results from the collapse of the block to a less swollen structure and to the formation of crystallites with a higher stability.

REFERENCES

1. M. Szwarc, M. Levy and R. Milkovich, J. Am. Chem. Soc. 78, 2656 (1956).
2. H. Mark, Textile Res. J. 23, 294 (1953).
3. R. J. Angelo, R. M. Ikeda and M. L. Wallach, Polymer 6, 141 (1965).
4. D. J. Meier, J. Polymer Sci. Part C, 26, 81 (1969).
5. H. Hendus, K. H. Illers and E. Ropte, Kolloid-Z. 216-217, 110 (1967).
6. J. F. Beecher, L. Marker, R. D. Bradford and S. L. Aggarwal, J. Polymer Sci. Part C, 26, 117 (1969).
7. E. E. Bostick, "Synthesis of Poly(diphenylsiloxane-block-dimethylsiloxane-block-diphenylsiloxane) Copolymers," paper presented at the Symposium on Block Copolymers, Division of Polymer Chemistry, American Chemical Society, New York, September, 1969.

8. A. K. Fritzsche and F. P. Price, "Structure and Morphology of Poly(dimethylsiloxane-block-dimethylsiloxane-block-diphenylsiloxane) Copolymers," paper presented at the Symposium on Block Copolymers, Division of Polymer Chemistry, American Chemical Society, New York, September, 1969.
9. G. Gee, Trans. Faraday Soc. 38, 418 (1944).
10. P. J. Flory, Ind. Eng. Chem. 38, 417 (1946).
11. H. Staudinger, W. Heuer and E. Husemann, Trans. Faraday Soc. 32, 323 (1936).
12. G. S. Whitby, J. C. McNally and W. Gallay, Colloid Symposium Monograph 6, 230 (1928).
13. J. E. Jordon, VAPOR PRESSURE OF ORGANIC COMPOUNDS, Interscience, New York, 1954.
14. "Physical Properties of Chemical Compounds," Advances in Chemistry Series, ACS.
15. G. M. Bristow and W. F. Watson, Trans. Faraday Soc. 54, 1733 (1958).
16. J. H. Hildebrand, SOLUBILITY OF NONELECTROLYTES, Reinhold, New York, 1950.
17. J. J. Hermans, J. Polymer Sci. 59, 171 (1962).
18. P. J. Flory, L. Mandelkern, J. Kinsinger and W. B. Schultz, J. Am. Chem. Soc. 74, 3364 (1952).

STRUCTURE AND FUNCTION OF BIOLOGICAL MEMBRANES

David E. Green and Garret Vanderkooi

Institute for Enzyme Research

University of Wisconsin, Madison, Wisconsin 53706

I. STRUCTURE

Types of Membranes

All living cells contain a multiplicity of membrane systems. One type of membrane forms the boundaries of cells (plasma membranes in mammalian systems); a derivative of a plasma membrane forms the multilayered insulating myelin sheath which surrounds nerves. Many subcellular organelles are both surrounded by membranes, and contain membranes within them which are made up of enzymes that carry out complicated concerted reactions, or function in energy transduction and conservation. For example, the mitochondrion is a system of two membranes, an outer membrane, and an inner membrane which contains the proteins that couple electron transfer reactions to the synthesis of adenosine triphosphate. The light-sensitive rhodopsin molecules found in the retinal rods of the eye are the principal protein components of the membranes which are concerned in the photoreception. The list could be extended, but the point to be made is that membrane systems, while having the same general structure and characteristics, have exceedingly diverse functions.

It is possible to categorize membranes into two classes, simple and complex, depending upon their protein composition. In simple membranes, the association between protein molecules is random, whereas in complex membranes, repeating sets of proteins are required for function, and proteins in a set have an invariant geometric relation one to another. The retinal rod disks are a good example of simple membranes; the rhodopsin molecules are spread out as individual spheroidal particles in the plane of the membrane (1). The mitochondrial membrane, on the other hand, is a complex membrane. The numerous protein species found here are arranged in several multimeric clusters, or repeating units (2), each of which carries out a concerted set of enzyme reactions.

Membrane Composition

Biological membranes are composed of lipids and proteins. There is great variation with respect to the nature and proportions of the lipid components, but all membranes contain a high proportion of one or another kind of amphiphilic lipid. Amphiphilic lipids usually have a long nonpolar aliphatic "tail" (16-24 carbon atoms) and a polar "head." Phospholipids, such as phosphatidyl ethanolamine and phosphatidyl choline (lecithin) are the most common examples of such amphiphilic lipids. Many membranes also contain a greater or lesser amount of neutral lipids, chiefly cholesterol; these lipids modulate the membrane properties, but are not essential for the structural integrity of the membrane. Membrane proteins account for from 50-70% of the dry weight in most membranes. Contrary to what was formerly believed, the proteins appear to be globular (i.e., compact or spheroidal), just as are the majority of well known soluble proteins, rather than fibrous or extended structures. Membrane proteins have molecular weights ranging from 12,000 to 50,000 and diameters of 30-50 Å.

General Structure of Membranes

A membrane may be described as a two-dimensional continuum of lipids and proteins. The thickness varies from 60 to 90 Å, depending upon the source. The lipids in a membrane are in the bilayer arrangement; that is, the long axes of the lipid molecules are normal to the plane of the membrane, the polar lipid heads are on the two surfaces, and the nonpolar lipid tails extend from the two surfaces toward the center of the membrane, where they meet in the middle. On account of the sizes of the protein molecules as compared to the thickness of a lipid bilayer, part of each protein molecule (one hemisphere) extends out of the lipid layer into the aqueous medium which surrounds the membrane. (In biological systems, the fluid surrounding membranes is always an aqueous solution of salts and other solutes.) Figure 1 shows a cross-section of a membrane, and illustrates the arrangement of the lipids and proteins (3). This diagram is purely schematic; there is no necessity for the protein molecules to be ordered in pairs in the manner shown; rather, all that is intended by the model is that there are two protein layers, and lipid bilayer between the protein

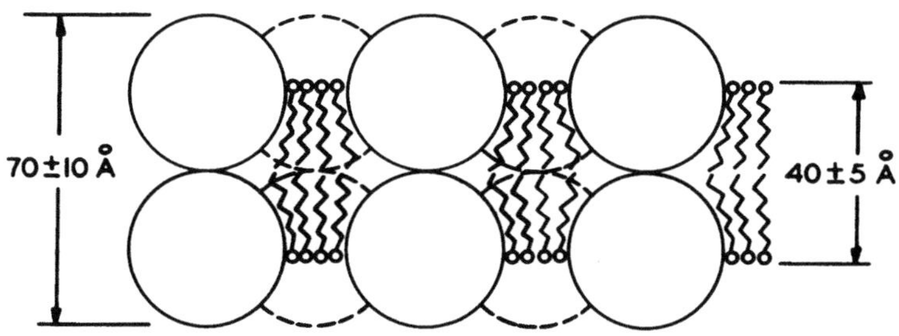

Figure 1. Membrane cross-section. The large circles are the proteins; the dashed circles are understood to be proteins behind the plane of the section. The small circles are the polar lipid heads, and the wavy lines are the nonpolar lipid tails.

molecules. Also, the value of 70 Å given for the thickness is only a typical value.

The membrane shown in Figure 1 has a two-fold plane of symmetry, but real membranes which have a true two-fold symmetry plane are probably rare. Indeed, several functions of membranes, such as ion translocation or energy transduction, appear to depend explicitly on the asymmetry of the membrane. In such membranes, an approximate two-fold plane of symmetry is retained, in that there are two layers of proteins and lipids, but the proteins in one layer are not the same as the proteins in the other layer.

Crystalline versus Liquid Crystalline Membranes

X-ray studies have established that the membranes of the rod outer segments have a high degree of fluidity in contrast to the more crystalline character of membranes such as those in the mitochondrion. The key to this spread in the degree of fluidity in biological membranes is the fluctuating character of the associative bonds between proteins. In the more crystalline membranes each molecule of protein is more less permanently associated with two to four other molecules in the same tier and with another molecule in the apposed tier. In the more fluid membranes, the number of associative bonds between protein molecules and their nearest neighbors is not only reduced but the direction of the bonds is fluctuating. While at all times the proteins in the membrane are attached to one another, the points of attachment are fluctuating. In the limit, the lipid in such fluid membranes form as much a continuum as does the protein because there is continual merging of pockets of lipid as the bonds between proteins fluctuate.

Geometry of Membranes

An invariant geometric principle is that the surface of all biological membranes is continuous. That is to say there are never open ends in undamaged membranes. The geometric form of membranes can be like that of a hollow sphere (plasma and nuclear membrane) or that of a hollow tube (microvilli of the intestinal wall). The space enclosed by such membranes is filled with an aqueous medium. Tubular membranes may be flattened or rounded or alternately constricted and expanded. Occasionally membranes are arranged in concentric layers, wrapped around a central core (myelin sheath).

Electron Microscopy of Fixed Membranes

The electron microscope has been a powerful tool for demonstrating that membranes generally have a universal structural pattern. The thickness, shape and organization of membranes are easily studied by standard techniques, using positive staining (e.g., osmium tetroxide), negative staining (e.g., phosphotungstic acid), or freeze etching. At high resolution, membranes fixed with glutaraldehyde, stained with osmium tetroxide and sectioned with a diamond knife, show a bilayer of spherical protein molecules (30-50 Å in diameter) with spaces or cavities between the proteins which can be filled with bilayers of phospholipid (see Figure 2). According to the electron microscopic data, the proteins which make up the membrane continuum should be spherical and relatively small, i.e., they should have molecular weights in the range between 12,000 and 50,000. Evidence is multiplying that the isolated and fully depolymerized proteins of membranes indeed fall within that range of molecular size.

Experimental Approaches to Membrane Structure

Diffraction of X-rays has been used for certain membrane systems (viz., the myelin sheath and the retinal rod) where many layers of identical membranes are stacked in an orderly manner. The X-ray diffraction from these systems can be analyzed to give the electron density profile across one of the membranes. The result obtained is a pair of high electron density regions separated by a low density central region (1,4). The high density regions correspond to the positions of the lipid head groups and the centers of the proteins, while the low density region is the center of the membrane, which is made up mostly of lipid tails, since the spheroidal proteins only touch tangentially in that area.

Conventional biochemical techniques have been used to obtain the chemical composition of membranes. The detailed lipid analyses and the percentages of lipid versus protein for a large number of membranes are known (5,6). The proteins have proved to be more intractable than the lipids, however; it is not yet possible to give a detailed list of the protein species found in any membrane except possibly for some very simple, single- protein microbial systems. The main reason for this difficulty is that membrane proteins tend to aggregate to an insoluble mass when the membrane is disrupted and the lipid removed. This is undoubtedly not an unsurmountable problem, but it has held up the careful study of membrane proteins for many years.

The application of several other physical measurement methods to membrane systems, and the information gained from them, have been reviewed by Steim (7). These include optical rotatory dispersion, circular dichroism, infrared spectroscopy, nuclear magnetic resonance spectroscopy, and scanning calorimetry. Electron spin resonance studies have also been employed in some instances. Each of these techniques has contributed in its own special way to the better understanding of membranes.

Role of the Lipids in Membranes

The polar lipids, being amphiphilic or bimodal molecules, naturally form a bilayer by themselves when placed in water under the proper conditions, since in this arrangement polar head groups line the two bilayer-water interfaces, and the nonpolar tails are out of contact with the water. In a lipoprotein membrane, the same forces stabilize the two-dimensional continuum. The membrane proteins will enhance the stability of the two-dimensional continuum if their surfaces are partially polar (charged) and partially nonpolar (uncharged) in an asymmetric manner. The part of the protein bearing the majority of the charged groups will naturally be exposed to the water, while the less polar part of the protein will be inside the membrane, where it can interact favorably with the lipid tails in a nonaqueous environment. Every membrane surface is, therefore, a boundary between an aqueous phase and a nonaqueous phase; the proteins lie in this surface, with part of the protein being in each phase. This serves to orient the proteins in a constant manner relative to the plane of the membrane, and presumably also makes possible many types of reactions which could not take place in either a pure aqueous or nonaqueous phase.

Intermolecular Forces

The stability of membranes depends upon noncovalent intermolecular forces among the lipid and protein components. The types of forces of interest are hydrophobic, nonbonded (i.e., dispersion forces), and electrostatic (ionic and dipole-dipole)(8). The

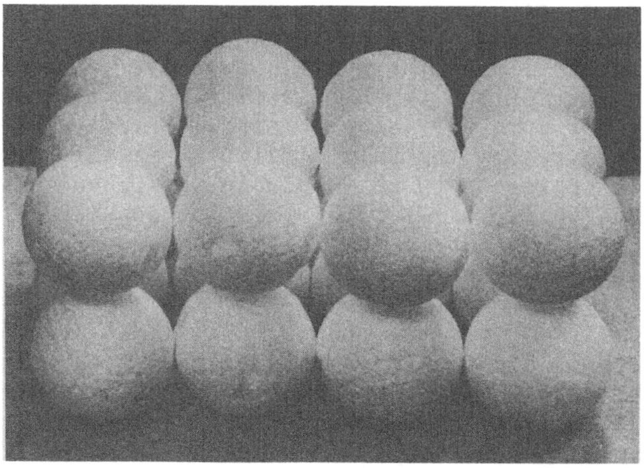

Figure 2. Photograph of a model showing the double layer arrangement of proteins in membranes. The lipid molecules fill the pores between the proteins which are normal to the surface, as shown in Figure 1.

hydrophobic forces arise in consequence of the properties of liquid water; the contact of nonpolar groups with water tends to be minimized, while the contact of water with polar groups tends to be maximized by the hydrophobic forces. This effect immediately accounts for the arrangement of lipids in membranes, the nonpolar tails being in the interior, out of contact with water, while the polar heads are on the surface. Since these hydrophobic forces depend upon the presence of water, removal of a membrane from an aqueous medium will abolish this primary determinant of structure. One would not expect the two-dimensional continuum to be stable in a nonaqueous solvent. Likewise, the addition of reagents to an aqueous medium which affect water structure would also be expected to perturb the properties of a membrane. These expectations are in agreement with observation (9).

The hydrophobic forces tend to keep the nonpolar parts of the lipids and proteins out of water and in the interior of the membrane, but they do not serve to organize the molecules in any special way within the nonaqueous phase. The short range nonbonded and longer range electrostatic forces do serve this function, however. The lipid tails are attracted to one another by the nonbonded forces; this attraction is opposed by the entropy of motion of the tails, the result being that at some temperature there is a thermal phase transition from a frozen tail state to a liquid tail state. In the frozen state, the tails line up so as to minimize the nonbonded energy, while in the liquid state they are free to move, maximizing the entropy. Steim (7) and others have studied this phase transition in membranes by calorimetry. In the liquid phase, the nonbonded energy of attraction among the lipid tails, and between the lipid tails and the proteins, is probably weak, on account of the severe distance dependence (inverse sixth power) of this type of energy. Nonbonded interactions will be present between the larger, and therefore more sluggish, proteins, however.

Electrostatic forces which act through the water surface of the membrane will be diminished by the shielding effect of the water and dissolved ions, but interactions among polar groups on the parts of the proteins which are in the interior of the membrane will not be susceptible to this shielding effect, and consequently such polar interactions as are operative in this nonpolar environment will be strong. Specific interactions of this type may serve to hold the proteins of membranes enzyme complexes together.

According to the protein crystal model of the membrane which we have presented, the link of protein and phospholipid is indeed hydrophobic but not all the phospholipid is committed to interaction with protein. If we may think of the lipid in terms of cores of phospholipid bilayers fitting into the interprotein cavities, then only the phospholipid molecules at the periphery of the core can interact with protein. Although limited in extent, this hydrophobic interaction of phospholipid to protein is nonetheless crucial in determining the stability of the bilayer within the interprotein cavity. Detergents which weaken the protein-lipid interactions can effectively displace phospholipid from the membrane.

Membranes can be depolymerized either into "soluble" complexes or into proteins and micellar lipids by bile acids and detergents. The complex membranes which are built up of repeating units are readily depolymerized into soluble lipoprotein complexes. Only at much higher concentrations of bile acids or detergents does depolymerization extend to the individual protein stage with release of micellar lipid. Membranes can be readily reconstituted from the soluble lipoprotein complexes by removal of the depolymerizing reagents. But this reconstitution of membranes from the ultimate proteins and micellar phospholipid is a process that has not yet been studied experimentally.

II. FUNCTION

Repeating Units of Membranes

As we have mentioned earlier, membranes may be simple or complex in respect to structure. In the simple membranes, there is no repeat structure, i.e., discontinuities. In the complex membranes, the continuum is built up of repeating units, each of which represents a segment of the membrane in which sets of proteins are arranged in precise pattern. Two kinds of repeating units have been recognized the macrotripartite repeating units of the mitochondrion and the chloroplast, and the microtripartite repeating units found in all membranes which carry out active transport (see Figure 3). These repeating units consist of three bonded parts which are separable by chemical means -- basepiece, stalk and headpiece. The basepiece is the sector which makes up the membrane continuum. The headpiece is a projection connected to the basepiece by the cylindrical stalk. The macro and microtripartite repeating units are geometrically indistinguishable except for size. The mass of the macro unit (lipid + protein) is about 800,000 Daltons while that of the micro unit is some 200,000 (10). It is only in the basepiece that lipid is found associated with protein. The headpiece and stalks are entirely protein in nature.

The precise function of each of the three sectors of the mitochondrial tripartite repeating unit has been established. The electron transfer chain is localized in the basepiece and the ATP-synthesizing or hydrolyzing system in the headpiece. The stalk is the physical and functional link between these two energy-coupled systems.

The repeating units are the molecular instruments for the energy-transducing functions of their respective membranes. Thus in the mitochondrion the macrotripartite repeating units couple electron transfer to synthesis of ATP; in the sarcoplasmic reticulum the microtripartite repeating units couple ATP hydrolysis to active transport of Ca^{++}.

Conformational Model

In the mitochondrion the coupling of electron transfer to synthesis of ATP by union of ADP and Pi (inorganic phosphate) appears to be achieved by a conformational mechanism. Thus electron transfer induces a conformational transition which leads to an excited state of the tripartite repeating unit. In this excited state ADP and Pi interact isoenergetically to form ATP and concomitantly the repeating unit relaxes to its nonenergized or ground state conformation.

The Membrane Potential

By virtue of the asymmetry of the membrane surface which the repeat structure in the membrane imposes, if for no other reason, there is likely to be an asymmetry of charge across the membrane. This charge asymmetry will lead to a membrane potential. The conformational cycle has as one of its sequellae the generation of even larger charge asymmetries and hence an increased potential.

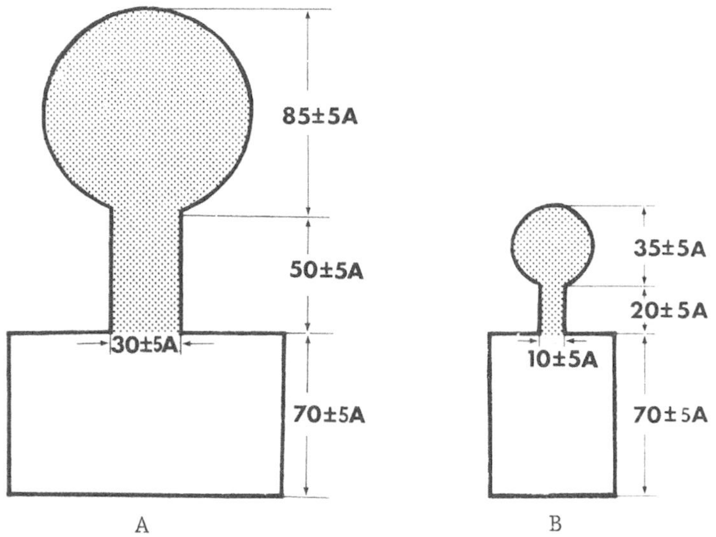

Figure 3. Diagrammatic representation of the macro and micro-tripartite repeating units.

There are in fact multiple devices in membrane systems for inducing a membrane potential. The asymmetric binding of charged ions such as Ca^{++} or Mg^{++} or phosphate to the membrane surface is one such device. The membrane potential plays a primary role in regulating ion and water movements through the membrane and is undoubtedly crucial in the mechanism of active transport as well as in the generation of electrical currents in nerve membranes.

Permeability of Membranes

Solute molecules can move across membranes only through the cavities or spaces between the protein molecules in the membrane continuum. Passage through the protein molecules themselves is completely excluded. When the cavities are filled with bilayers of phospholipid as is predominantly the case, the problem of permeability reduces to either the solubility of solutes in the lipid phase or the ability of small molecules or ions to maneuver within the fluctuating water channels that can provide an alternative passage through the lipid phase. Very few solute ions or molecules are small enough to enter by way of the transitory water channels. Water, H^+, glycol and glycerol are among the few ions or molecules that can apparently traverse membranes through aqueous channels in the lipid phase. All the rest must traverse the membrane by alternative means. Molecules which are nonpolar can dissolve in the lipid phase of the membrane. These molecules in fact do not necessarily traverse the membrane. Given high solubility these will accumulate in the lipid phase. Some membranes have water-filled cavities and such cavities provide a royal highway through the membrane for polar solute molecules which are too large to be accommodated within the aqueous channels. There are two other known devices for insuring the passage of solute ions and molecules through the membrane -- ionophores and permeases. Ionophores are cage-like molecules in which monovalent ions like Na^+ or K^+ can be hydrogen bonded. These ionophores which are lipid soluble even when bonded to the cations can ferry monovalent ions through the membrane with great effectiveness. Permeases are proteins which are intrinsic to the membrane and which can bind specifically to some ion such as sulfate or to a molecule such as glucose. How the complex of permease and its binding ion or molecule actually traverses the membrane is still an unsolved problem.

Molecules which can dissociate, such as phosphoric acid, present yet another problem of permeability. In the fully protonated forms these molecules may be quite soluble in the lipid phase, but not in the ionized forms. Given the possibility of complete protonation, a considerable number of polyacids such as phosphate and sulfate can readily traverse a biological membrane. Anions can also serve as coions for neutral ionophores which require an anion in addition to the cation which is specifically bound.

Active Transport and Conformational Change in Repeating Units

A large proportion of biological membranes have the capacity to carry out ATP-energized concentration of ions within the aqueous medium enclosed by the membrane. There is a net flux of ions from the external to the internal medium and this flux can go on even when the internal concentration of the ion in question is many times higher than the external concentration. Some insight into the mechanism of active transport of monovalent ions has been gained in the study of the translocation of K^+ by the inner membrane of the mitochondrion. There appear to be two critical events in the translocation -- first the binding of weak acid anions to appropriate sites in the membrane, and second the conformationally-dependent transition of the domain of the binding sites from a hydrophobic to an aqueous environment. The cations which serve as coions of the membrane-bonded weak acid anions are concentrated together with the anions. The anions can move directly into the membrane in their fully protonated form whereas the cations are brought in by means of appropriate ionophores.

Binding of Ions to Membranes

As one would expect, a consequence of the polyelectrolyte nature of membranes is that ions are electrostatically bound to the surface to give local charge neutrality. This includes both nonspecific binding of counterions, and also specific binding of Ca^{++} and Mg^{++} to sites on the lipids and proteins. In the case of the mitochondrion, there also appear to be specific binding sites for K^{+}, as well as for phosphate ions and the organic anions involved in mitochondrial function. There is some evidence which would indicate that the specific site binding in the membrane of K^{+} or Na^{+} plays an important role in the energy coupling process.

Summary and Conclusion

In most general terms, a biological membrane is an extensive two-dimensional structure of relatively small molecules which are held together by noncovalent bonds, and is stable in an aqueous environment. The interior of the membrane is mainly nonpolar, while the surface is mainly polar. This effect reflects the central role of amphiphilic molecules in the lipid of the membrane. The proteins, which are considerably larger than the lipid molecules, must also be partially nonpolar, in order to interact favorably both with the lipid and with the water. Hydrophobic and nonbonded forces appear to be responsible for maintaining the integrity of the membrane as a two-dimensional continuum, but Coulombic and dipole-dipole forces probably play an important role in the arrangement of the molecules within the membrane.

Given the fundamental properties of biological membranes as outlined above, it ought to be possible to design and build synthetic membranes using the same principles, but with materials which are more stable or easier to make than are the natural lipids and proteins. The possible practical uses for such synthetic membranes have not been looked into, but considering the great many uses to which membranes are put in living tissue, one would suppose that a large number of uses could be found for tailor-made membranes as well.

ACKNOWLEDGEMENT

This work was supported in part by project grant number GM-12,847 from the National Institute of General Medical Sciences (USPHS).

REFERENCES

1. Vanderkooi, G. and Sundaralingam, M., Proc. Nat. Acad. Sci. (U.S.) 67, September (1970).

2. Green, D.E. and Perdue, J.F., Proc. Nat. Acad. Sci. (U.S.) 55, 1295 (1966).

3. Vanderkooi, G. and Green, D.E., Proc. Nat. Acad. Sci. (U.S.) 66, 615 (1970).

4. Blaurock, A.E. and Wilkins, M.H.F., Nature 223, 906 (1969).

5. Korn, E.D., Science 153, 1491 (1966); Federation Proc. 28, 6 (1969).

6. Poincelot, R.P. and Abrahamson, E.W., Biochim. Biophys. Acta. 202, 382 (1970).

7. Steim, J.M., Advances in Chemistry Series, No. 84, "Molecular Association in Biological and Related Systems," 259 (1968).

8. Scheraga, H.A., The Harvey Lectures, Series 63, Academic Press, New York, 99 (1969).

9. Hatefi, Y. and Hanstein, W.G., Proc. Nat. Acad. Sci. (U.S.) 62, 1129 (1969).

10. Green, D.E. and Baum, H., Energy and the Mitochondrion, Academic Press, New York (1970).

THE EFFECT OF AN OXIDATIVE-CAUSTIC ENVIRONMENT ON GRAFT COPOLYMER MEMBRANES*

Edward F. Cuddihy, Jovan Moacanin, Donald E. Walmsley,

and Ho Yet Tom

Jet Propulsion Laboratory, Pasadena, Calif.

This paper discusses the accumulated results of a continuing study on the long-term effect of the environment in a Ag-Zn battery on a separator material prepared from thin films of crosslinked polyethylene grafted with poly (potassium acrylate), the acrylate content being usually approximately 40 wt %. These materials have been prepared for JPL by Southwest Research Institute, San Antonio, Texas (Ref. 1). This type of separator membrane is considered for spacecraft applications because of its relative stability and the ability to withstand heat sterilization at 135°C.

Separator membranes are required in Ag-Zn batteries to prevent silver ions, dissolved from the silver electrode, from reaching and poisoning the zinc electrode. In addition, the separator membranes are used to prevent physical contact between the electrodes, and must also have the property of allowing ready transport of the charge carrier ions. This current class of graft copolymer membrane satisfies these basic requirements.

However, silver dissolved (30 mg Ag^+/liter) in the aqueous 40 wt % KOH battery electrolyte results in the formation of a strong oxidative-caustic solution. As a result, the separator membrane performs its function in a caustic oxidative environment (Ref. 2), where it was observed that in use these membranes become black

---

*This paper presents the results of one phase of research carried out at the Jet Propulsion Laboratory, California Institute of Technology, under Contract No. NAS 7-100, sponsored by the National Aeronautics and Space Administration.

due to the accumulation of deposited silver products. This observation initiated a study of the cause of the accumulation as well as a general study of the interaction between the membrane and the battery environment.

ENVIRONMENT EFFECTS

To determine the effect of the environment, the membrane materials were first characterized in terms of their chemical and physical properties (Ref. 3), and then re-analyzed after exposure. The results of this effort demonstrated that exposure of the separator to this strongly oxidative-caustic environment resulted in a loss of poly (potassium acrylate) PKA from the films, along with a corresponding accumulation of deposits of metallic silver and silver oxide. Total PKA losses were observed to occur by two mechanisms, one a portion extractable by KOH and the remainder from oxidative degrafting which also resulted in the accumulation of the reduced metallic silver. Silver oxide deposition is not completely explained, but appears related to the level of unsaturation in polyethylene.

The first portion of this paper presents the result of a chemical study investigating the PKA losses and silver products accumulation, while the second portion consists of the result of a morphology study employing scanning electron microscopy, transmission electron microscopy, and an electron microprobe. The latter technique was successfully employed in monitoring the PKA concentration gradients which are established across the membranes as a result of exposure to the battery environment.

CHEMICAL STUDY

Experiment

Samples of battery separators were immersed at 95, 80, 65, and 50°C in 45% KOH saturated with Ag_2O, and the concentration in the film of silver, silver oxide, and acrylate were monitored as a function of time. For comparison, samples were also exposed to KOH solution not containing Ag_2O. The acrylate contents were determined by an analytical method described previously (Ref. 3). Silver and silver oxide were analyzed by a potassium cyanide (KCN) method, which relies on the solubility of silver oxide in KCN solution. Metallic silver must first be oxidized to a silver ion which is then soluble in KCN solution (Ref. 4). Thus, a sample of a battery separator membrane is exposed to KCN under conditions which prevent oxidation of metallic silver. Silver oxide is extracted and the

Table 1. Loss of Potassium Acrylate from Grafted Membranes Exposed at 95°C to a 45% KOH Solution

Unsterilized		Sterilized[a]	
Time, hr	% Loss of Initial	Time, hr	% Loss of Initial
21	3	272	2
162	12	679	5
812	17	1296	7
1796	20		
2695	22		
3712	24		

[a]Sterilized by exposure to 45% KOH solution at 135°C for 120 hours which resulted in a loss of 19 wt % of initial potassium acrylate. Values shown are subsequent losses from the continued exposure at the lower temperature of 95°C. Actual total losses from the combined exposures are the values shown plus 19%.

loss of weight of the membrane sample gives the silver oxide content. The resultant sample is then further exposed to KCN under oxidizing condition. The additional weight loss gives the amount of metallic silver. Because of the number of steps in weighings in these analyses, the cumulative relative errors are rather high.

Results and Discussion

GX-152 Membranes. Table 1 and Fig. 1 give the results for an unsterilized grafted crosslinked membrane (GX-152) exposed at 95°C to a 45% KOH solution containing no silver oxide. After 3712 hr, this exposure results in a loss of approximately 24% of potassium acrylate from the membranes. For sterilized GX-152 samples, equivalent losses are also observed. Sterilization for 120 hr at 135°C in 45% KOH solutions results in a loss of 19 wt % PKA, and after an additional exposure of 1296 hr at 95°C, another 7 wt % PKA is

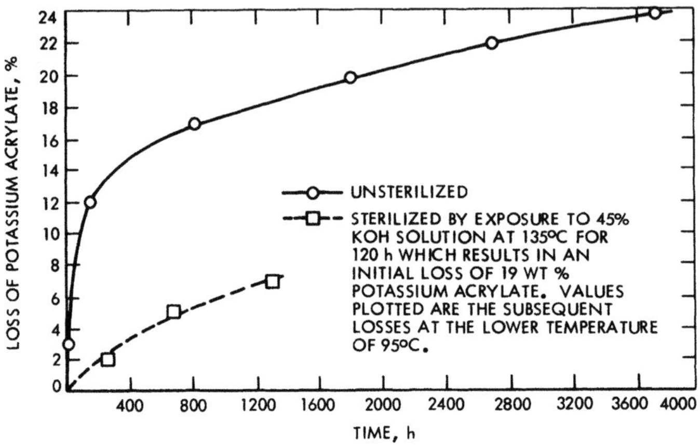

Fig. 1. Loss of Potassium Acrylate from GX-152 Membranes Exposed at 95°C to a 45% KOH Solution

removed from the membrane for a total of 26 wt % PKA lost by KOH extraction. Since the extraction curves given in Fig. 1 appear to be approaching an asymptotic limit, this suggests that the KOH-extractable PKA may be residual homopolymer rather than the result of being chemically degraded by KOH. There is no evidence that KOH alone can degrade the grafted poly (potassium acrylate).

PKA losses from sterilized and unsterilized separator samples exposed at 95°C to silver oxide saturated 45% KOH solutions are given in Table 2, with Fig. 2 being a plot of the PKA losses from the sterilized samples. The PKA losses given for the sterilized samples are percentages based on the PKA remaining after sterilization. The intent is that, since sterilization removes the bulk of the KOH-extractable PKA, these percentages should reflect only the PKA loss by the oxidation reaction with the silver ion. The semi-log plot in Fig. 2 suggests the possibility of a first-order kinetic oxidation reaction between silver and PKA, with the curvature most likely attributable to some diffusion control of the silver ion into the core of the films. The latter point will be discussed more fully in the section on morphology.

For the unsterilized samples, the initial rates of PKA loss (up to 316 hr) are higher above the sterilized samples by an amount corresponding to the additional KOH-extractable PKA present in the unsterilized samples. Using the information given in Table 1 to account for the presence of the latter, losses caused by oxidative

Table 2. Results of Exposure of Grafted Membranes to Silver Oxide Saturated 45% KOH Solution at 95°C

Item	Sterilized Membranes							Unsterilized Membranes				
Exposure Time, hr	20	57	69	316	800	1728	3150	69	316	336	1460	3476
Potassium Acrylate, Percent Loss of Initial	20	22	15	22	35	52	60	22	33	41	44	46
Total Gain of Deposited Silver Products, Expressed as wt % of Initial Potassium Acrylate	14	16	21	104	57	68	185	27	49	50	93	142
Percent as Ag^{o}	72	92	66	95	83	92	94	68	92	85	86	91
Percent as Ag^{+}	28	8	34	5	17	8	6	32	8	15	14	9
Equivalence Ratio, Potassium Acrylate Lost Ag^{o} Deposited	0.62	1.5	1.0	0.21	0.61	0.62	0.35	1.2	0.74	1.0	0.53	0.34
gm Ag^{+} Deposited/ gm Polyethylene	0.027	0.021	0.059	0.018	0.037	0.040	0.030	0.059	0.033	0.039	0.044	0.073

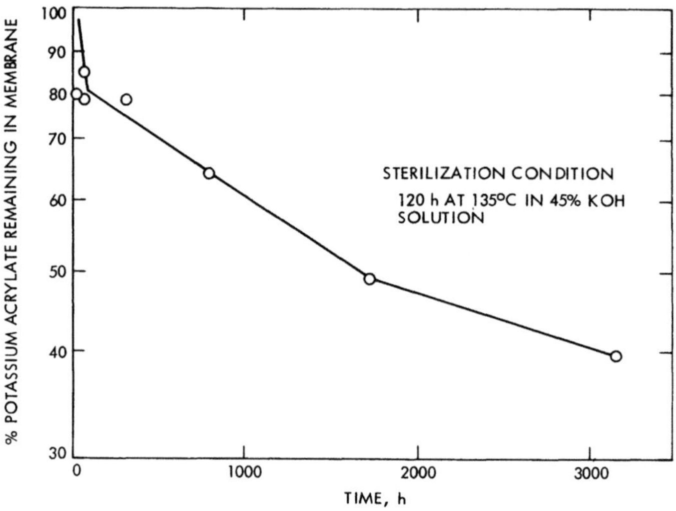

Fig. 2. Loss of Potassium Acrylate from Sterilized Membranes Exposed to a Silver Oxide Saturated 45% KOH Solution at 95°C

degradation of 17% after 69 hr and 22% after 316 hr can be calculated. These calculations are in reasonable agreement with the results for the sterilized sample. Above 316 hr, the rate of PKA loss from the unsterilized sample slows significantly as compared to the sterilized sample, an effect not yet explained.

Information on silver products deposition is also given in Table 2. Silver oxide dissolved in the KOH solution yields a strong oxidizing medium which attacks the membranes, resulting in degradation and loss of acrylate from the membrane. Simultaneously, the silver oxide is reduced to metallic silver and deposited within the membrane. Damage to membrane separators by soluble silver has been previously reported (Ref. 2). Table 2 gives the equivalence ratio of potassium acrylate lost to metallic silver deposited. The equivalent weight ratio of acrylate to Ag, being close to unity, implies that, on the average, one acrylate moiety is lost for each atom of deposited neutral silver.

Data on exposure of unsterilized membranes to silver oxide saturated 45% KOH solution at 80, 65, and 50°C are given in Table 3. The rate of PKA lost decreases as expected with decreasing temperature. An interesting observation is that the percentage of deposits occurring as metallic silver decreases with decreasing temperatures; however, this may just be an artifact resulting from a slower oxidation reaction between silver ion and PKA.

Silver Oxide. It is observed at all temperatures that the quantities of deposited metallic silver increase with increasing time, and in correlation with the decrease in PKA content. Silver oxide, on the other hand, appears to deposit into the membrane very early in the exposure and rapidly achieves a steady state concentration. This phenomenon can be observed in Tables 2 and 3. (Silver oxide deposits are expressed as weight ratio of gm Ag^+ deposited per gm of polyethylene. It has been observed (Ref. 3) that weight of polyethylene is unchanged during exposure.)

An attempt to understand the mechanism for silver oxide deposition was initiated, and discussed below are some results from a study investigating the influence on silver oxide deposition from unsaturation in polyethylene and the chemical type of polyethylene crosslinker.

Polyethylene. Two samples of polyethylene were exposed at 95°C to a silver-oxide saturated 45% KOH solution. They are (1) DOW 400, which is completely saturated, and (2) Petrothane 301, which contained 0.12 mM/gm of unsaturated pendent vinylidene groups.

The saturated DOW 400 accumulated less silver products than the unsaturated Petrothane (Table 4). This indicates that, unsaturated bonds being easily oxidized, they promote accumulation of deposits, particularly metallic silver.

The amounts of silver oxide detected in these films are of the same order of magnitude as those found in the grafted membranes when the results are expressed on a polyethylene basis (Tables 2 and 3). This observation suggests that most of the silver oxide detected in the grafted membranes is trapped in the hydrophobic polyethylene phase.

1, 7-Octadiene Crosslinked Membrane. The GX-152 membranes under discussion were crosslinked with divinylbenzene. The results for a grafted membrane crosslinked with an aliphatic crosslinker, 1, 7-octadiene, are given in Table 5 for exposure at 95°C. It was suspected that the crosslinker divinylbenzene could attract and increase the deposition of silver oxide by introducing aromatic groups into the membrane. However, the results obtained on the membrane were not significantly different from the divinylbenzene-containing membranes. (Compare the silver oxide weight ratios with the base polyethylene, Petrothane 301, as well as the silver oxide weight ratios given in Tables 2 and 3.)

Acrylate losses after 316 hr compare with the GX-152 separator, but the initial rate for the first 69 hr is higher (Tables 2 and 5). Also, there is a noticeable reduction in the amount of deposited

Table 3. Results of Exposure of Unsterilized Grafted Membrane to Silver Oxide Saturated 45% KOH Solution at 80, 65, and 50°C

Item	80°C		65°C		50°C
Exposure Time, hr	263	1224	912	1643	528
Potassium Acrylate, Percent Loss of Initial	25	39	1.4	26	23
Total Gain of Deposited Silver Products, Expressed as wt % of Initial Potassium Acrylate	20	31	24	19	7
Percent as Ag^{o}	69	77	57	50	60
Percent as Ag^{+}	31	23	43	50	40
Equivalence Ratio, Potassium Acrylate Loss/Ag^{o} Deposited	1.8	1.6	0.09	2.8	5.5
gm Ag^{+} Deposited/gm Polyethylene	0.027	0.053	0.066	0.075	0.025

Table 4. Results of Exposure of Polyethylene Films to Silver Oxide Saturated 45% KOH Solution for 1000 hr at 95°C

Polyethylene Type	Dow 400	Petrothane 301
Silver Products, wt % Deposited	2	4
Percent of Total as Ag^{o}	86	73
Percent of Total as Ag^{+}	14	27
Weight Ratio, gm Ag^{+} Deposited/gm Polyethylene	0.0023	0.010

Table 5. Results of Exposure of a 1, 7-Octadiene Cross-Linked Membrane to Silver Oxide Saturated 45% KOH Solution at 95°C

Exposure Time	69 hr	316 hr
Percent Loss of Initial Potassium Acrylate	32	34
Total Gain of Deposited Silver Products, Expressed as wt % of Initial Potassium Acrylate	15	20
Percent of Total as Ag^{o}	60	60
Percent of Total as Ag^{+}	40	40
Equivalence Ratio, Potassium Acrylate Lost/Ag^{o} Deposited	3.5	2.8
Weight Ratio, gm Ag^{+} Deposited/gm Polyethylene	0.033	0.041

silver products as compared to the divinylbenzene-containing GX-152 membranes (Table 2). However, available data are insufficient to relate these observations to the differences in chemical features of the crosslinkers.

MORPHOLOGY

The morphology of the graft copolymer membranes resulting from exposure to the oxidative-caustic environment was investigated employing a series of microscopic techniques. Surface topology was viewed employing scanning electron microscopy (SEM), while views of the cross-section, and the concentration gradients of PKA and silver deposits through the films were studied employing transmission electron microscopy (TEM) and electron microprobe analysis (EMA).

Experiment

For the latter two techniques, film samples were imbedded in an epoxy matrix and thin slices (between 600-900 Å) obtained on an LKB ultramicrotome. The pattern of slicing was carried out in such a way to ensure that the cross-sections of the films were contained within the thin slices. These slices were then positioned on a 300 mesh copper grid and the film cross-section viewed with a Phillips 200 transmission electron microscope.

For microprobe analysis, the samples consisted of the epoxy slugs which remained after the microtoming. These samples had smooth surfaces which resulted from the microtoming and included the exposed cross-section of the graft copolymer film.

To increase the ability of the EM to see the acrylate domains within the membrane films, the grafted poly (potassium acrylate) was converted to poly (cesium acrylate) (Ref. 3). Cesium is more opaque to electrons than potassium (Ref. 5) and thus an enhanced black-white contrast between acrylate and polyethylene would occur.

The EM was employed to achieve an overall view of the cross-section while EMA was employed to obtain the concentration gradients of acrylate. To accomplish this, the electron microprobe (Applied Research Lab. Model EMX) was stepped across the cross-section at 3μ intervals, analyzing for cesium concentration at each point. As the film thicknesses can vary from 25-35μ, between eight to nine point determinations were made to obtain the concentration gradients. Since it has been previously demonstrated (Ref. 3) that 100% conversion of the acrylate to the cesium salt is obtainable, the level of cesium concentration detected at a given point by the microprobe

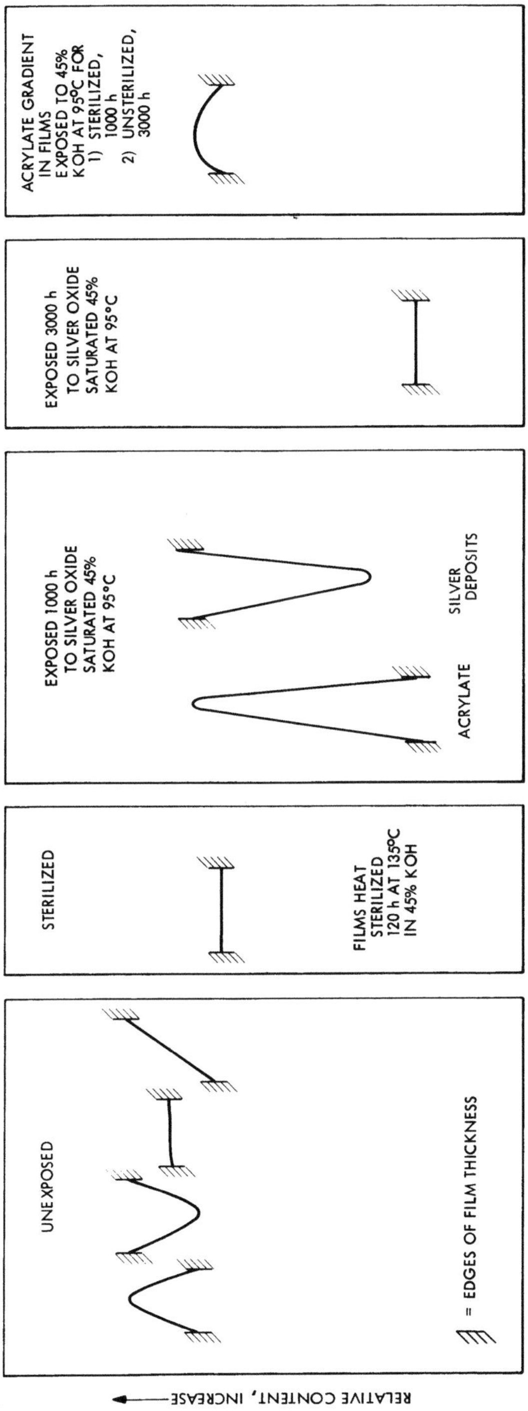

Fig. 3. Relative Acrylate Content Across Graft Copolymer Membranes

was taken to be directly proportional to the acrylate content. Further, no attempt was made to calibrate the microprobe response counts; therefore, the results show only the relative acrylate distribution across the thickness of a given sample. Inasmuch as the same instrument settings were used for all analyses, qualitative comparisons between samples could be made.

Acrylate Gradients

It was previously shown (Ref. 3) that the concentration of grafted PKA in the films was extremely non-uniform, with values as low as 8 and as high as 50 wt % PKA found within the same film samples. This same non-uniformity also applied to the acrylate concentration across the thickness of the starting film samples. Figure 3a shows the various PKA concentration gradients observed across different sections of grafted copolymer films. These sections range from parabolic-up and parabolic-down to linear and graded. No characteristic could be attributed to the starting films. However, after heat sterilization (120 hr at 135°C in 45% KOH), which results in an average loss of 19 wt % PKA, the concentration generally became uniform as shown in Fig. 3b. Since it is suspected that heat sterilization removes excess PKA homopolymer trapped in the membrane, perhaps the non-uniformity of the PKA concentration in the starting materials results from the random dispersion of PKA homopolymer.

The PKA gradient established after ~1000 hr at 95°C in silver oxide saturated 45% KOH is shown in Fig. 3c. In all cases studied, a dominant parabolic-up gradient was observed obviously resulting from the preferential attack of silver on the PKA closer to the surface. The silver deposit gradient (Fig. 3c) is just the reverse, parabolic-down, indicating, as expected, that the heavier deposits of reduced silver occurred at the sites where the most PKA was removed.

After 1000 hr of exposure, the silver begins to attack the PKA in the core of the film, thus reducing the PKA concentration to a near-linear distribution which is observed after ~3000 hr (Fig. 3d). This process requires diffusion of the silver ion to the center of the film, and thus diffusion control of the oxidation kinetics should ensue. This shows up in the curvature which occurs in the PKA loss curves (Fig. 2) above 1000 hr.

Finally, PKA gradients for samples exposed to 45% KOH only at 95°C are shown in Fig. 3e. In all cases, a definite parabolic-up distribution was observed; however, the time to achieve it varied depending on whether the samples were sterilized or unsterilized prior to exposure.

For sterilized specimens, the starting concentration was essentially uniform (Fig. 3b) and the parabolic-up distribution was achieved after ~1000 hr. This demonstrates that the initial PKA extraction occurred preferentially from the surfaces, to be followed then by a slower, diffusion controlled removal of PKA from the core.

For the unsterilized samples, about 2000 to 3000 hr at 95°C are required to achieve a definite parabolic-up distribution. Gradient curves taken for times less than 2000 to 3000 hr were erratic, presumably because large amounts of randomly dispersed homopolymer were still present in the film. Once this is essentially removed, the dominant parabolic-up distribution emerges. Thereafter, the continued extraction of PKA proceeds as previously discussed for the sterilized case.

A point not resolved in this study is whether KOH is degrading the grafted PKA, or whether the observed PKA losses results from diffusion-controlled extraction of homopolymer.

Electron Microscopy

Electron microscopic examinations of the cross-sections did not reveal any evidence of specific acrylate domains within the polyethylene matrix. Inspection at magnifications up to 13,500 X were carried out on microtomed slices which were 600-900 Å thick. Figures 4a and 4b are photograph cross-sections of starting film, and are characterized by a dark uniform appearance caused by the opaqueness to electrons of the cesium ion. This appearance was typical of all the cross-sections which were viewed. Cross-sections of films containing only poly (acrylic acid) were light and clear. This confirmed that the presence of cesium in the film provided the dark contrast.

In only one case where there was a known extreme gradient variation in acrylate content did the EM also observe a variation in the intensity of the dark appearance. After 1000 hr at 95°C in silver oxide saturated 45% KOH, the microprobe revealed a dominant parabolic-up PKA gradient (Fig. 3c) which resulted from extensive surface degradation of PKA. Figure 5 shows a PKA rich dark core which falls off on either side to a light clear appearance as the PKA-depleted edges of the film are approached.

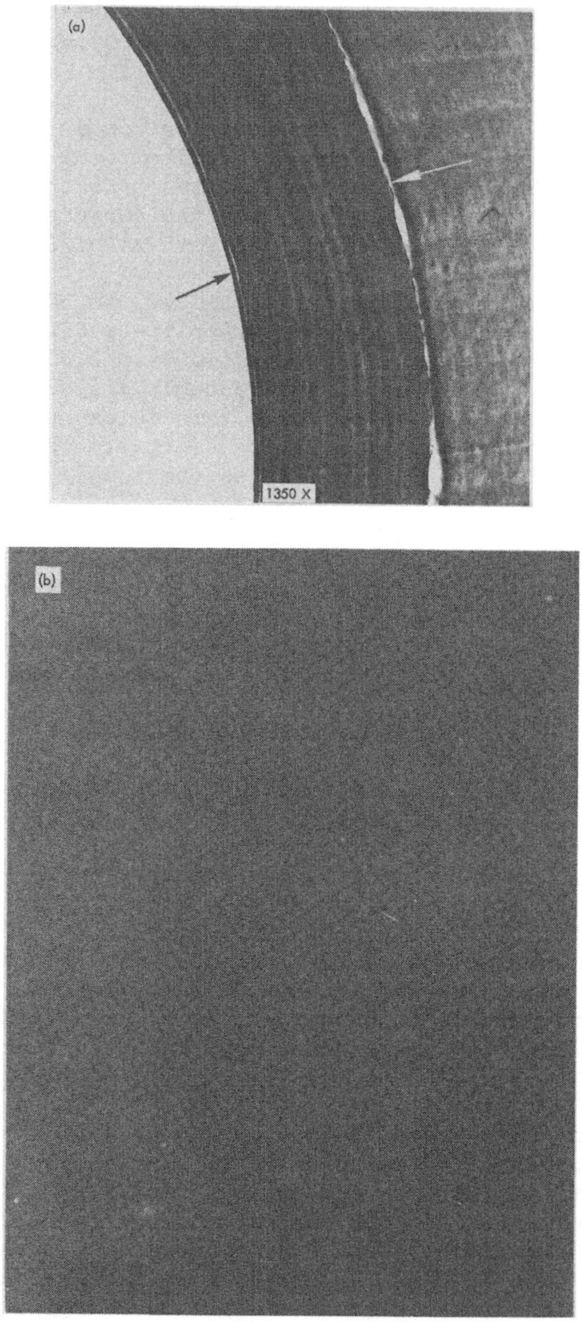

Fig. 4. Electron Microphotograph of an Unexposed Graft Copolymer Membrane (Arrows Point Out Edges of Film Sample)

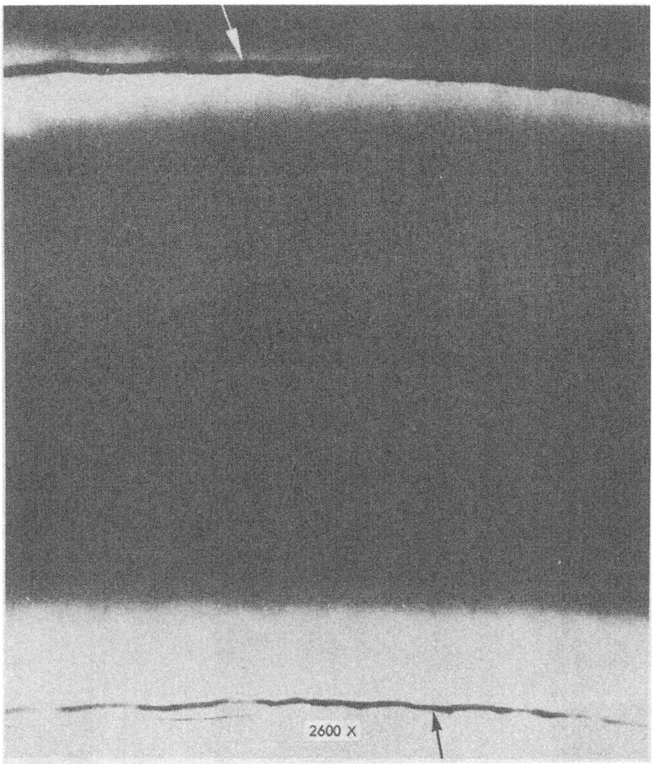

Fig. 5. Electron Microphotograph of Graft Copolymer Membrane after 1000 hr at 95°C in Silver Oxide Saturated 45% KOH. Note Reduction in Dark Field at Edges Resulting from Surface Removal of Acrylate by Silver Oxidation. Compare with Acrylate Gradient Shown in Fig. 3 (c). (Arrows Point Out Edges of Film Sample.)

Scanning Electron Microscopy

The surfaces of the films were viewed with a Cambridge Stereoscanning Electron microscope. Figure 6a shows the surfaces of the ungrafted polyethylene at two levels of magnification. The surfaces were observed to be uniformly smooth. After grafting, some surfaces of the films were observed to have bubbles (Fig. 6b); however, this effect was random. Other areas of grafted film (also Fig. 6b) were found to be as smooth as the starting polyethylene.

After heat sterilization, however, all surfaces were covered with the bubbles and with a layer which apparently consists of fused fragments of bubbles (Fig. 6c). Figure 6d, 6e, and 6f are

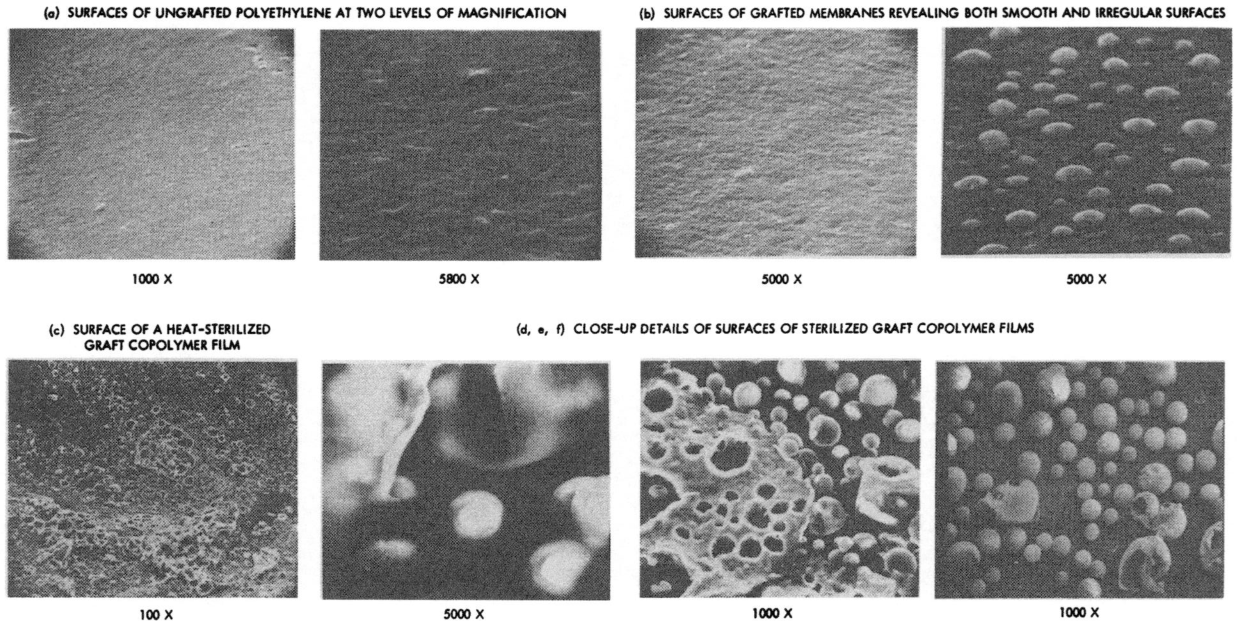

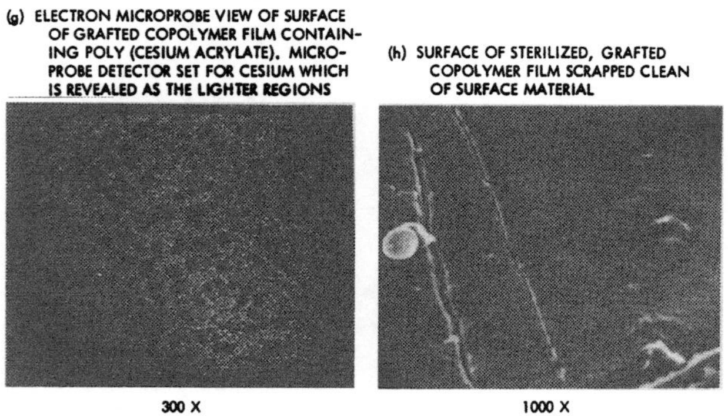

Fig. 6. Scanning Electron Microphotograph of the Surfaces of the Graft Copolymer Membranes

close-up views of the surfaces of heat-sterilized grafted film. Figure 6f is particularly interesting in revealing that the surface bubbles are hollow.

Figure 6g shows the result of an Electron Microprobe view of the surface of a sterilized, grafted film which contained poly (cesium acrylate), with the detector set to detect cesium (which shows up light in the photograph). The result reveals that the bubbles did not contain cesium and, therefore, demonstrates that the bubbles are not grafted acrylate. Further, it was found that the surface matter could be easily scraped off with the edge of a razor blade (Fig. 6h). Sufficient matter was removed and an IR analysis revealed the material to be polyethylene.

The process by which the grafted films form bubbles of polyethylene on the surface is not understood. Perhaps, in the case of unsterilized film (Fig. 6b), the appearance of bubbles is related in some way to the local concentration of grafted acrylate. Perhaps while sterilized films are heated above 110°C the polyethylene melting point, an internal equilibrium is achieved between the relative concentration of PKA and polyethylene in such a way that an excess of polyethylene exists and then is forced out onto the surface in the form of bubbles. This phenomenon warrants additional study.

References

1. Adams, L. M., Harlowe, Jr., W. W., and Lawrason, G. C., Fabrication and Testing of Battery Separator Material from Modified Polyethylene, Final Report for Project 01-1842, JPL Purchase Order CA-384805, Southwest Research Institute, San Antonio, Tex., June 7, 1966.

2. Lander, T. T., Degradation by Soluble Silver. Chapter 10: Characteristic of Separators for Alkaline Silver Oxide Zinc Secondary Batteries, 773-10-216 publication No. 300. Edited by J. E. Cooper and Arthur Fleisher, Air Force Aero Propulsion Laboratory, Wright-Patterson AFB, Dayton, Ohio, Sept. 1965.

3. Cuddihy, E. F., Moacanin, J., and Walmsley, D. E., "Chemical and Physical Analysis of Grafted Membranes for Sterilizable Silver-Zinc Batteries," Proc. of the Symposium on Battery Separation, Columbus Section of the Electrochemical Society, p. 316, Feb. 1970.

4. Latimer, W., and Hildebrand, J., Reference Book of Inorganic Chemistry, Third Edition, The Macmillan Company, New York, 1951.

5. Hall, C. E., Introduction to Electron Microscopy, Second Edition, McGraw-Hill Book Co., New York, 1966.

MICROPHASE SEPARATION IN ORGANIC POLYMERS CONTAINING IONS

W. J. MacKnight

Chemistry Department, University of Massachusetts

Amherst, Massachusetts 01002

INTRODUCTION

In recent years there has been a considerable amount of interest in a new class of polymeric materials consisting of copolymers of olefins and carboxylic acids, the hydrocarbon being the major constituent. It has been found that the properties of these materials undergo profound alterations when the carboxylic acid groups are neutralized to form various salts. In particular, the optical clarity, tensile strength, impact resistance and melt viscosity greatly increase upon neutralization. Studies carried out on such materials in these laboratories[1,2] as well as others[3] indicate that the phenomenon of microphase separation occurs with the dispersed phase consisting of ionic aggregates or domains, probably less than 100A in size. A related system, consisting of polyethylene modified by the inclusion of phosphonic acid side groups has been recently studied[4,5] and apparently exhibits microphase separation at high phosphonic acid contents even without ionization. In this case the dispersed phase consists of hydrogen bonded aggregates of phosphonic acid groups.

All of the above copolymers contain a random distribution of functional groups and it is of interest that microphase separation occurs inasmuch as it is usually associated with an organized chain structure such as that of a block copolymer. In this paper it is intended to review the experimental evidence for microphase separation in organic polymers containing ions as well as theoretical attempts that have been made to explain the occurrence of this phenomenon.

Discussion will be restricted to random copolymers of ethylene and carboxylic and phosphonic acids. In all cases the acid side groups are present to less than 10 mole per cent. This restriction is not intended to imply that other hydrocarbon polymers containing polar side groups do not undergo microphase separation.

It is the contention of the author that microphase separation occurs in any material in which polar groups are present in a medium of sufficiently low dielectric constant. The phenomenon must obviously depend on the concentration of the polar groups in the medium. Thus, at extremely low concentrations the principle species present are presumably isolated groups and hydrogen bonded dimers in the case of substituents such as carboxylic acids and isolated ion pairs and dimers consisting of two ion pairs in the case of salt groups. At intermediate concentrations, higher aggregates start to form and finally, when the limiting cases of poly acids and polyelectrolytes are achieved, the structure is probably akin to the three dimensional networks found in some inorganic glasses. The actual concentration range in which microphase separation occurs must depend on the polarity of the substituent. The exact nature of this relationship has not yet been investigated.

The ethylene copolymers treated here possess an additional complication in that the hydrocarbon matrix may undergo partial crystallization when the polar substituents are present in concentrations less than about 8 mole per cent. The crystals formed appear to consist entirely of polyethylene segments with the polar substituents excluded and are similar in properties to the crystals in ethylene homopolymers. The spherulitic superstructure and the kinetics of the crystallization process undergo considerable modification as a result of the presence of the polar groups.

Space limitations prevent treatment of all systems studied and extensive data are available on the ethylene copolymers. Among others, carboxylic rubbers ionized with various cations have been stated to undergo microphase separation[6] and a recent review contains a more complete exposition of the subject than that presented here.[7]

MORPHOLOGICAL INVESTIGATIONS

X-Ray investigations have been carried out on ethylene-methacrylic acid copolymers and their salts[8,9]. It has been found that the characteristic polyethylene diffraction pattern is discernible up to an acid content of about 8 mole per cent and that neutralization produces relatively little change in the degree of crystallinity in well annealed samples. Thermal history has a large effect which generally increases with increasing salt content. It is the low angle region, here defined as angles below about $2\theta=5^{\circ}$, which yields information concerning structural reorganization as a result of neutralization. Two peaks have been observed in this region. The first, which occurs at an angle corresponding to a spacing of the order of 20A[8], is absent in the acid material and is present in all salts, regardless of cation size or valence. Fig. 1 compares the acid form of a 15 weight per cent ethylene-methacrylic acid copolymer with the sodium salt of the same material. The low angle peak is discernible in the salt form. The angular position of this peak varies somewhat with cation concentration. Further, the peak is present at temperatures up to 300°C and is essentially destroyed by absorbed water. These results lead to the postulate that this peak reflects some ordered structure within the ionic domains. At still lower angles a peak has been observed in a cesium salt of an ethylene copolymer containing 4.1 mole per cent methacrylic acid which corresponds to a spacing of about 80A[9]. Fig. 2 illustrates this behavior for a 4.1 mole per cent ethylene-methacrylic acid copolymer and its 59% cesium salt. This peak is sensitive to the nature of the cation, being entirely absent with light metal salts such as lithium. No information is available regarding the effect of water or temperature on this peak. In view of the dependence of the 80A peak on the scattering power of the metal, it is probable that this peak arises from electron dense regions caused by ionic aggregations. The available x-ray evidence thus leads to a three phase structural model consisting of a polyethylene crystal phase, and a polyethylene amorphous phase throughout which ionic clusters are dispersed.

Electron microscopic investigations have also been carried out on ethylene-methacrylic acid copolymers and their salts[10,11] with somewhat inconclusive results. It appears that the lamellar structure characteristic of polyethylene and present in the acid material is largely destroyed on ionization being replaced by a grainy structure which has been said[10] to indicate phase separation. Caution must be taken in interpreting these results inasmuch as it has been shown that the presence or absence of the grainy structure in the ionized materials depends on the method of sample preparation[11].

PHYSICAL PROPERTY STUDIES

The concept of ionic aggregate formation and the occurrence of a microphase separation in organic polymers containing ions provides a framework within which the mechanical, dielectric and optical properties as well as the melt rheology of these materials may be understood. In general, four relaxation regions have been discerned in bulk ethylene-carboxylic acid copolymers and their salts in the temperature region from -180° to 100°C. These are labeled γ, β, β' and α in order of increasing temperature. The dispersions which

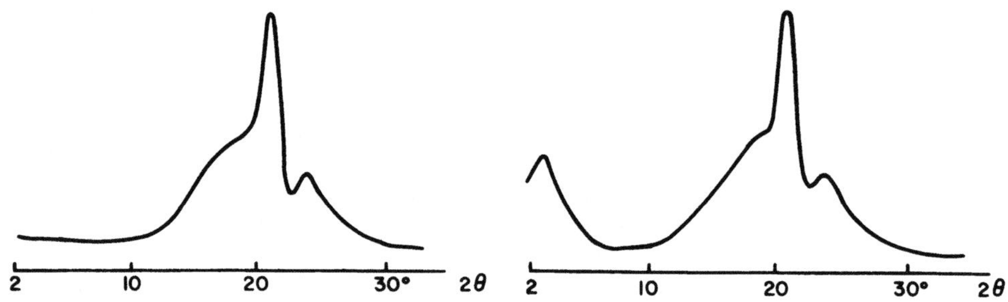

1. X-Ray diffraction patterns from a 15 weight % ethylene-methacrylic acid copolymer and its sodium salt (90% ionization). [After Wilson, Longworth and Vaughan. (8)].

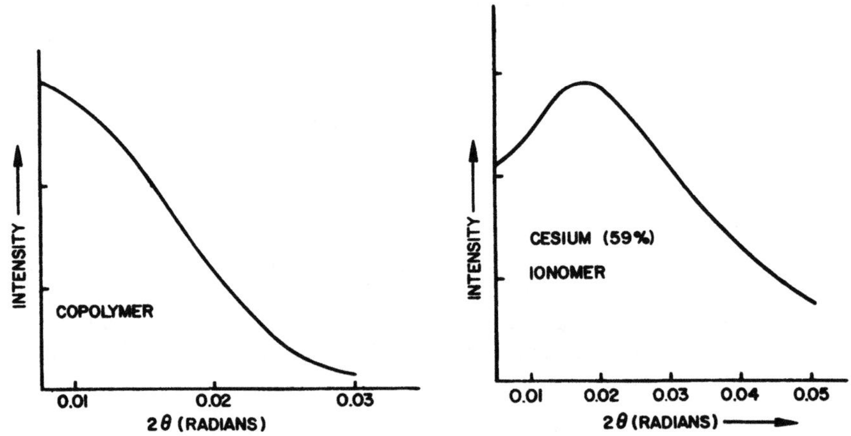

2. Low angle x-ray scattering from an ethylene-methacrylic acid copolymer (4.1 mole per cent acid) and its cesium salt (59% ionization). [After Delf and MacKnight (9)].

particularly bear upon the question of microphase separation are the β, β' and α peaks. In the acid material containing 4.1 mole per cent carboxylic acid groups the β and α relaxations are absent and the β' relaxation occurs in the vicinity of room temperature at frequencies of 100 cycles per second. This relaxation has been studied as a function of acid content[12] and its behavior has been shown to be consistent with a mechanism involving the main glass transition. The temperature dependence of the β' relaxation on acid content appears to be due to the presence of hydrogen bonded species acting as crosslinks[12]. Upon complete neutralization the β' relaxation disappears, being replaced by two new relaxations, the β, occurring at -10°C for the 4.1 mole per cent copolymer and the α, occurring at 50° - 100°C for the 4.1 mole per cent copolymer depending on the nature of the cation. The β relaxation is essentially independent of the cation at comparable levels of neutralization[1] for well annealed samples, both with regard to temperature location and magnitude. Fig. 3 is a plot of the temperature dependence of the loss modulus E" in the β relaxation region[1]. The behavior of the sodium, lithium and calcium salts are all very similar. The α relaxation, on the other hand shows both temperature and magnitude dependence on the concentration and nature of the cations present. Figs. 4 and 5 show the effect of the concentration of salt groups on the dielectric α relaxation. The magnitude of the α dielectric loss, (tan δ), increases from about 5×10^{-3} for the 53% sodium salt of the 4.1 mole per cent copolymer to in excess of 15×10^{-3} for the 90% sodium salt of the 8 mole per cent copolymer. Similarly the peak temperature increases from 50°C for the former to in excess of 90°C for the latter. In addition this relaxation is sensitive to water, moving to lower temperatures with increasing water content. This behavior as well as the very great dielectric activity of the α relaxation[2] led to the conclusion that the α relaxation reflects motion within the ionic aggregates. The relative insensitivity of the β relaxation to the nature of the cation, on the other hand, suggests that this relaxation has to do with the hydrocarbon matrix. In view of its temperature location being close to that of the relaxation associated with the glass transition in branched polyethylene, it has been assigned to the glass transition of the matrix. It should be pointed out that the temperature of the β relaxation depends to some extent on the concentration of cations present[13]. This has led to the suggestion that the salt groups are not microphase separated but are uniformly dispersed throughout the amorphous phase. However, in view of the fact that a qualitative change in the relaxation spectrum takes place upon neutralization, it is difficult to understand how the neutralization process could involve merely the replacement of one kind of polar group with another.

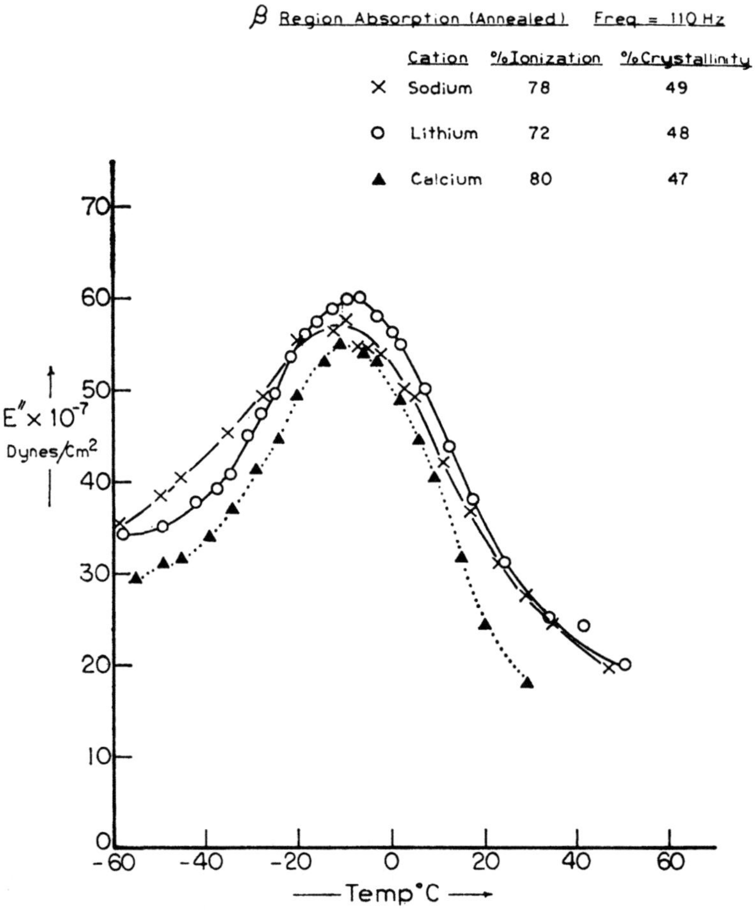

Figure 3

3. Temperature dependence of E" in the β relaxation region for annealed sodium, lithium and calcium salts of a 4.1 mole per cent ethylene-methacrylic acid copolymer at 110 Hz. (After MacKnight, Kajiyama and McKenna (1)].

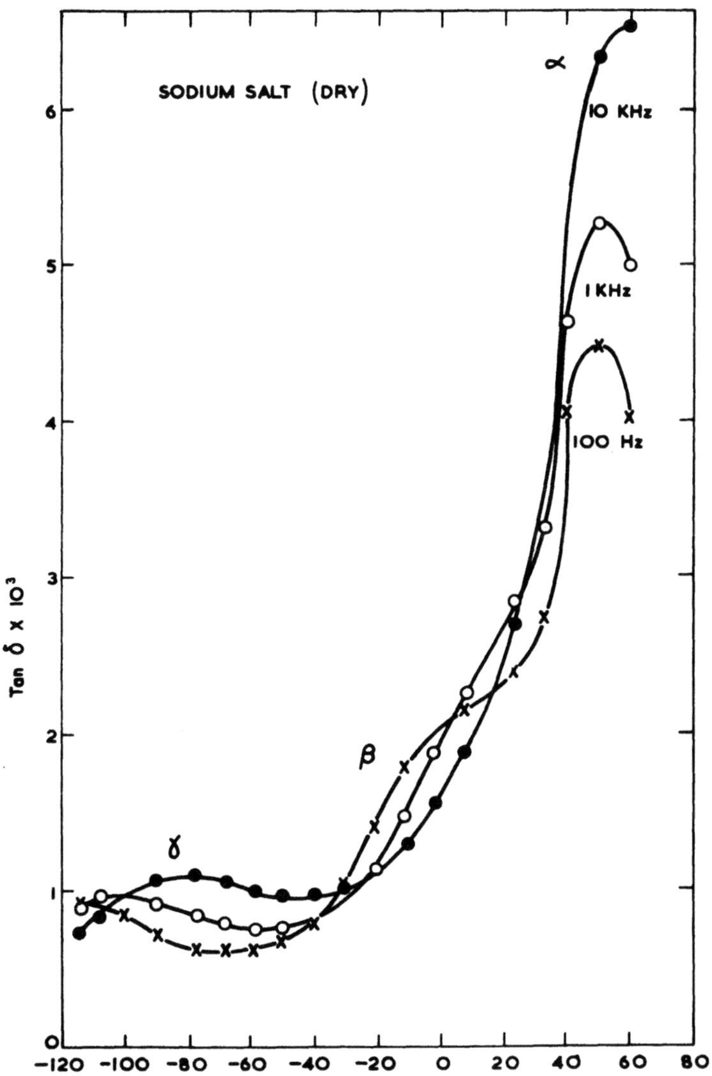

Figure 4

4. Temperature dependence of the dielectric loss tangent at various frequencies for the 53% sodium salt of a 4.1 mole per cent ethylene-methacrylic acid copolymer. (After Read, Carter, Connor and MacKnight (2)].

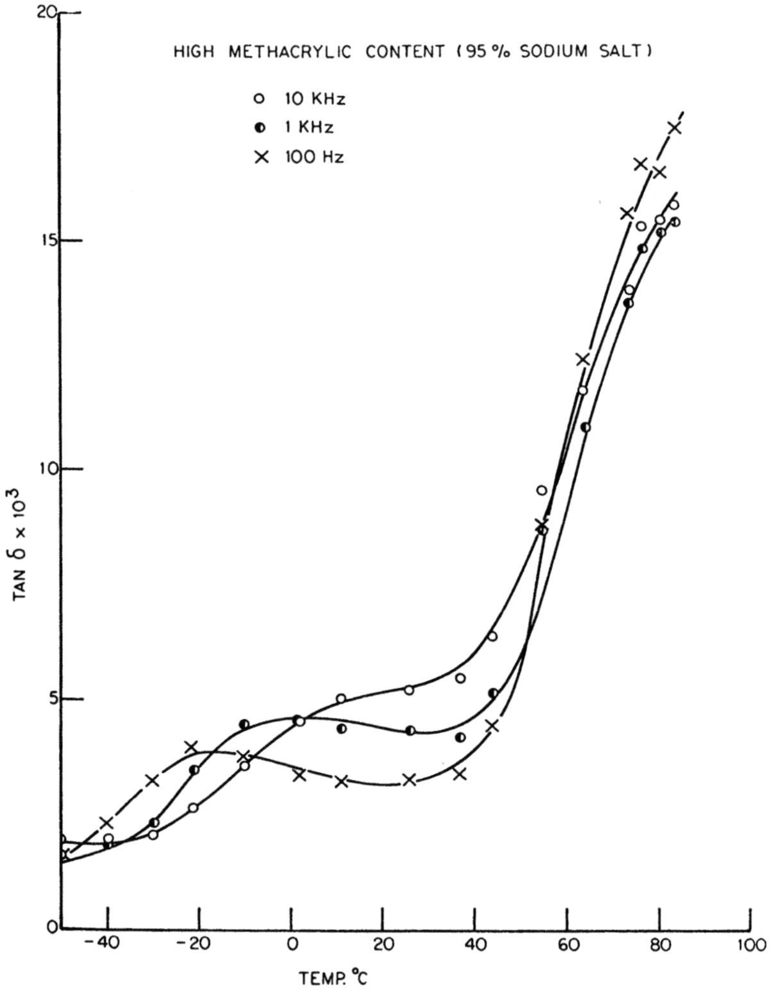

Figure 5

5. Temperature dependence of the dielectric loss tangent at various frequencies for the 90% sodium salt of an 8.3 mole per cent ethylene-methacrylic acid copolymer. (O) 1 KHz; (O) 10 Kz; (x) 100 Hz. (After Phillips and MacKnight [J. Polym. Sci. A2(8), 727 (1970)].

Melt rheology studies indicate that the ionic domains retain their integrity in the melt to high temperatures[14]. This is consistent with the x-ray results previously mentioned. This conclusion is largely based on the finding that while the acid copolymer shows good correlation between the frequency dependence of the complex viscosity and the shear rate dependence of the apparent viscosity this is not true for the salts investigated. Fig. 6 illustrates the frequency dependence of the complex viscosity and the shear rate dependence of the apparent viscosity for the 4.1 mole per cent methacrylic acid copolymer and its 59% sodium salt. It can be seen that in the acid material the two types of viscosity data yield identical plots with the exception of the 100°C points which exhibit deviations due to the fact that 100°C is just the melting point of the crystalline regions of the copolymer. In the 59% sodium salt the apparent viscosity is significantly larger than the corresponding complex viscosity at all temperatures. This difference in rheological behavior between the acid and salt forms may be rationalized on the basis that the complex viscosity reflects short range motions occurring in the hydrocarbon matrix which is relatively fluid at the temperatures investigated, while the apparent viscosity includes longer range motions which are considerably influenced by the ionic domains and must reflect the interchange of ionic groups from one domain to another.

Optical studies serve to further reinforce the microphase separated structural model. In particular, the strain optical coefficient of a 4.1 mole per cent methacylic acid copolymer and its partially neutralized salts has been measured both statically and dynamically as a function of temperature and cation concentration[15]. An α optical loss peak corresponding to the mechanical and dielectric α relaxation previously discussed is found to be present in the salts but not in the acid material. This peak reflects orientational relaxations occurring in the ionic domains.

Apparently the phenomenon of microphase separation is not limited to the presence of salt groups. It has recently been found[5] that in a series of random ethylene-phosphonic acid copolymers, new mechanical and dielectric relaxation peaks occur in a copolymer containing 7.4 mole per cent phosphonic acid groups. Fig. 7 contrasts the temperature dependence of the loss modulus of a branched polyethylene homopolymer with the 7.4 mole per cent copolymer. This copolymer is based on the polyethylene shown in the figure. (See experimental section). The peak at 50°C in the copolymer is not the polyethylene α relaxation since this is known to originate in the crystal phase and the copolymer is totally amorphous. The behavior is very analogous to that of the salts of ethylene-carboxylic acid copolymers discussed above. It is probable that clusters or domains of phosphonic acid groups exist in these materials, being stabilized by hydrogen bonding. The

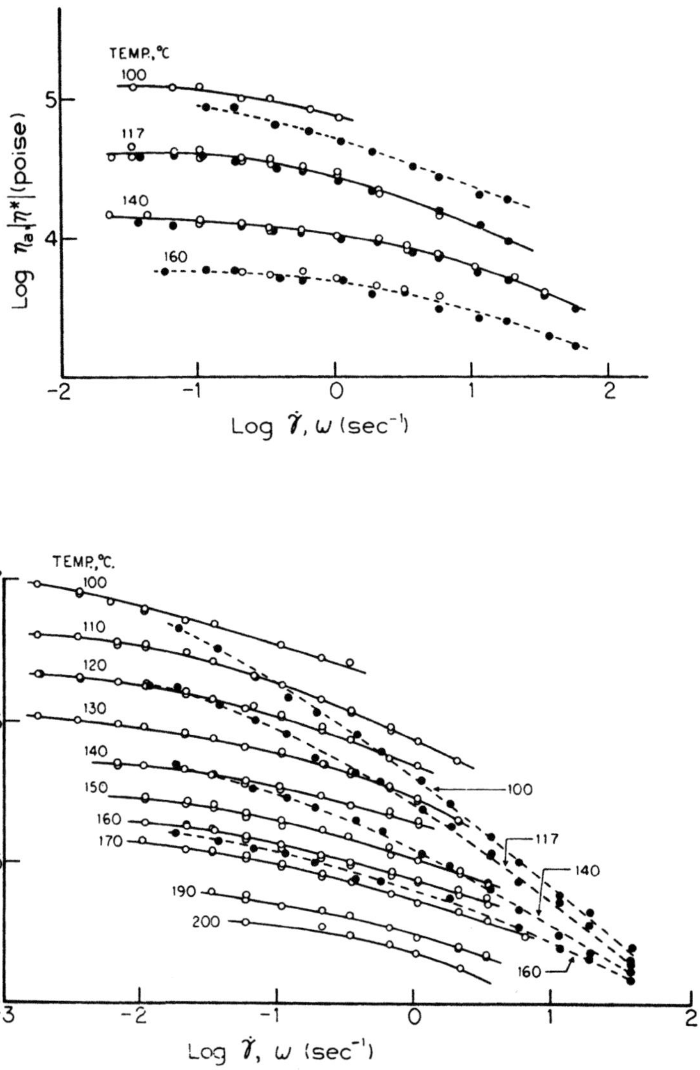

Figure 6

6. Upper Graph: Apparent viscosity $\eta_a(0)$ vs. shear rate $\dot{\gamma}$ and complex viscosity (η^*) vs. angular frequency ω for ethylene-methacrylic acid copolymer containing 4.1 mole % methacrylic acid. Lower Graph: Apparent viscosity $\eta_a(0)$ vs. shear rate $\dot{\gamma}$ and complex viscosity (η^*) vs. angular frequency ω for the 59% sodium salt of the 4.1 mole per cent methacrylic acid copolymer illustrated in the upper graph. [After Sakamoto, Porter and MacKnight (14)].

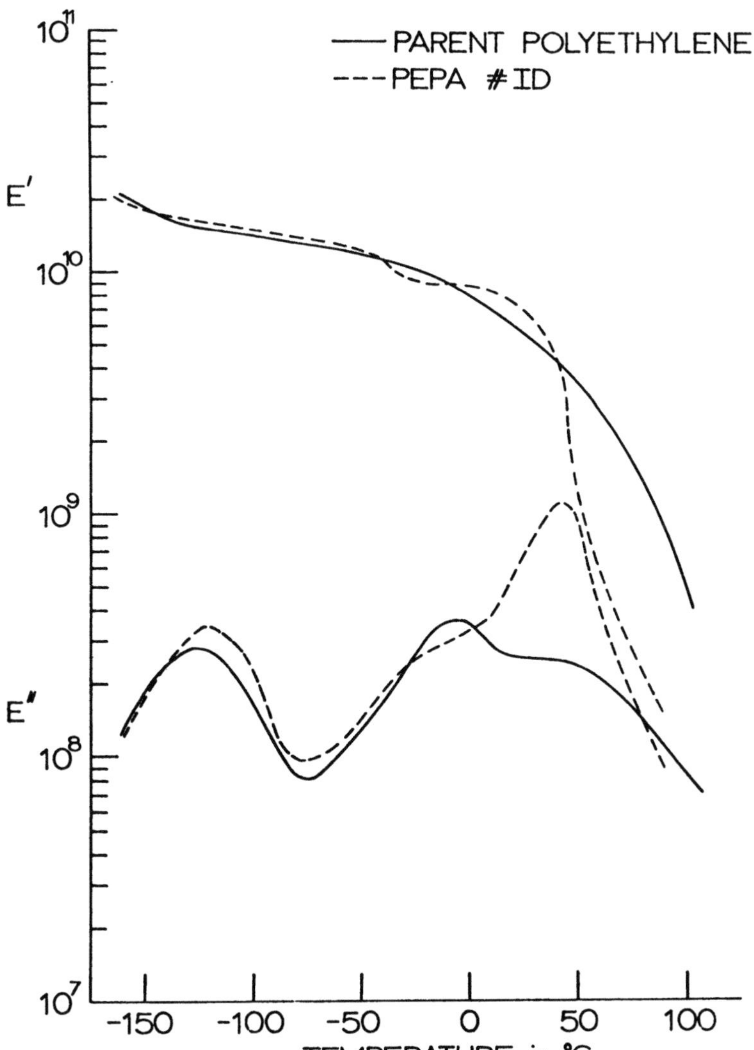

Figure 7

7. Temperature dependence of E' and E" for the parent polyethylene and an ethylene-phosphonic acid copolymer containing 7.4 mole per cent phosphonic acid groups. [After Phillips, Emerson and MacKnight (5)].

morphology of these materials is currently being investigated and positive identification of the occurrence of microphase separation must await these results.

THEORETICAL CONSIDERATIONS

It is difficult to construct an adequate theoretical model for the occurrence of microphase separation in organic polymers containing ions in view of the paucity of experimental data presently available concerning the geometry and structure of the dispersed ionic phase. Thus, in the materials discussed above, the geometry of the domains is unknown. They are unlikely to be spherical, however, in view of the random placement of the groups along the chain and steric requirements. It may be postulated that lamellar structures are present, perhaps analogous to micellar structures formed from salts of fatty acids in non-polar media. The x-ray results indicate some degree of order within the domains although they appear to be non-crystalline. It is certain that the domains cannot consist entirely of salt groups but must include considerable amounts of entrapped hydrocarbon segments.

A thermodynamic approach to the problem has recently been undertaken by Eisenberg[16]. It is first shown that ions in organic media of low dielectric constant exist most probably as pairs or higher multiplets. At low temperatures these multiplets aggregate to form clusters or domains. The driving force for this process is electrostatic in nature and is opposed by the reduction in entropy suffered by the chains being forced to assume strained conformations. It is assumed that this elastic effect and the electrostatic energy involved in cluster formation become equal at some critical temperature, above which the aggregate becomes unstable, breaking down into smaller species. It is tempting to identify this critical temperature with the α relaxation discussed above. It should be realized, however, that Eisenberg's model is an equilibrium model and involves the assumption of specific cluster geometries so that it is premature to attempt to apply it to the ethylene-carboxylate salt experimental results.

EXPERIMENTAL

The ethylene-carboxylic acid copolymers are generally prepared under high pressure, free radical conditions, adjusted to yield essentially random placement of the carboxylic acid side groups. Ionization of the resulting materials may be carried out either in the bulk by milling at elevated temperatures with an appropriate salt such as an acetate salt[3] or by solution techniques[1,13].

The ethylene-phosphonic acid copolymers are prepared by first dissolving polyethylene in phosphorous trichloroide and then bubbling oxygen through the material. This results in the random placement of phosphonyl chloride side groups along the chain according to[4]

$$\sim CH_2\ CH_2\sim\ +\ PCl_3\ +\ \frac{1}{2}\ O_2\ \rightarrow\ \sim\underset{\displaystyle POCl_2}{\underset{|}{CH}}\ CH_2\sim\ +\ HCl$$

Hydrolysis then yields the phosphonic acid form.

$$\sim CH\ CH_2\sim\ +\ 2H_2O\ \rightarrow\ \sim\underset{\displaystyle PO(OH)_2}{\underset{|}{CH}}\ CH_2\sim\ +\ 2HCl$$

ACKNOWLEDGEMENTS

The author is grateful to the National Science Foundation for partial support of this research.

REFERENCES

1. W. J. MacKnight, L. W. McKenna and T. Kajiyama, Polym. Eng. and Sci. 8, 267 (1968).
2. B. E. Read, E. A. Carter, T. M. Connor and W. J. MacKnight, Br. Poly. Journal 1, 123 (1969).
3. R. Longworth and D. J. Vaughan, Nature 218, 85 (1968).
4. P. J. Phillips and W. J. MacKnight, J. Polym. Sci., B8, 87 (1970).
5. P. J. Phillips, F. A. Emerson and W. J. MacKnight, Macromol., in press.
6. A. V. Tobolsky, P. T. Lyons and N. Hata, Macromolecules 1, 515 (1968).
7. E. P. Otacka, Reviews in Macromolecular Chemistry, in press.
8. F. C. Wilson, R. Longworth and D. J. Vaughan, Polymer Preprints 9(1), 505 (1968).
9. B. W. Delf and W. J. MacKnight, Macromol. 2, 309 (1969).
10. H. A. Davis, R. Longworth and D. J. Vaughan, Polymer Preprints 9(1), 515 (1968).
11. W. J. MacKnight and H. K. Plummer, unpublished results.
12. E. P. Otacka and T. K. Kwei, Macromol. 1, 244 (1968).
13. E. P. Otacka and T. K. Kwei, ibid, 1, 401 (1968).
14. K. Sakamoto, W. J. MacKnight and R. S. Porter, J. Polym. Sci., A2, 277 (1970).
15. T. Kajiyama, R. S. Stein and W. J. MacKnight, to be published.
16. A. Eisenberg, Macromol., in press.

IONOMER GRAFT COPOLYMERS OF POLYSTYRENE-POLYBUTYLACRYLATE

B. Vollmert, W. Schoene

Polymer-Institut, Universität Karlsruhe, Germany

Nearly all systems of two or more different polymers are incompatible, meaning that solutions of the two polymers above a certain concentration as well as the polymers themselves are not mixible together in the sense of a molecular dispersed mixture. This also applies to blends of polystyrene and polybutylacrylate.

We have investigated the influence of the presence of 5, 10 or 15 weight % acrylic acid as a copolymer component in the polymer chains on the compatibility of the polystyrene polybutylacrylate system, where the ionomer components are stepwise neutralised with different alkali hydroxides.

These experiments could also be seen as an attempt to prepare ionomer graft copolymers (see fig.5), where the connexion between the different coils is effected by dipol-dipol-association of COOM-groups. In such crosslinked graft systems, formed by secondary valence bonds, it is not to be avoided, contrary to covalent grafted polymer systems[1] that the same mutual interaction is also effective between the macromolecules of the same type.

METHOD OF PREPARATION

A statistic styrene acrylic acid copolymer (5, 10 or 15 % by weight of acrylic acid) was dissolved together with a butylacrylate acrylic acid copolymer in a mixture of THF/ CH_3OH = 4/1. The solutions of about 20 % are neutralised with methanolic alkali hydroxide solutions (LiOH, KOH, CsOH) up to the desired degree of neutralisation. The solvent was evaporated under a stream of nitrogen on plates of polyethylen at 50° C and the films were dried for 24 hours in vacuum at a temperature of 70° C.

It is unimportant for the quality of the films whether only one of the components is neutralised in the solution and than mixed with the other component or whether the mixture is prepared first and than neutralised up to the desired degree of neutralisation. Cation exchange takes place.

TESTING OF THE PROPERTIES OF THE PREPARED SAMPLES

The testing of the compatibility was carried out by studying the so obtained films in the following ways:

1. Transparency measurements by a photometric method.
2. Phase contrast microscopy.
3. Determination of damping maxima by torsion pendulum (SCHMIEDER-WOLF)
4. DTA-measurements.

Furthermore some mechanical properties of the prepared ionomer mixtures were tested:
Hardness of the films (pendulum hardness according to DIN 53157).
Flexibility and toughness of the films by slow deformation according to ERICHSEN (DIN 53156).
Impact strength according to DIN 53453.
The solutions of the single components in tetrahydrofurane (THF) and THF/CH_3OH = 4/1 mixtures were tested by viscosity measurements and conductimetry[5].

Measurement of the Relative Transparency

The transparency of our ionomer blends is a good test of the degree of compatibility and its improvement by neutralisation with alkali hydroxides, because the two polymer components, polystyrene and polybutylacrylate, have considerably different refractive indices.

Films of 0,2 mm thickness were compared in a ZEISS Spectral Photometer at 600 μm with a polystyrene film of the same thickness.

Table 1 shows the result of such measurements for an ionomer blend of polystyrene (S_9) and polybutylacrylate (B_9) (S_9/B_9 = 8/2), both containing 9 weight % acrylic acid.

It is remarkable that the Li-salts and the tetraethylammonium salts do not effect any improvement of the compatibility and that K-salts were found to give the maximum of compatibility among the salts of the alkali group.

Mixtures with other blend ratios and ionomer mixtures with higher and lower acrylic acid contents give quite similar results. 9/1-mixtures of K-ionomers show a distinct maximum of transparency at a neutralisation degree of 0,6.

Table 1
Relative light transparency of polystyrene (S_9) butylacrylate (B_9) ionomer mixtures with different cations. S_9/B_9 = 8/2
Reference standard: polystyrene bulk polymer
$\bar{M}_W(S_9)$ = 150 000 $\bar{M}_W(B_9)$ = 210 000
Acrylic acid content (in both components): 9 % by weight

Degree of Neutral.	Relative Transparency (% of a Polystyrene Standard)			
	Li-	Na-	K-	Cs-
0	1,1	1,1	1,1	1,1
0,2	0,5	0,7	1,0	0,6
0,4	0,3	0,8	0,8	0,6
0,6	0,1	0,9	56,2	0,5
0,8	0,05	0,4	87,5	13,0
1,0	0,3	90,5	95,0	89,5

Phase Contrast Micrographs

As photographic objects we have used thin films (10 to 15 μm) of the ionomer mixtures, which were prepared by pouring out mechanically a 20 % solution of the copolymers in THF/ CH_3OH = 4/1 directly on the object holder.

The phase contrast pictures make sure, what we could expect because of the change of transparency with increasing degree of neutralisation, namely that the size of the particles decreases with increasing degree of neutralisation. In this case, to, lithium salts are an exception. Fig. 1 shows as an exemple the phase contrast pictures of a K-salt ionomer mixture with different degrees of neutralisation.

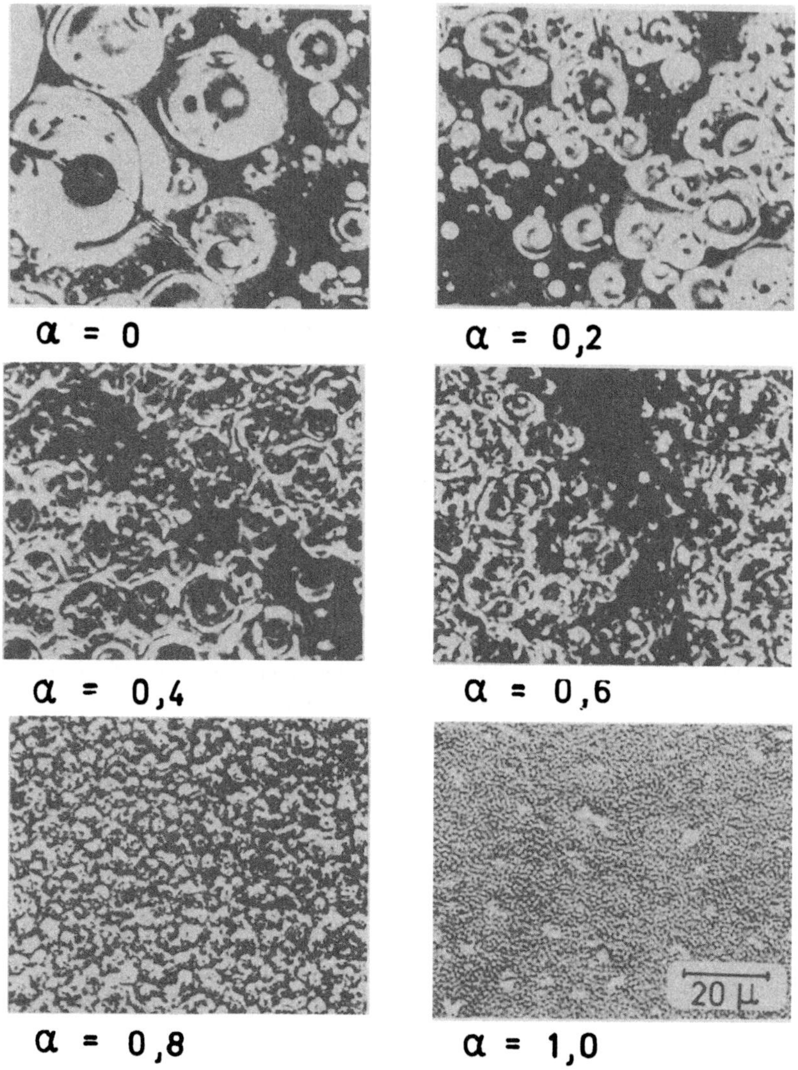

Fig. 1

Phase contrast micrographs of a polystyrene (S_5)-polybutylacrylate (B_5)-ionomer-mixture in the ratio $S_5/B_5 = 8/2$
Acrylic acid content (in both components): 5 weight %
Base for neutralisation: KOH (α = neutralisation degree)
Styrene copolymer: $\bar{M}_W = 200\ 000$
Butylacrylate copolymer: $\bar{M}_W = 170\ 000$

At an analogous ionomer blend with 9 % acrylic acid we have found at $\alpha = 0,6$ already the same distribution degree as in Fig. 1 at $\alpha = 1$. The transparent ionomer mixture films ($\alpha = 0,8$ and $\alpha = 1,0$, see Table 1) no more permit recognizing separate particles in the phase contrast micrographs. Li-ionomer mixtures do not show an essential change of particle size with increasing degree of neutralisation.

As for the details of the morphological structure, the phase contrast micrographs of ionomer mixtures with low degrees of neutralisation show mostly spherical light particles, which can be identified as polybutylacrylate phase because of the refractive indices of the single components. With increasing degrees of neutralisation the particles show irregular shapes, so that we can speak no more of a continuous and a dispersed phase. Especially in samples with very fine grained structures we can see patterns of spaghetti like intertwined chains of beads.

The pictures according to Fig. 1 do not show microtomic slices but surfaces of thin films. Therefore the microstructure of the particles is to be perceived but indistinctly and is difficult to be discussed.

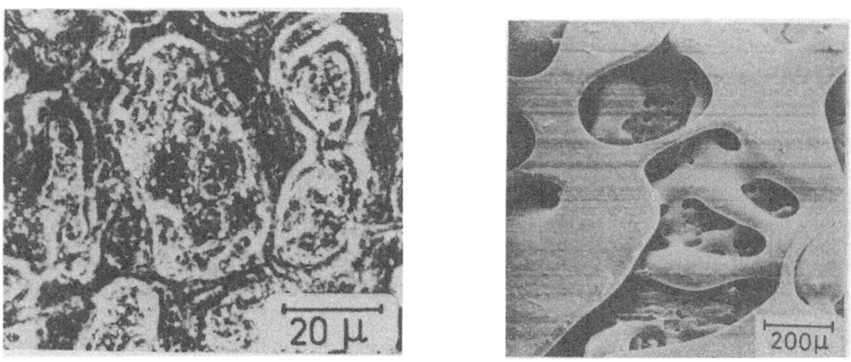

Fig. 2 Phasecontrast picture of a polystyrene-polybutylacrylate-ionomer blend film and a stereoscan picture of the same film after removing the polybutylacrylate-phase by extraction with water. Styrene copolymer: 15 % acrylic acid, $\bar{M}_W$ = 250 000
Butylacrylate copolymer: 15 % acrylic acid, $\bar{M}_W$ = 250 000
Degree of neutralisation: 0,6 Base: KOH

Due to the water solubility of the polyacrylic phase in some samples with relativly high acrylic acid content, this phase could be removed by extented extraction with water, so that only the polystyrene skeleton remained in a highly swollen state (see Fig.2).

Relaxation Measurements (SCHMIEDER-WOLF Torsion Pendulum)

These measurements should show how and to what extent the transition temperatures (glass temperatures) of the copolymers and the ionomer blends are influenced by neutralisation. We are indebted to Dr. SCHMIEDER (BASF), who was kind enough to carry out these experiments in his laboratory.

Table 2
The influence of neutralisation on the damping maxima of styrene acrylic acid copolymers (S_9) and butylacrylate acrylic acid copolymer (B_9) with 9 % acrylic acid in each case

Ionomer Sample	Degree of Neutral.	Base MOH	Damping Maxima °C
S_9	0	--	135
S_9	1,0	KOH	155
S_9	1,0	CsOH	150
B_9	0	--	-15
B_9	1,0	KOH	-20
B_9	1,0	CsOH	-18

Table 2 shows that the maxima of polystyrene increase with neutralisation and those of polybutylacrylate decrease slightly. Probably at low temperatures hydrogen bonds are more effective than dipol-dipol-interaction between the COOM-groups, whereas at higher temperatures the contrary is true.

The investigation of the ionomer blends showed a rather surprising result: in the case of potassium ionomer blends already above the degree of neutralisation 0,4 the damping maxima of the polybutylacrylate component disappear completely (Fig. 3). Only the lithium ionomer blends show a weak maximum at aproximately 10° C.

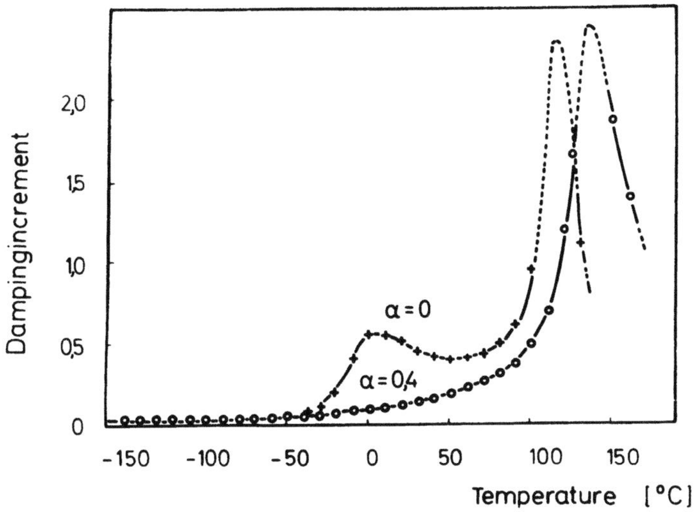

Fig.3 Damping maxima of ionomer blends S_9/B_9 = 8/2 (see table 2) with α= 0 and α= 0,4

So fare we cannot give a substantial explanation for the disappearance of the polybutylacrylate maxima. We would like to presume, however, that the ionomer blend structure is much more complicated than the phase contrast pictures show, so that the polybutylacrylate ionomer phase - even at medium degrees of neutralisation - forms a crosslinked graft system by dipol-dipol-interaction with the polystyrene ionomer macromolecules. As may be easyly understood in such an ionomer network system the polyacrylate molecules have lost their mobility to an extent that the chain segements are not able to respond to the frequency of 1 Hz in the SCHMIEDER-WOLF experiment.

DTA-Measurements

An explanation of the phenomena represented in Fig. 4 is not made easier if we have to take notice of the fact that the DTA-measurements show clearly two steps of a heat effect for the transition of both ionomer components, even in the case of the fully transparent K- or Cs-blends. As an example Fig.4 represents the behaviour of a potassium ionomer serie 8/2.

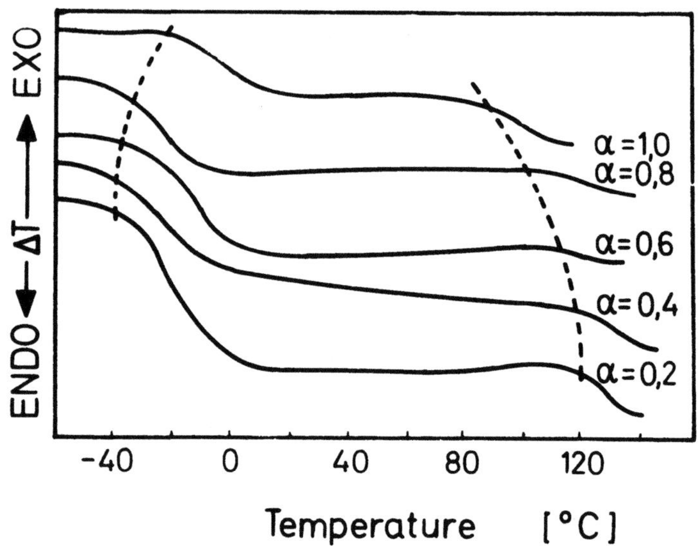

Fig. 4 DTA-thermogramms of polystyrene-polybutylacrylate-ionomer blends. Blend ratio: S_9/B_9 = 8/2 Acrylic acid content = 9 % weight % Base: KOH

It should be noted that with increasing degree of neutralisation α the glass temperature of the two components in the blend runs in the opposite direction as it does in the case of the individual ionomer components (table 3). This is another evident indication of interaction between both of the different components in the ionomer blends.

Table 3

Glass temperatures (determined by DTA) of polystyrene ionomer S_9 and polybutylacrylate ionomer B_9, as free acids and as salts (α = 1)

Ionomer Sample	α	Base MOH	Tg°C (DTA)
S_9	0	--	123
S_9	1,0	LiOH	131
S_9	1,0	KOH	129
S_9	1,0	CsOH	127

Ionomer Sample	α	Base MOH	T_g °C (DTA)
B_9	0	--	-39
B_9	1,0	LiOH	-38
B_9	1,0	KOH	-45
B_9	1,0	CsOH	-48

Although the heat effect at the transition from the glassy state to the rubbery state - even in the transparent ionomer blends - appears seperate for each of the ionomer components as indicated by DTA-measurements, this does not allow to draw conclusions with regard to the particle sizes in the ionomer blends. The disappearance of the second damping maxima in the torsion pendelum experiment shows that in the samples with higher degrees of neutralisation there must exist a very intensive mixture of the polybutylacrylate phase in the polystyrene phase (or vice versa), which can possibly go to the extent of a dispersion of individual polymer coils or chains of coils so that there is a certain justification to speak of an ionomer graft polymer (Fig. 5).

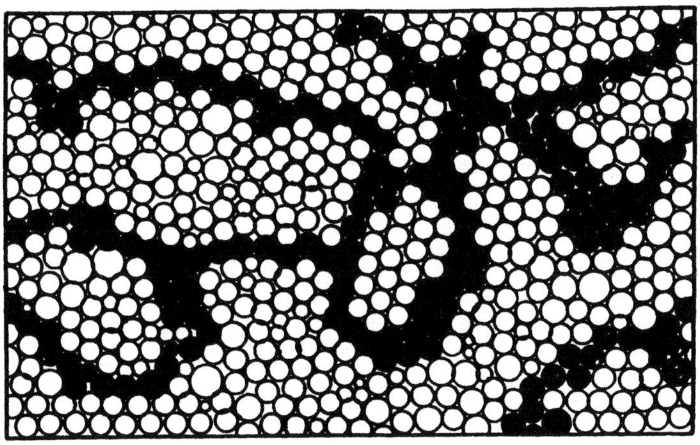

Fig. 5 Hypothetic scheme of the structure of polystyrene-polybutyl acrylate ionomer graft polymer
● polybutyl acrylate ionomer coil
○ polystyrene ionomer coil

Mechanical Properties

The change of the structure of the ionomer blends with increasing degree of neutralisation effects a striking change of the mechanical properties. For example the flexibility and the toughness of the ionomer blends, without change in hardness, increase steeply as soon as the samples become transparent, whereas the inhomogeneous blends (Li-salts) retain the brittleness of polystyrene. Table 4 shows the ERICHSON-values (as a measure for toughness and flexibility) as a function of an encreasing degree of neutralisation for different alkali salt ionomer blends (the value of 11 mm is the highest measurable value). For comparison reference is made to the measured values of the polystyrene ionomer components alone and to those of some well-known polymer materials. The transparent ionomer blends are tough and extensible only if they are slowly deformed. Upon rapid deformation they behave as brittle materials. This strikingly different behavior at slow and rapid deformation is in close relation with the disappearance of the damping maxima in the torsion pendulum experiment. It is interesting that graft copolymers with the same polymer components (polystyrene and polybutylacrylate), but with covalent graft links (esterlinks or carbonamid links) are excellent aging-resistant impact-resistant polystyrenes[1].

Table 4

Toughness and hardness of ionomer blends at different degrees of neutralisation and of different cations for polystyrene S_9 and for ionomer blends S_9/B_9 = 8/2

	ERICHSEN DATA mm				PENDULUM-HARDNESS				IMPACT STRENGTH
		Ionomer blends				Ionomer blends			
	S_9	S_9/B_9			S_9	S_9/B_9			S_9/B_9
	K^+	Li^+	K^+	Cs^+	K^+	Li^+	K^+	Cs^+	K^+
0	1,5	1,0	1,0	1,0	148				
0,2	2,0	1,2	0,6	0,9	153	44	126	97	4,0
0,4	1,9	1,1	2,6	1,7	157	63	135	125	5,2
0,6	1,0	2,5	9,4	8,3	160	140	137	128	8,4
0,8	1,0	2,7	9,6	9,8	162	158	142	123	9,2
1,0	0,6	1,4	10,0	10,0	159	160	145	128	5,4

Comparative Data:

Polymer material	Erichsen Data	Pendulum-Hardness	Impact Strength
Polystyrene	0,7	131	22
Impact Polystyrene	2,5	115	70
PPO	10,0	132	>100

S_9 = Styrene acrylic copolymer with 9% by weight acrylic acid

B_9 = Butylacrylate acrylic acid copolymer with 9 % by weight acrylic acid

..... =border line of transparency

SHORT SUMMARY AND DISCUSSION OF THE RESULTS

The experiments have shown that copolymers of styrene and acrylic acid and copolymers of butylacrylate and acrylic acid are incompatible and that they become compatible if the carboxyl groups are partially or completely neutralised with NaOH, KOH or CsOH.

No improvement of compatibility could be achieved by neutralisation with LiOH. One can suppose that the improvement is effected by a dipol-dipol interaction between the COOM-groups[2]. X-ray structure studies with sodium acetate[3] and UR-spectroscopic studies[4] seem to indicate a symmetric structure of the $C\lesssim^{O}_{O}\ominus$ M^+ group. This could be confirmed by our own UR-spectroscopic measurements with Cs-salts of polystyrene ionomers.

The presence of dipols, however, will not be effective enough for the occurence of interactions between the two different polymer components. The dipol-groups in both components also have to be available to the same extent. This became evident by exchange of the butylacrylate component by methylacrylate and by dodecylacrylate. In the methylacrylate-acrylic acid copolymers the COOM-group is sterically much less obstructed than in the styrene-acrylic acid copolymer whereas

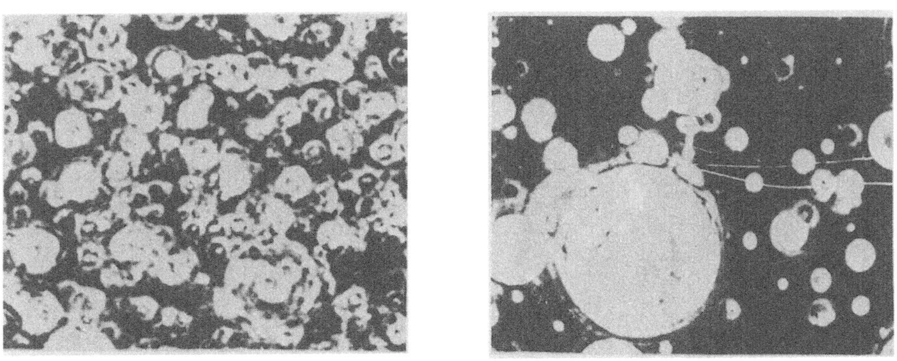

a b

Fig.6 Phase contrast micrographs of ionomer blends of
a) polystyrene (S_9)-polymethylacrylate
b) polystyrene (S_9)-polydodecylacrylate
Blend ratio 8/2 (9 % acrylic acid in both components)
Degree of neutralisation $\alpha = 1$ Base: KOH

the obstruction in the dodecylacrylate-acrylic acid copolymers is more complete than in styrene copolymers. In both cases no compatibility of the ionomer blends could be reached, as shown in Fig. 6.

In order to get an improvement of compatibility it seems to be necessary that the interaction between polymer coils of the same type and of the other type have the same order of magnitude. In every case, the intermolecular interaction ("crosslinking") is connected with an intramolecular interaction of chain segments inside the coils. More details about that and a possible influence of dissociation phenomena can be obtained by viscosity measurements[5].

From a thermodynamical point of view, the improvement of compatibility is effected by a decrease of the positive values of the mixing enthalpy or - in extreme cases - by a change of the sign to negative values of ΔH. In our case the decrease of the entire mixing enthalpy is caused by the exothermic association of the COOM-dipole-groups. As the experiments have shown, it depends on the number of COOM-groups (i.e. on the degree of neutralisation) for which particle size ΔH and $T \cdot \Delta S$ become equal. At a given degree of neutralisation, therefore, the compatibility should become better with increasing mixing temperature. That could be confirmed by experiments (see Fig. 7).

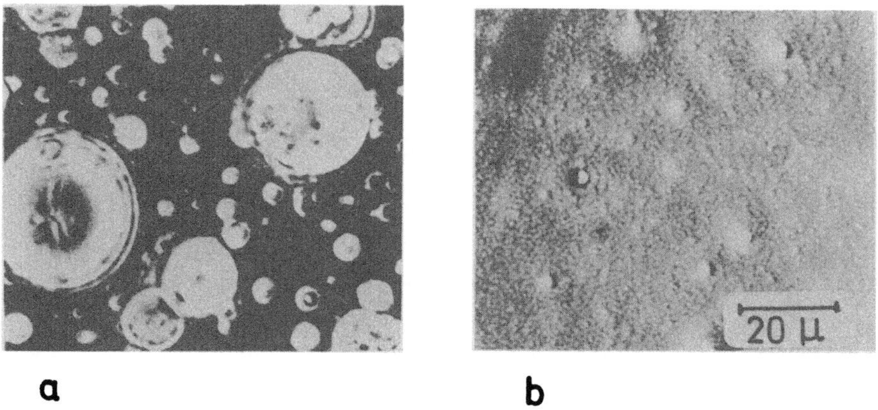

Fig. 7 Phase contrast micrographs of S_9/B_9-ionomer blends 8/2 with 9 % acrylic acid in both components
a) prepared at 70° C b) prepared at 140° C
Degree of neutralisation in both samples $\alpha = 0,2$ Base:KOH

LITERATURE

1) B.Vollmert, Die Angewandte Makromolekulare Chemie 3 (1968) 1- 27
2) E.P.Otocka and F.R.Eirich, Polymer Preprints 8/1 (1967) 66
E.P.Otocka and F.R.Eirich, J. Polymer Sci. A2,6 (qi68) 921
3) W.H. Zachariasen, J.Am.Chem.Soc. 62 (1940) 1011
4) K.Nakamoto, J.Morimoto and A.E.Martell, J.Am.Chem.Soc. 83 (1961) 4528
5) B.Vollmert, W. Schoene, Die Angewandte Makromolekulare Chemie (in print)

INFRARED DICHROISM OF SEGMENTED POLYURETHANE ELASTOMERS

G. M. Estes, R. W. Seymour, S. J. Borchert,
and S. L. Cooper

DEPARTMENT OF CHEMICAL ENGINEERING

UNIVERSITY OF WISCONSIN — MADISON, WISCONSIN 53706

INTRODUCTION

The mechanical and optical properties of block polymers in uniaxial extension have been well-documented (1). Characteristic of these materials are high moduli, high extensibility, and high resilience relative to random copolymers of equivalent composition. Also common to block polymers is the phenomenon of stress-softening (2, 3), whereby the stress and modulus of the material are substantially lowered by the cyclic application of strain. It is generally agreed that these properties are the product of a heterophase microstructure, which arises from the segregation of unlike blocks into separate domains (1). Higher modulus domains serve both as filler particles and as physical, multifunctional cross-link points to reinforce the elastomer. Stress-softening is postulated to result from the deformation, orientation, or partial destruction of the reinforcing domains.

In order to understand the morphological changes which occur during straining, it is necessary to have information describing how each of the phases of the block polymer deform. Several investigators have used strain-birefringence techniques to study segmented polyurethane elastomers (4-6), ter-block polymers based on polystyrene-diene rubber (7-10), and other block polymer systems (1, 11). The birefringence techniques are limited however, because the experimental quantity "birefringence" is a measure of the average orientation in the sample. Although it has not been proven rigorously correct, it has been generally assumed that the total birefringence in a heterophase system is equal to the sum of component contributions from the different phases plus a form birefringence which arises from interaction and distortion of the light wave at the interface between the phases. The latter component is not well defined, and although recent studies (9, 10) have suggested it to be important, most investigators (7, 8, 11) have assumed it to be negligible. Infrared dichroism is a more powerful technique since it allows the orientation of specific polymer chains to be monitored. By analysis of appropriate infrared absorption bands, the orientation of polymer segments in the different domains may be determined.

It is the purpose of this investigation to determine morphological changes which occur in polyether- and polyester-urethanes by measuring the infrared dichroism of samples in uniaxial extension. The strain-birefringence responses of these same materials have been described previously (4, 5).

INFRARED DICHROISM (12)

Absorption of infrared radiation can occur only when a vibrational mode produces a changing dipole moment, characterized by a transition moment vector $\underset{\sim}{M}$. If the incident radiation has a component in the direction of $\underset{\sim}{M}$ absorption will occur. The phenomenon of infrared dichroism arises from this fact. If plane polarized radiation is used, maximum absorption will occur when $\underset{\sim}{M}$ and the direction of the polarization $\underset{\sim}{P}$ are coincident. If $\underset{\sim}{M}$ and $\underset{\sim}{P}$ are orthogonal no energy will be absorbed. The utility of dichroic effects to study orientation is thus apparent, for differences in absorption intensity with polarization direction should occur if the transition moment vectors are preferentially oriented.

The dichroic ratio D for a particular absorption band in an oriented specimen is defined by Equation (1)

$$D = A_{\parallel}/A_{\perp} \tag{1}$$

where $A_{\parallel}$ and $A_{\perp}$ represent the absorbancies of infrared radiation polarized parallel and perpendicular to the stretching direction (uniaxial tension). Using a baseline technique the absorbance A at the frequency of maximum absorption (ν_0) is given by Equation (2),

$$A = \log_{10} \frac{I_0(\nu_0)}{I(\nu_0)} \tag{2}$$

where I_0 is the baseline intensity and I is the intensity of the transmitted radiation. The dichroic ratio is unity for an unstrained sample, and depending upon the nature of the vibrational mode may become either greater than or less than unity with increasing orientation.

In order to quantitatively characterize the orientation of absorbing groups, the definition of an orientation function f is necessary. Ideally this would be a function describing the complete distribution of chain orientations. However, since the measured dichroic ratio represents an average over all orientations which are present, such a distribution function is not possible. In view of this difficulty, the procedure followed is to assume a simple model which defines an orientation parameter to characterize the oriented sample, and to calculate the expected dichroic ratio. Experimental dichroic ratios are then correlated with the orientation parameter through the resulting formula.

Two models (13), both equally applicable, have been employed to attain this goal. One model assumes that given any state of strain, the polymer chains make an angle θ with the stretching direction. The second model assumes that a fraction f of the polymer chains are perfectly aligned into the stretching direction, while the remaining $(1 - f)$ chains are randomly oriented. The two orientation parameters may be related by defining the standard orientation function (14) in Equation (3),

$$f = \frac{3\,\overline{\cos^2\theta} - 1}{2} \tag{3}$$

where f is equivalent to the previous orientation parameter. Fortunately, the two models give rise to the same result given in Equation (4),

$$f = \frac{(D_0 + 2)}{(D_0 - 1)} \cdot \frac{(D - 1)}{(D + 2)} \tag{4}$$

where D is the dichroic ratio defined above and D_0 the dichroic ratio for perfect alignment. D_0 is related to the angle α which the transition moment vector makes with the polymer backbone by Equation (5),

$$D_0 = 2\cot^2\alpha \quad ; \qquad 0° \leqslant \alpha \leqslant 90° \tag{5}$$

Using Equation (4), infrared dichroism measurements are capable of yielding information about molecular orientation, providing suitable absorption bands exist. If, as in the unstrained state, the polymer chains are randomly oriented, then $D = 1$ and $f = 0$. When the polymer chains align preferentially into the direction of strain, then $D \neq 1$ and $f > 0$ with $f = 1$ characterizing complete alignment.

EXPERIMENTAL

The materials used in this study were a polyether-urethane (designated ET-38) and a polyester-urethane (ES-38), both described in Figure 1. Both polyurethanes were block polymers composed of alternating blocks of soft (polyether or polyester) segments and hard (urethane) segments. The soft segment prepolymers (~~~~~) were polytetramethylene oxide in ET-38 and polytetramethylene adipate in ES-38. The molecular weight of the prepolymers was 1000. The hard segments were based on p,p'-diphenylmethane diisocyanate (U) chain extended with 1,4-butanediol (G). On the average each hard segment contained three diisocyanate residues, so the average molecular weight of the hard segments (UGUGU) was approximately 930. Both materials contained 38 weight per cent diisocyanate.

Two experiments have been performed. In the first type (Type I) the polymer films were stretched to a nominal strain level and the polarized IR spectra scanned. Then the samples were relaxed at zero load for five minutes after which the spectra were again obtained. Each sample yielded only one data point on an orientation function-strain curve. The second type of experiment (Type II) involved extending one sample by step increments to 360% strain and scanning the polarized IR spectra at each step. In this case one sample yielded the entire orientation function-strain curve. Type II experiments have the advantage that only one sample is used, thereby eliminating the variability between samples as a source of experimental error. The Type II experiments were performed in duplicate.

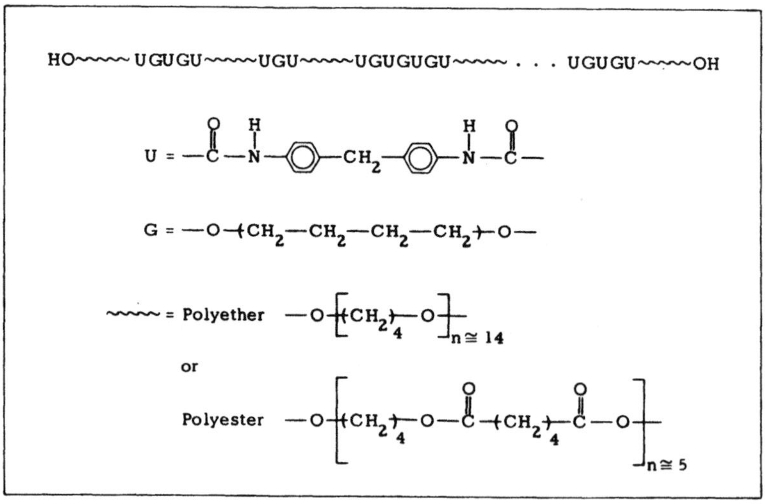

Figure 1. Schematic Representation of Segmented Polyurethane Structure.

Details of sample preparation, techniques used for infrared measurements, and complete infrared spectra of ET-38 and ES-38 are published elsewhere (15).

RESULTS AND DISCUSSION

Two-phase microstructures have been postulated to explain the observed mechanical and optical behavior of block polymers. Direct evidence of discrete domains has been provided by electron microscopic studies of ter-block polymers (1) and recently of the segmented polyurethane elastomers investigated in this report (16). To more fully understand the role of the domain in determining mechanical properties, it is both desirable and necessary to describe molecular orientation in the different domains. Infrared dichroism has served this purpose.

The applicability of infrared dichroism to the study of polyurethane elastomers is enhanced by the availability of at least one well-characterized absorption band in each phase. It has been established that the soft domains are comprised of prepolymer (polyether or polyester) segments, while the hard domains contain urethane segments extended with 1,4-butane diol. The NH stretching vibration which is located in the urethane linkage has been employed to characterize the orientation of the higher modulus or hard domains. This N-H absorption appears at 3320 cm^{-1} in both ET-38 and ES-38. The asymmetric C-H stretching absorption occurs at 2935 cm^{-1} in ET-38 and at 2960 cm^{-1} in ES-38, and has been used to describe the orientation of prepolymer segments which comprise the rubbery or soft domains. Although the CH_2 linkage also is present in the chain extending diol in the hard domains, the fraction of CH_2 in the soft segments is sufficiently greater (84% in ET-38 and 78% in ES-38) that the asymmetric C-H stretch can be confidently employed for soft segment characterization.

Both the N-H and C-H stretching modes give rise to transition moment vectors which lie perpendicular ($\alpha \simeq 90^\circ$) to the polymer backbone (12, 13). For this specific case Equation (4) reduces to Equation (6),

$$f = -2 \cdot \frac{D - 1}{D + 2} \qquad (6)$$

which has been used to compute the orientation functions presented in this paper.

As described in the Experimental section, two simple experiments have been performed. Figures 2 and 3 display the orientation functions for both phases in ET-38 and ES-38, respectively, during

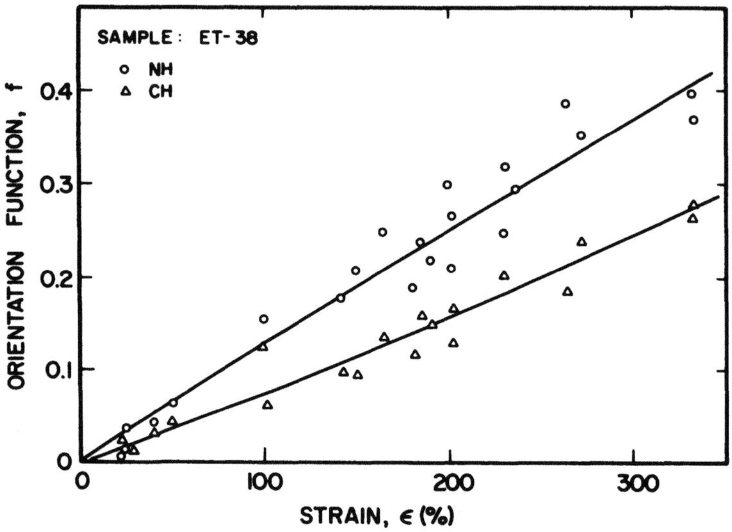

Figure 2. Orientation Function Versus Strain for Polyether-Urethane, ET-38 (Type I Experiment).

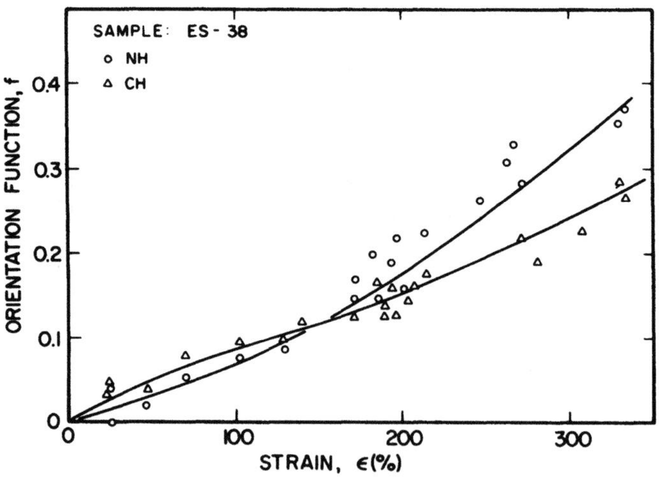

Figure 3. Orientation Function Versus Strain for Polyester-Urethane, ES-38 (Type I Experiment).

the initial extension of Type I experiments. The orientation of hard and soft segments is of similar magnitude, with the hard segments aligning to a higher degree at all strain levels in ET-38 and above 150% in ES-38. In ES-38, the CH orientation is at low strain levels greater than the NH orientation, with the two functions crossing at about 150%.

The orientation which remains after 5 minutes relaxation is represented by the residual orientation functions plotted in Figures 4 and 5 for ET-38 and ES-38, respectively. The abscissa in these figures is the strain level to which the samples were extended prior to relaxation; hence, it is labelled as prestrain. Although both the soft and hard domains orient to nearly the same degree upon initial straining, when relaxed the hard domains (NH orientation) exhibit substantially larger orientation functions than the soft domains. Both ES-38 and ET-38 show this behavior. Since the elastomers creep considerably (65-70% following straining to 300%), it might be thought that the residual orientation is due to the residual strain. This could be true in part, but cannot explain the entire residual hard segment orientation for at 70% initial strain the orientation function for NH alignment is only about 0.07. Moreover, if creep were entirely responsible, then the residual orientation in the two phases should be approximately equal, but lower than those in Figures 2 and 3. It appears that the differences in the residual orientation functions are due to different retractive stresses

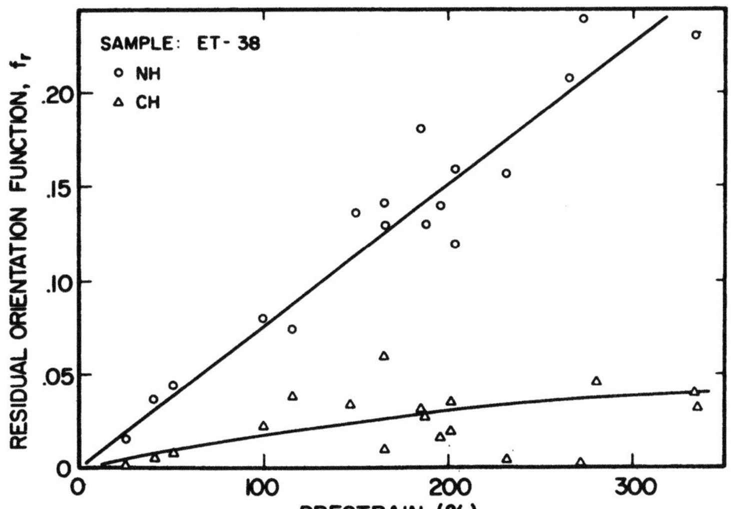

Figure 4. Residual Orientation Function Versus Strain for Polyether-Urethane, ET-38 (Type I Experiment).

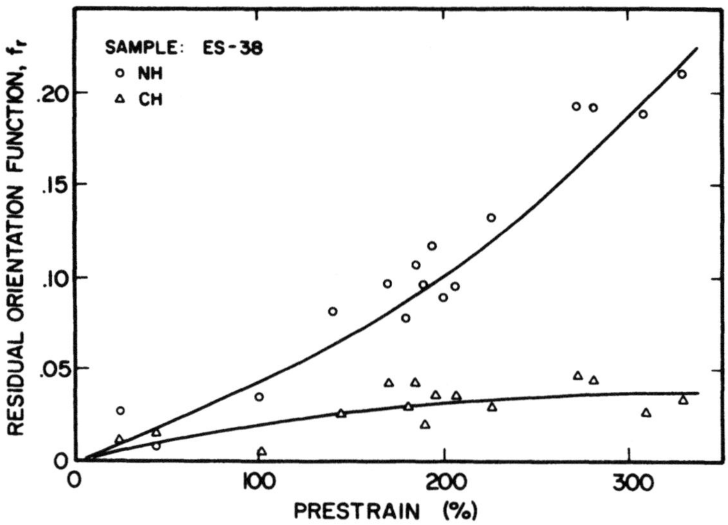

Figure 5. Residual Orientation Function Versus Strain for Polyester-Urethane, ES-38 (Type I Experiment).

acting on the separate phases. The high residual orientation of the hard domains is related to the stress-softening in the stress-strain-birefringence results for ES-38 and ET-38 which have been presented previously (4, 5).

To further investigate the connection between strain history and residual orientation, films of ET-38 and ES-38 were subjected to Type II experiments both without prestrain and also after pre-straining to 200 per cent elongation. The results of these experiments are shown in Figures 6 and 7. Despite the slightly different time scales of the two types of experiments, the non-prestrained curves in these figures compare favorably with those in Figures 2 and 3 for the Type I experiments.

For both ET-38 and ES-38, the orientation of the soft domains is relatively insensitive to prestrain, while the orientation of the hard domains is very dependent upon strain history. At strain levels lower than that of the prestrain (200%), the hard segments orient to a noticeably higher degree than is observed for non-prestrained samples. However, at extensions above the prestrain level, the NH orientation functions are essentially independent of prestrain. This behavior exactly parallels the stress-softening which is prevalent in these materials.

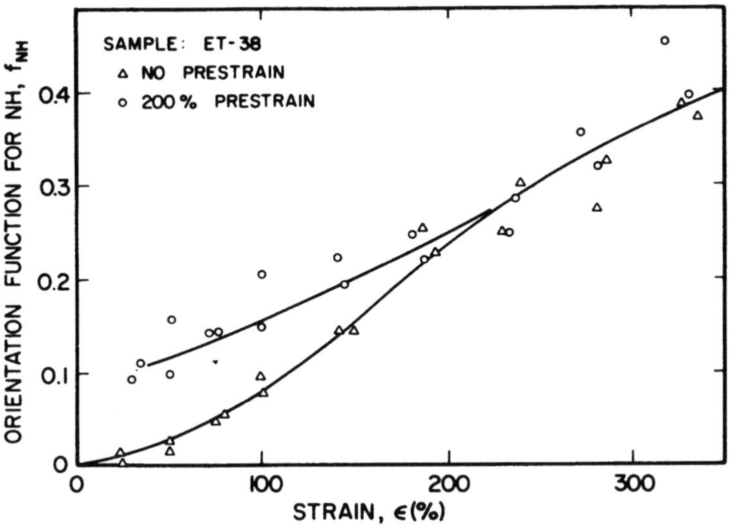

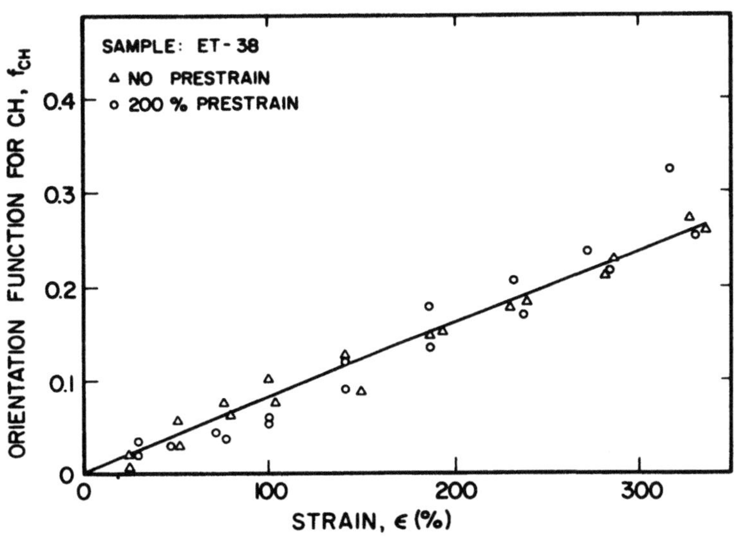

Figure 6. Orientation Functions for ET-38 as a Function of Prestrain (Type II Experiment).

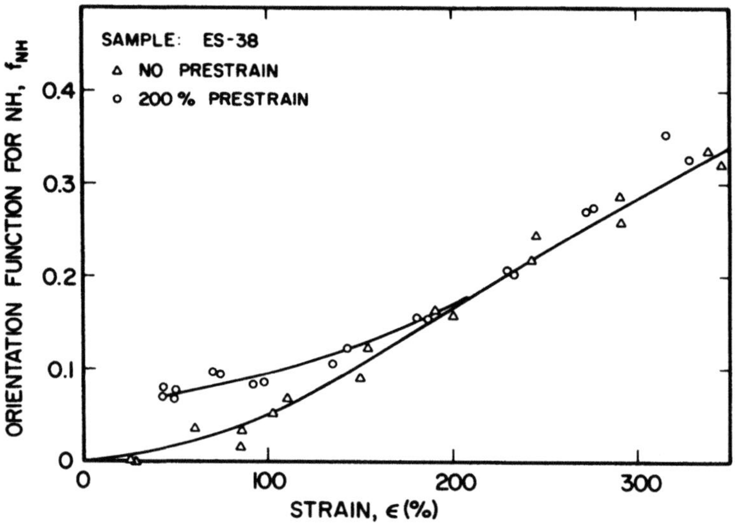

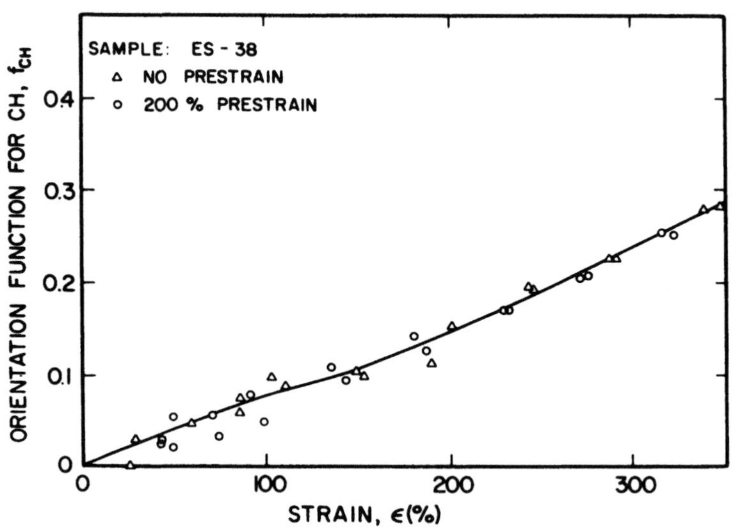

Figure 7. Orientation Functions for ES-38 as a Function of Prestrain (Type II Experiment).

CONCLUSIONS

The infrared dichroism data presented in this report lends further support to the existence of a previously inferred two-phase microstructure in polyurethane block polymers. By monitoring the NH and CH infrared stretching absorptions, it has been shown that the orientation of the different backbone segments depends strongly upon strain history. Upon initial straining, both segments orient similarly, but after relaxation, chains containing the NH group remain oriented to a substantially higher degree than those containing the CH group. If the sample is again strained, at strain levels lower than those previously applied, the NH containing chains are more highly oriented than during initial extension. The orientation of backbones containing CH groups is unaffected by the prestrain.

The distinct difference between the orientation functions for the two types of chains (in both ES-38 and ET-38) provides positive evidence that the different types of polymer chains exist in different domains, and react somewhat independently to the application and removal of stress. Hence, it may be inferred that one phase (previously referred to as the soft phase) is composed of prepolymer which contain the predominance of CH_2 groups. A second type of domain consists of ordered bundles of urethane linkages (previously referred to as the hard phase), and contains the NH group.

In an unstrained sample the ordered hard domains, although individually anisotropic, are randomly arranged so that the sample appears mechanically and optically isotropic. Initially as strain is applied, the prepolymer segments in the soft domains are extended and straightened out, but large-scale motion is impossible since each prepolymer segment is constrained by two hard domains. At higher strain levels the soft segments become taut so that further extension tends to orient the hard domains as well as the soft. Polymer segments in both phases then orient similarly up to high levels of strain.

When the load is removed, the soft segments are able to return to an almost isotropic state. However, there is little restoring force to disorient the hard domains, so they retain a large degree of their orientation. Upon subsequent straining, at low strain levels only the chains in the soft domains must be oriented so that lower stress levels are required.

A model consistent with the data envisions a dispersion of irregularly shaped soft and hard domains with neither type of domain more discrete nor dispersed than the other. This model is similar to one proposed by Bonart (17) for polyurethane elastomers which may be crystalline in both phases. Bonart's model is, however, not quantitatively applicable to the elastomers studied in this report, since wide angle X-ray investigations (18) on ET-38 and ES-38 have shown that neither domain is crystalline nor crystallizes during extension up to 300% strain. Using this general model, the mechanical properties and specifically the stress-softening in amorphous polyurethane elastomers may be qualitatively explained.

ACKNOWLEDGEMENT

The authors wish to thank Dr. Edward A. Collins of the B. F. Goodrich Chemical Company who kindly prepared laboratory size samples of well characterized segmented polyurethanes for our research. We are also grateful to the National Science Foundation for support of this research through grant GK-4554.

REFERENCES

1. Estes, G. M., S. L. Cooper, and A. V. Tobolsky, J. Macromol. Sci., C4 (1), 167 (1970).
2. Trick, G. S., J. Appl. Polymer Sci., 3, 252 (1960).
3. S. L. Cooper, D. S. Huh, and W. J. Morris, Ind. Engr. Chem., Prod. Res. Development, 7, 248 (1968).
4. Estes, G. M., R. W. Seymour, D. S. Huh, and S. L. Cooper, Polymer Engr. Sci., 9, 383 (1969).
5. Estes, G. M., D. S. Huh, and S. L. Cooper, Block Polymers, (S. L. Aggarwal, Ed.), Plenum Press, New York (1970), p. 225.
6. Puett, D., J. Polymer Sci., A-2, 5, 839 (1967).
7. Henderson, J. F., K. H. Grundy, and E. Fischer, J. Polymer Sci., C16, 3121 (1968).
8. Fischer, E. and J. F. Henderson, Rubber Chem. and Technol., 40 (5), 1373 (1967).
9. Wilkes, G. L. and R. S. Stein, J. Polymer Sci., A-2, 7, 1525 (1969).
10. Nishioka, A., J. Furakawa, S. Yamashita, and T. Kotani, J. Appl. Polymer Sci., 14, 799 (1970).
11. LeGrand, D. G., Polymer Letters, 7, 579 (1970).
12. Zbinden, R., The Infrared Spectroscopy of High Polymers, Academic Press, New York, 1964.
13. Fraser, R. D. B., J. Chem. Phys., 21 (9), 1511 (1953); 28 (6), 1113 (1958).
14. Stein, R. S., J. Polymer Sci., 31, 327 (1958).

15. R. W. Seymour, G. M. Estes and S. L. Cooper, Macromolecules, in press.

16. Koutsky, J. A., N. V. Hien, and S. L. Cooper, J. Polymer Sci., B, in press.

17. Bonart, R., J. Macromol. Sci., B2 (1), 115 (1968).

18. Huh, D. S., unpublished observations.

EMULSIFYING EFFECT OF BLOCK AND GRAFT COPOLYMERS - OIL IN OIL EMULSIONS

Gérard RIESS, Jacques PERIARD, Albert BANDERET

Ecole Supérieure de Chimie

3, rue Alfred Werner 68 - MULHOUSE (France)

Summary

Block and graft copolymers act as oil in oil emulsifier for the corresponding incompatible homopolymers and for two non miscible liquids, each of them being a selective solvent of one of the sequences of the copolymer.

This emulsifying effect of polystyrene-polyisoprene block and graft copolymers has been studied for DMF-hexane mixtures, in function of the solvent volume ratio, the molecular weight, composition and structure of the copolymer.

Viscosity measurements on these oil in oil emulsions and determination of the particle size of the dispersed phase which have been performed, show that the type of emulsion, as well as the mean diameter of the dispersed phase is, a function of these parameters. It appears also that the continuous phase of the emulsion is preferentially formed by the solvent having the best solubility for the emulsifier.

Information on the emulsifying effect of block and graft copolymers may be of value for the design of polymeric composite systems, like ABS resins on High Impact Polystyrene.

It has been shown previously that block and graft copolymers act as oil in oil emulsifier for the corresponding incompatible homopolymers (1-4). Especially in polymeric composite systems, like ABS resins or High Impact Polystyrenes, it has been possible to show that block and graft copolymers regulate the particle size of the dispersed rubber phase and contribute to the adhesion at the boundery of the two phases (4-5).

In order to show that this emulsifying effect of A-B block and graft copolymers is a general behaviour, we have studied the simpler case of two non miscible liquids a and b, where we have not the inconvenience of high viscosity as for polymeric systems. The two non miscible liquids are choosen such that one of the liquids, say a, is a good solvent for the sequence A of the copolymer and a non solvent for the other sequence B. Reversely, the liquid b is a solvent of sequence B and a non-solvent for sequence A of the copolymer.

These requirements are met for the system acetonitrile-cyclohexane as immiscible solvents in presence of polystyrene-polymethylmethacrylate block or graft copolymers or also for the system DMF-n hexane in presence of polystyrene-polyisoprene (PS-PI) block or graft copolymers.

A preliminary report has already shown that these systems give stable oil-in-oil emulsions whereas random copolymers or mixtures of corresponding homopolymers have no emulsifying effect (6).

In order to complete these previous results, we have studied the system DMF-hexane-copolymers polystyrene-polyisoprene, in function of the molecular characteristics, e.g. molecular weight, composition and structure of the copolymer.

I. Experimental

The blockcopolymers polystyrene-polyisoprene (Cop SI) are two blockcopolymers prepared by anionic polymerization. The graft copolymers polystyrene-g-isoprene (Cop G SI) and polyisoprene-g-styrene (Cop G IS) are also prepared by anionic polymerization after preliminary lithiation of the polymer, either polystyrene (PS) or polyisoprene (PI) with BuLi-NN', NN' tetramethylethylene diamine (7-8).

Molecular weight ($\overline{M}_n$) of the copolymers is determined by osmometry in toluene and their composition is obtained by selective degradation of the PI sequence according to KOLTHOFF (9).

These characteristics are given in Table 1. For a certain number of copolymers we have also indicated the $\overline{M}_n$ of the corresponding PS sequence.

Table 1

Characteristics of the copolymers

Cop	$\overline{M}_n$	$\overline{M}_n$ polystyrene sequence	polystyrene weight fraction	type of copolymer
SI 2	200,000	87,000	0.48	two-block
SI 3	33,000	13,000	0.44	"
SI 8	350,000	190,000	0.50	"
SI 9	44,000	36,000	0.72	"
SI 10	36,000	25,000	0.60	"
SI 11	25,000	14,000	0.55	"
SI 19	50,000	25,000	0.40	"
SI 21	20,000	12,500	0.53	"
G IS 5	70,500	/	0.154	graft : PI backbone
G SI 7	81,000	/	0.153	graft : PS backbone

After having shown that under given shaking conditions, we obtain in a few minutes oil in oil emulsions with reproductible characteristics, we have determined their viscosity and the mean diameter of the particles of the dispersed phase.

In fact, it has been established that viscosity is a characteristic property of water in oil or oil in water emulsions (10). In this case the viscosity is a function of

- the volume of the dispersed phase
- the physical properties of the two phases (viscosity, density...)
- the emulsifier (concentration, distribution between interface and the two phases ...)
- the size and the distribution of the emulsified particles.

We have found that viscosity is also a convenient method to characterize oil in oil emulsions and to determine the phase inversion point.

These viscosities have been determined in a Ubbelohde viscosimeter having following characteristics :

flow time of n hexane at 25° C : 70.7 sec
flow time of DMF at 25° C : 120.7 sec

The mean diameter of the dispersed particles is determined on phase contrast microphotographs.

II. Results

1) Molecular weight of Cop SI

In order to study the influence of the molecular weight of the copolymers on the characteristics of such oil in oil emulsions, we have selected Cop SI_2, Cop SI_3 and Cop SI_8 having same structure and practically same composition.

By plotting the viscosity of the emulsions, given by the flow time, in function of the volume ratio DMF-hexane, it appears a characteristic variation in function of the molecular weight of the Cop (Fig. 1).

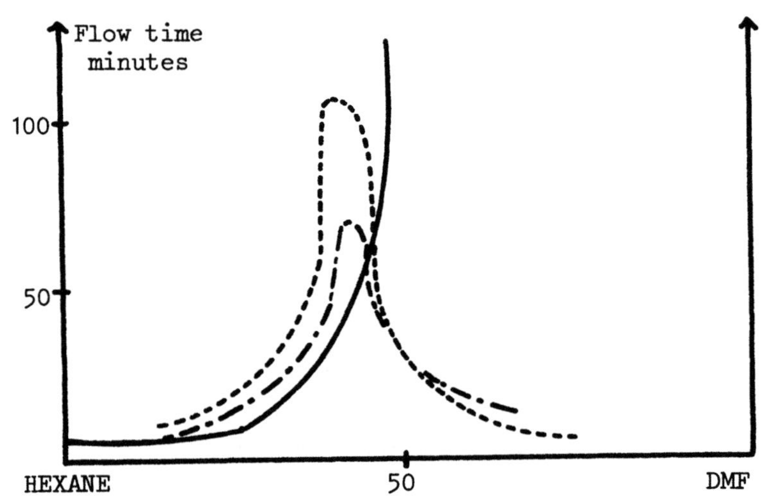

Fig. 1 : Influence of $\overline{M}_n$ on the viscosity of oil-in-oil emulsions
Copolymer concentration 3 g/100 cm^3 solvent mixture

----- SI_2

-·-·- SI_8

—— SI_3

So for Cop SI_3, having the lowest $\overline{M}_n$ and the best solubility in hexane, it is only possible to obtain stable emulsions of the type DMF in hexane, whereas those of hexane in DMF are unstable.

This behaviour is consistent with Bancroft's rule established for conventional oil-water emulsions, according to which to continuous phase of an emulsion is formed by the solvent having the best solubility for the emulsifier (11). Considering the solubility parameters in our case, we found that hexane is a better solvent for PI than DMF for PS. DMF is practically a Θ solvent for PS at 25° C.

By increasing the molecular weight of the copolymer (Cop SI_2), the two types of emulsions, DMF in hexane and hexane in DMF, are obtained. This can be attributed to the fact that Cop SI_2, having a higher molecular weight than Cop SI_3, has no preferential solubility neither in DMF nor in hexane. But hexane in DMF emulsions are still less stable than those of DMF in hexane.

For Cop SI_8 finally, we have the same behaviour as previously, but the viscosities are quite lower. At the same copolymer concentration, it appears that the emulsifying effect decreases by increasing its molecular weight. This can also be shown be considering the mean diameter of the dispersed particles for these emulsions (Table 2).

Table 2

Influence of the molecular weight of the copolymer on the emulsion characteristics

Cop	$\overline{M}_n$	Mean diameter µ m	type of emulsion
SI_3	33,000	1	DMF/hexane
SI_2	200,00	2	DMF/hexane and hexane/DMF
SI_8	350,000	5	DMF/hexane and hexane/DMF (greater polydispersity)

Note : the mean diameter of the dispersed particles is determined for an initial volume ratio DMF/hexane = 1 and at a concentration of 3 g Cop/100 cm^3 solvent mixture.

Phase inversion for emulsions of Cop SI_2 and Cop SI_8 corresponds to a mixture of 42.5 % vol. DMF - 57.5 % vol. hexane, which represents an initial ratio of 1 for the phase volumes, by taking in to account the slight mutual solubility of the solvents. In contrary for Cop SI_3 the domain of stable emulsions is greater, and phase inversion, leading in this case to unstable emulsions, corresponds to about 60 % vol. DMF in the initial mixture, or to 75 % vol. of dispersed phase, which is the well known theoritical limit for emulsions.

As shown previously, phase inversion can also be controlled by using a selective dyestuff of DMF (6).

2) Composition of the copolymers

By studying the emulsions obtained from Cop SI_{19}, Cop SI_3, Cop SI_{21}, Cop SI_{11}, Cop SI_{10} and Cop SI_9, having practically the same $\bar{M}_n$ but styrene contents ranging from 40 to 72 %, we obtain the viscosity curves given by Fig. 2 and the particle sizes given by Table 3.

Table 3

Influence of the copolymer composition on the emulsion characteristics

Cop	composition PS weight fraction	mean diameter μ m	type of emulsions	Remarks
SI_{19}	0.40	7	DMF/hexane	polydispersity
SI_3	0.44	1	"	monodispersity
SI_{21}	0.53	3	"	"
SI_{11}	0.55	6	"	polydispersity less stable emulsions
SI_{10}	0.60	> 10	DMF/hexane multiple emulsions	"
SI_9	0.72	> 10	DMF/hexane hexane/DMF	"

Note :

Cop SI_{19}, SI_3, SI_{21} copolymer concentration 3 g/100 cm^3 solvent mixture
Cop SI_{11} " " 4 g/100 cm^3 " "
Cop SI_{10}, SI_9 " " 5 g/100 cm^3 " "

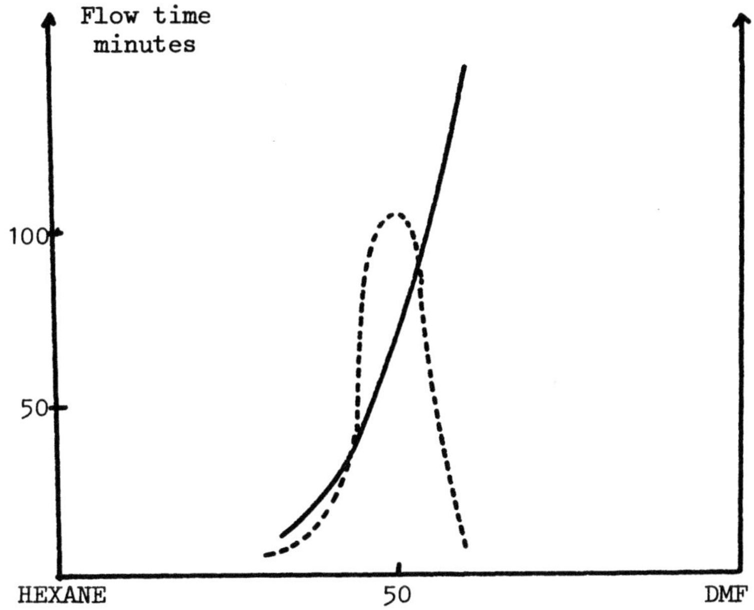

Fig. 2 : Influence of the copolymer composition on the viscosity of oil-in-oil emulsions

Copolymer concentration 5 g/100 cm^3 solvent mixture

——— SI_{10} ----- SI_9

From this table it appears that for this series of lower molecular weight copolymers, hexane in DMF emulsions can only be obtained for Cop SI_9, which has a PS content of 72 %.

Increasing the amount of PS in the Cop, corresponds to an increase of the total solubility of the Cop. in DMF, giving this way, according to Bancroft's rule, also emulsions where DMF can be the continuous phase. Nevertheless, the emulsifying effect of copolymers with high PS content is less, if we take into consideration the particle size and the stability of the emulsions.

This appears also in Table 4, where we have indicated the minimum amount of copolymer necessary to obtain a stable emulsion.

3) Structure

The two graft copolymers G IS_5 and G SI_7 have the same composition, similar M_n but very different structures. So, Cop G IS_5 has a PI backbone and PS grafts, whereas Cop G SI_7 has the reverse structure, e.g. a PS backbone and PI grafts.

Table 4

Minimum amount of copolymer for the formation of stable emulsions

Cop	composition PS weight fraction	minimum amount of copolymer % Cop for a mixture : 1 vol. hexane - 1 vol. DMF
SI_{19}	0.40	≃ 3
SI_{3}	0.44	3
SI_{21}	0.53	3
SI_{11}	0.55	4
SI_{10}	0.60	6
SI_{9}	0.72	8

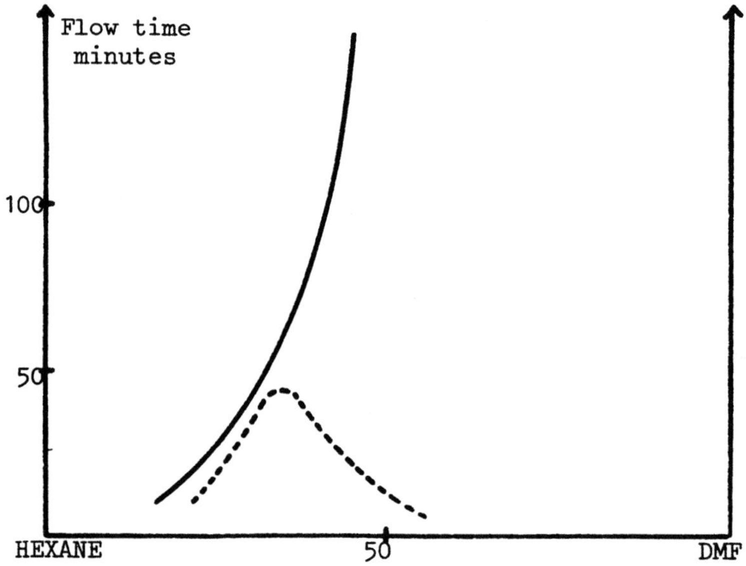

Fig. 3 : Influence of the copolymer structure on the viscosity of oil-in-oil emulsions
copolymer concentration 3 g/100 cm^3 solvent mixture ---G IS_5 —G SI_7

Fig. 3 shows that Cop G SI_7 gives only DMF in hexane emulsions, like two block copolymers of low PS content, whereas with Cop IS_5 it is possible to obtain the two types of emulsions hexane/DMF and DMF/hexane. So in a certain way, the type of the grafts commands the type of emulsion which can be obtained. Without considering the preferential solubility of PI in hexane, we notice that stable emulsions are obtained only if the continuous phase of the emulsion is a solvent of the grafts.

Conclusion

From this study it turns out that block and graft copolymers act as oil in oil emulsifier. The type of emulsion, the stability, as well as the mean diameter of the dispersed phase depends on the relative solubility of the copolymer in the two phases. This solubility being a function of molecular weight, composition and structure of the copolymer.

According to Bancroft's rule, established for conventional oil-water emulsions, it appears also for oil in oil emulsions that the continuous phase of the emulsion is preferentially formed by the solvent having the best solubility for the emulsifier.

References

1) G.E. MOLAU
J. Polymer Sci. A 3 1267 (1965)

2) G.E. MOLAU
J. Polymer Sci. A 3 4235 (1965)

3) G. RIESS, J. KOHLER, C.TOURNUT and A. BANDERET
Makromol. Chem. 101 58 (1967)

4) G. RIESS
IUPAC Meeting Budapest 1969

5) J. PERIARD, A. BANDERET and G. RIESS
Revue Gén. Caoutchouc - Plastiques in press

6) J. PERIARD, A. BANDERET and G. RIESS
Polymer Letters 8 109 (1970)

7) M.A. JAMPOLSKAYA, O. JU OKHLOBYSTIN, S.L. DAVYDOVA and N.A. PLATE
Vysokomol. Soedin. 8 771 (1966)

8) Y. MINOURA, K. SHIINA and H. HARADA
J. Polymer Sci. A 1 6 559 (1968)

9) I.M. KOLTHOFF, T.S. LEE, C.W. CARR
J. Polymer Sci. 1 429 (1946)

10) P. BECHER
Emulsions : Theory and Practice
New York Reinhold Publishing Corporation (1957)

11) W.D. BANCROFT
J. Phy. Chem. 17 501 (1913)
J. Phy. Chem. 19 275 (1915)

ABS POLYMER AS A COLLOIDAL DISPERSION

M. G. Huguet and T. R. Paxton

The B. F. Goodrich Research Center, Brecksville, Ohio

INTRODUCTION

ABS polymer consists of a dispersion of rubber particles in a matrix of styrene-acrylonitrile (SAN) copolymer resin. The rubber is usually polybutadiene. The initials ABS stand for the three monomers present, but this tells nothing about the morphology. Since the rubber and the SAN copolymer are not mutually soluble, there are two polymeric phases. The role of the rubber phase is to provide toughness to this structural thermoplastic. The plastic will be shaped by melt extrusion and molding. It is therefore important that the morphology which provides impact strength be able to survive melt processing.

ABS polymers of the graft type are made by copolymerizing styrene and acrylonitrile monomers in the presence of rubber. During the polymerization some of the SAN copolymer molecules become grafted to the rubber. This grafted SAN acts as a dispersing agent. It stabilizes the particles of rubber in the SAN polymer medium during melt flow.

The object of this paper is to point out the analogies between ABS dispersions and more conventional colloidal systems. Viewing ABS polymers as colloidal dispersions helps to understand many of the interesting properties of these graft polymer systems. To illustrate the colloidal properties of ABS polymers we have made some model polymers in which the amount of graft "dispersing agent'' is varied for several particle sizes of rubber. We have examined the behavior of these samples in solution and in melt flow. Electron micrographs confirm some of our conclusions.

EXPERIMENTAL DETAILS

We prepared the polymers by free-radical catalysis in emulsion. Two polybutadiene latices were made having different particle sizes; 880 and 2800Å. We measured particle sizes by standard soap titration techniques. Electron micrographs confirmed the particle diameters and also showed that the particle size was relatively narrow. A reasonably narrow distribution is necessary if particle size is to be an experimental parameter. The gel contents of the rubber samples were 80% for the small-particle latex and 10% for the large particle latex. However, since the rubber is extensively crosslinked during the grafting step this difference is not considered important for our present purposes.

Grafting was accomplished by polymerizing styrene and acrylonitrile monomers (30% acrylonitrile by weight) in the polybutadiene latex. By stopping this polymerization at various stages, we obtained a series of grafted rubber samples in which the graft content varied over a wide range. In a similar polymerization system we prepared some pure SAN copolymer – no rubber present. This SAN polymer was mixed with the grafted rubbers to give the desired rubber concentration in the final polymer mixture. We added 1.5 phr of a hindered phenol antioxidant emulsion to the latex mixture before coagulating and drying the ABS polymer.

The intrinsic viscosity of the SAN copolymer was 0.5 dl./g at 25°C in tetrahydrofurane. This was true for the free SAN copolymer extractable from the graft samples and for the SAN copolymer polymerized separately. The extent of grafting was determined from measurements of solubility in acetone. SAN polymer in this molecular weight range is readily soluble in acetone, whereas polybutadiene is not. Hence the grafted SAN is also insoluble. The insoluble portion can be separated from the acetone solution of ungrafted SAN by ultracentrifuging at about 10,000 times gravity.

Electron micrographs of the ABS polymers were made both of latex particles and of ultrathin sections of extrudates from the melt rheometer. We used the O_sO_4 staining techniques described by

Kato[1] to provide contrast between the rubber and SAN portions of the polymer sample. We measured melt viscosities in an Instron capillary rheometer. Measurements were made at a variety of temperatures and shear rates. However, for simplicity we report here only the results for a shear rate of 100 sec^{-1}. The melt viscosities at other shear rates show essentially the same trends as the ones we have selected to discuss. The rheometer capillary had an L/D ratio of 79, and the Rabinovitsch correction[2] was applied.

RESULTS

Table I describes our ABS samples. The degrees of grafting listed in column two are expressed as grams of grafted SAN per gram of polybutadiene substrate. The range covered is nearly a factor of ten in graft content. All of these ABS samples have been adjusted to 20% polybutadiene by addition of the proper amount of SAN copolymer.

TABLE I

EXPERIMENTAL ABS POLYMERS CONTAINING 20% RUBBER

Rubber Particle Size	Graft to Rubber Ratio	Behavior in Acetone	Melt Viscosity (100 sec^{-1}, 260°C) 10^3 Poise	Surface Conc. of Graft
880Å	0.15	Did not disperse	7.8	17Å
	.23	Did not disperse	6.8	28
	.31	Did not disperse	7.4	37
	.61	Partly dispersed	5.6	73
	.62	Partly dispersed	5.5	75
	.84	All dispersed	5.2	102
	.85	All dispersed	5.8	103
	1.30	All dispersed	7.7	157
2800Å	.13	Did not disperse	3.7	49
	.27	Partly dispersed	3.0	103
	.41	All dispersed	3.6	158
	.60	All dispersed	3.9	232
	.79	All dispersed	4.8	304

The first indication of the role which grafted SAN plays as a dispersing agent comes in measuring the graft content. When the samples are placed in acetone, those with more than a certain minimum amount of graft disperse to form a stable colloid. The others do not; the insoluble portion promptly settles out. It is the stable dispersion which makes an ultracentrifuge necessary for separating soluble from insoluble polymer. The critical graft level for dispersability is smaller for the larger particle size rubber. The fourth column of Table I lists the melt viscosities of these polymers at 100 sec^{-1} and 260 °C. As the degree of grafting increases the melt viscosity first decreases and then increases. The minimum in melt viscosity occurs at about the same critical level needed for dispersability.

SAN polymer and polybutadiene are not mutually miscible. Hence we can expect that polybutadiene latex particles with SAN copolymerized on them will be heterogeneous. Figure 1 shows electron micrographs of such particles at two stages of growth with SAN. The rubber, stained dark with O_sO_4, is in the core, while

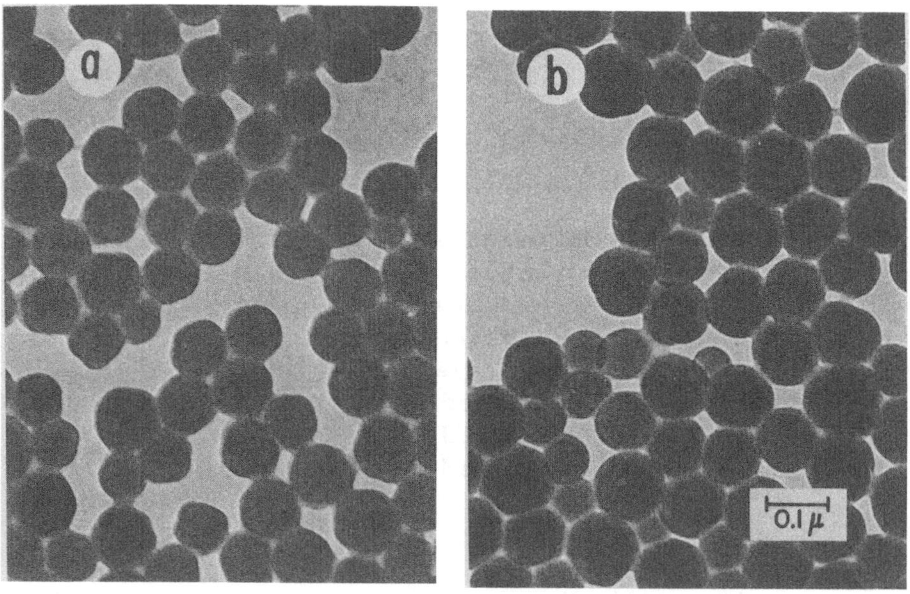

Figure 1. Electron micrographs of ABS latex particles showing the SAN polymer shell on rubber particles of 880Å diameter. The rubber is stained dark with O_sO_4. The graft ratios are 0.23 in (a) and 0.85 in (b).

the SAN forms a shell on the outside. Steinhofer,[3] Kato[4] and others have also reported such core-shell structure. Here we can clearly see that with further polymerization of SAN the shell becomes thicker. Of course only part of this SAN shell is grafted to the rubber. But since the growth of SAN occurs near the particle surface, the grafted SAN is probably also on the surface of the rubber. Since we regard the grafted SAN as a dispersing agent, a useful parameter would be the surface concentration of graft on the rubber. We therefore have calculated the quantity λ = (graft ratio) x $d_{BN}(\rho_{BN}/6\rho_{SAN})$, in which d_{BN} is the rubber particle size and ρ's are polymer densities. This is the volume of grafted SAN per unit area of polybutadiene surface. The quantity λ has the dimensions of length and represents the thickness of a hypothetical shell of pure grafted SAN. But the importance of λ is that it is a number which is proportional to the surface concentration of graft – a graft coverage factor. Values of λ are listed in Table I.

Figure 2 shows the variation of melt viscosity with graft content for the two particle sizes of rubber. The rubber concentra-

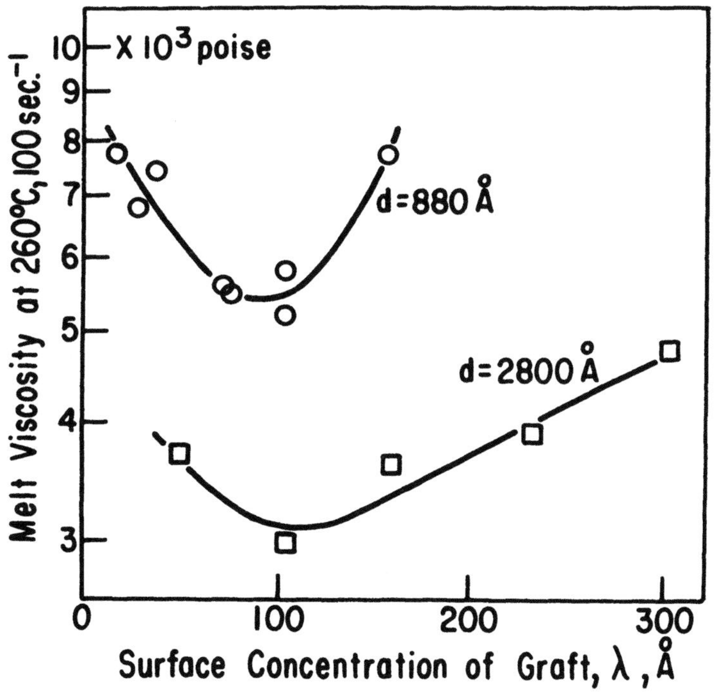

Figure 2. Melt viscosity as a function of the surface concentration of graft for ABS with small and large rubber particles.

tion in these polymers is constant at 20%. It is the proportion of grafted SAN which varies. Since the rubber and the graft attached to it constitute the disperse phase, we should expect the melt viscosity to increase with increasing graft ratio. We indeed see this trend in the range of high graft ratios. But at low graft ratios the melt viscosity is again high, creating a minimum in the curves. The melt viscosity minima for the two different rubber particle sizes occur at about the same surface concentration of graft, even though the weight graft ratios are quite different. The reason for the high melt viscosity at low surface concentration of graft is that the rubber particles flocculate in the melt. This parallels the behavior in acetone. Frazer[5] also has noted that graft helps to stabilize rheological properties in ABS polymers. The flocculation of rubber particles is easily seen in Figure 3. These pictures are transmission electron micrographs of ultrathin sections of extrudates from the melt rheometer. The sections were taken perpendicular to the melt flow direction. In Figure 3(a) the well-grafted rubber particles are separately dispersed even after melt shearing at 200 °C and 2000 sec^{-1}. Figure 3(b) shows a compression-molded sample (very little shear) of an ABS from the same rubber but with very little graft. It appears that the rubber particles may have flocculated somewhat during recovery of the polymer from its latex form. In Figures 3(c) and (d) we see that melt shearing in the rheometer at 200° and 290 °C and 2000 sec^{-1} produces further flocculation. Figures 3(e) and (f) show a similar contrast in the stability of the rubber dispersion for a larger particle size rubber with high and low graft ratios.

The melt flow behavior of well-grafted ABS polymers provides further illustrations of their analogy to classical colloidal dispersion. First, we saw in Figure 2 that the small-particle dispersion has a higher viscosity than the large-particle dispersion at the same rubber concentration for any graft ratio. The smaller particles are closer to each other and therefore interact more strongly in melt flow. Second, we have the dependence of viscosity on the volume fraction disperse phase. For this purpose, a portion of large particle size rubber, with a degree of grafting of 0.89 g./g., was mixed with SAN polymer so as to form final rubber contents of 5, 10, 15, 20, and 22%. The melt viscosities of these samples at four temperatures are shown in Figure 4. Here the concentration is expressed as the volume fraction rubber plus graft. The viscosity increases with concentration in the normal fashion for colloidal dispersions. By plotting log η as a function of $1/T$ we obtain the

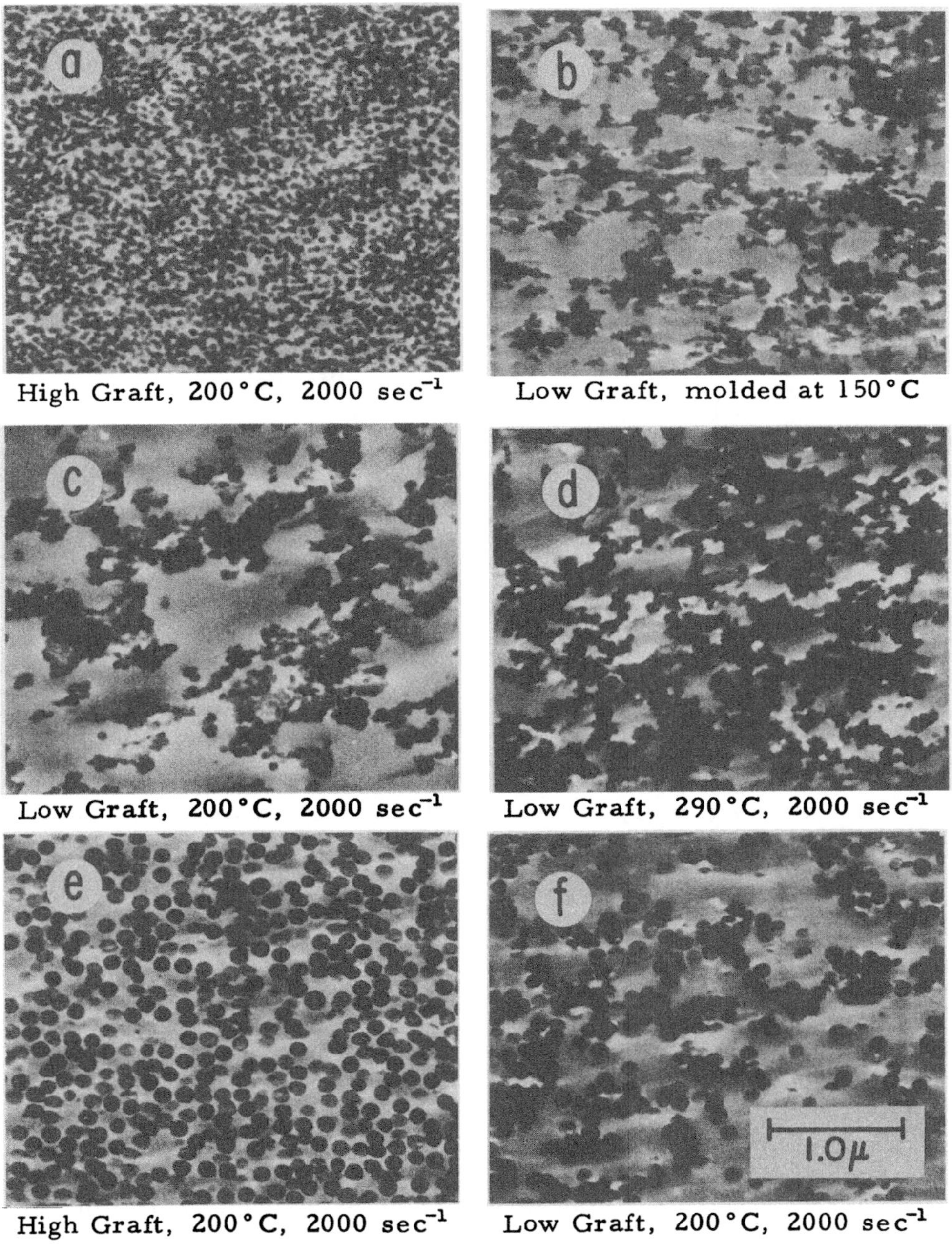

Figure 3. Electron micrographs of stained sections of extrudates of ABS containing 20% rubber: (a) 880Å diameter rubber, 0.85 graft ratio; (b), (c), (d), same rubber, 0.23 graft ratio; (e) 1900Å diameter rubber, 0.73 graft ratio; (f) same rubber, 0.05 graft ratio.

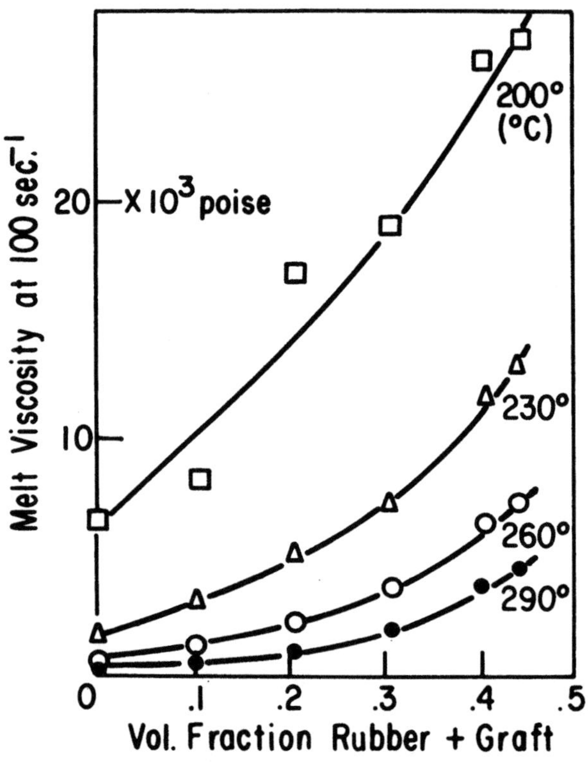

Figure 4. Melt viscosity of well-grafted ABS polymer as a function of the volume fraction of rubber plus graft, at various temperatures.

activation energies for melt flow, ΔH_{flow}. These are plotted in Figure 5 as a function of the volume fraction disperse phase. And we find that ΔH_{flow} decreases linearly with concentration and appears to be zero for pure disperse phase. In other words, the dispersed particles make no contribution to the variation of viscosity with temperature. The temperature coefficient is entirely due to the matrix of SAN polymer. Maron[6] found similar behavior for polystyrene latex. The latex particles made no contribution to the flow activation energy except to reduce the amount of water. We feel that is further evidence that the dispersed particles in ABS polymers flow as particles. Their motion is not a molecular process which can be aided by higher temperature.

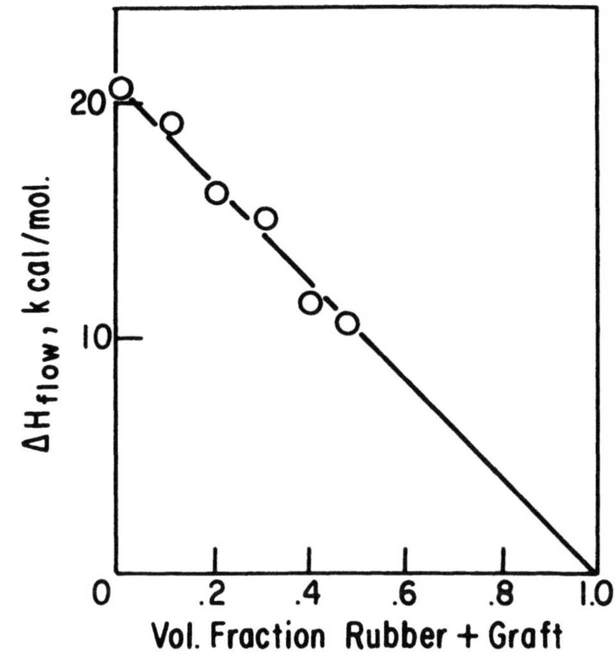

Figure 5. The activation energy for melt flow as a function of the volume fraction of rubber plus graft, for well-grafted ABS polymer.

CONCLUSIONS

The grafted SAN in ABS polymer serves as a dispersing agent for the particles of rubber. This is evident when the polymer is put into solvent, and also in the melt. If there is too little graft the rubber particles flocculate. The flocculation produces high melt viscosity. The amount of graft needed for a stable dispersion is proportional to the surface area of the dispersed rubber. In melt flow well-grafted rubber particles flow as particles, not molecules. They make very little contribution to the activation energy for melt flow.

ACKNOWLEDGEMENT

We wish to acknowledge gratefully the contribution of Mr. Ronald W. Smith in making the election micrographs, the assistance of Mr. F. W. Kunig in preparing polymers, and the permission of the B. F. Goodrich Co. to publish these results.

REFERENCES

1. Kato, K., Polym. Engng. Sci. 7, 38 (1967).
2. Rabinovitsch, B., Z. physik. Chem. 145A, 1 (1929).
3. Steinhofer, A., Kunststoffe 55, 551 (1965).
4. Kato, K., Koll. Z. & z.f. Polymere, 220, 24 (1967).
5. Frazer, W.J., Chem. & Ind., Aug. 13, 1966, p. 1399.
6. Maron, S.H., and S.M. Fok, J. Colloid Sci., 10, 482 (1955).

FORMATION OF HIGH IMPACT POLYSTYRENE FROM BLENDS OF POLYSTYRENE WITH BUTADIENE/STYRENE BLOCK COPOLYMERS

C. W. Childers, G. Kraus, J. T. Gruver and Earl Clark

Research and Development Department, Phillips

Petroleum Company, Bartlesville, Oklahoma 74004

I. INTRODUCTION

Two processes have been described for preparing impact polystyrene, graft polymerization of rubber solutions in styrene(1,2,3) and mechanical mixing of polystyrene with rubber(4). In both processes, toughness results from formation of dispersed rubber particles which are crosslinked and adhere to the continuous polystyrene phase(5,6,7). Size of the rubber particles in the graft process is determined by intensity of agitation during precipitation of the rubber(3) while antioxidant content and mixing time control the rubber dispersion of blends(4). Predominance of the graft process(8) in current production of impact polystyrene results from more efficient use of rubber in producing toughness.

We wish to describe a mixing process which utilizes rubber efficiently in producing impact polystyrene(9). Rubbery butadiene-styrene block copolymers are blended with polystyrene in the presence of free radical precursors. The rubbery blocks are crosslinked, and the polystyrene blocks attached to the crosslinked rubber particles act as anchors to the polystyrene matrix to produce a structure akin to the resins made by graft polymerization.

II. EXPERIMENTAL

Mixing

All compounds were mixed under nitrogen in a C. W. Brabender Company Plastograph equipped with cam style mixing blades. The mixing chamber was heated by hot oil circulated from a bath set at 185°C. Polystyrene was first fluxed at 5 rpm followed by addition of rubber. Mixing speed was then raised to 100 rpm for ten minutes. Peroxide, when included in the blend, was added three minutes after the start of mixing at 100 rpm. Torque and temperature were recorded continuously. One percent by weight of antioxidant was added two minutes before completion of mixing.

Testing

Specimens for measuring tensile strength and flexural modulus were cut from 1/16-inch thick compression-molded slabs and impact specimens from 1/8-inch thick slabs. "Dog-bone" tensile specimens were cut from strips 1/2" x 5 3/4" x 1/16" with a Tensil Kut milling machine. For polymers with high elongation, tensile strength is the maximum in stress which occurs around two percent elongation. Flexural modulus was measured according to ASTM D790-61, and notched Izod impact strength by ASTM D256-56. Melt flow was measured with a 5-kg weight at 200°C.

Dynamic properties were measured at 110 Hz with a Vibron Direct Reading Viscoelastometer (Toyo Measuring Instruments Company, Tokyo, Japan).

Materials

Elastomers listed in Table I were polymerized in hydrocarbon solution with alkyllithium initiator by methods described previously(10). The set includes polybutadiene (A) and two random copolymers containing 25 percent styrene, one with constant composition along the chain (E) and one with a small polystyrene block (F). Four block copolymers (B,C,D and K) with different molecular weights contain a pure styrene block and a random copolymer block. Polymers G, H, I and J are pure block copolymers with varying block lengths.

Molecular weights of the elastomers were calculated from gel permeation chromatograms. Values for block and random copolymers are only approximate since calibration curves intermediate between pure polybutadiene and pure polystyrene were used. No distinction was made between random and block polymers even though hydrodynamic volumes would be different for the two cases. Block lengths were estimated by multiplying the apparent molecular weight of the block polymer by the fraction of block styrene in the polymer.

TABLE I

ELASTOMERS USED IN BLENDS

	$M_w \times 10^{-3}$	Heterogeneity Index	Styrene, %	Block Styrene, %	$M \times 10^{-3}$ of Styrene Block	Polybutadiene Structure, %		
						cis	trans	Vinyl
A	96	1.16	0	0	-	60.7	33.3	6.0
B	148	1.17	25.5	20.4	30	60.8	32.9	6.3
C	98	1.15	25.3	20.2	18	49.5	42.6	7.9
D	50	1.15	25.5	18.6	9	57.3	35.9	6.8
E	86	1.12	25.8	0	0	59.5	31.6	8.9
F	86	1.13	25.7	0.9	0.8	58.8	33.8	7.4
G	91	1.10	10.5	9.2	8	40.4	50.2	9.4
H	93	1.17	18.1	17.3	16	55.8	37.3	6.9
I	93	1.18	25.5	24.9	23	49.9	42.1	8.0
J	155	1.20	24.9	23.4	36	51.7	40.5	7.8
K	83	1.2	25.0	18	15	38	53	9

Block polystyrene(11) and microstructure of the polybutadiene portion(12) were determined by previously described methods.

Polystyrene molecular weights were calculated from dilute solution viscosity measurements in toluene using $[\eta] = 10^{-4} \times M^{0.72}$ as given in reference 13. Unless specified, M_v of the polystyrene was 3.6×10^{-5} and polystyrene/elastomer ratio was 75/25.

In all blends where peroxide was included, 0.1 percent of recrystallized dicumylperoxide was used.

III. RESULTS AND DISCUSSION

As pointed out by earlier workers, elastomers are relatively inefficient in raising impact strength of polystyrene in simple blends. This is attributed to poor adhesion between rubber and polystyrene. An elastomeric butadiene-styrene block copolymer might be expected to improve this compatibility, with polystyrene blocks functioning similarly to grafted polystyrene chains in impact polystyrenes. In Table II, physical properties of 75/25 polystyrene blends with polybutadiene, 75/25 butadiene/styrene random copolymer and 75/25 butadiene/styrene block copolymer are compared. Molecular weights of the three elastomers are approximately equal. Although the block copolymer improves toughness as measured by the notched Izod impact test, flexural modulus and tensile strength are lowest with this elastomer.

TABLE II

POLYSTYRENE ELASTOMER BLENDS

Elastomer	Melt Flow, 5 Kg, 200°C	Impact Strength, Ft Lb/In	Flexural Modulus, PSIx10^{-3}	Tensile, PSI	Elong. %
A Polybutadiene	4.5	0.4	216	2350	3
E Random Copolymer	7.2	0.6	217	2120	2
C Block Copolymer	2.8	3.8	204	1780	4

A remarkable strengthening and touchening of these blends is produced by addition of 0.1 percent dicumyl peroxide which generates free radicals during mixing. As shown in Table III, blends of all three elastomers are toughened by peroxide and in the case of the block copolymer, strength and hardness are improved as well. Separate experiments in which 0.05 and 0.2 percent peroxide were added produced essentially the same properties.

TABLE III

POLYSTYRENE ELASTOMER BLENDS WITH PEROXIDE

Elastomer	Melt Flow, 5 Kg, 200°C	Impact Strength Ft Lb/In	Flexural Modulus, PSIx10^{-3}	Tensile, PSI	Elong. %
A Polybutadiene	0.6	0.8	213	2310	6
E Random Copolymer	0.8	1.2	224	2500	21
C Block Copolymer	1.1	4.3	232	3180	47

Plots of Brabender torque and mixing temperature as a function of mixing time are displayed in Figure 1. Initially polystyrene and rubber are fluxed at slow speed and then mixed for three minutes at 100 rpm. During this period temperature rises to about 185 C and torque diminishes from heat generated by mixing. About 15 seconds after peroxide is added, the torque begins to rise, reaching a maximum about one minute later. Decomposition half-life of dicumyl peroxide is near 15 seconds at this temperature. Beyond the torque maximum, both torque and temperature approach constant values. Similar results were found with both block and random copolymers, as well as with polybutadiene.

Bucknall(4) reports that without peroxide the rubber in blends tends to form layers resulting in a soft waxy material, and that peroxide causes a more complete dispersion of the rubber into larger particles. These same changes are apparently associated with block copolymer elastomer blends, although it will be shown that the rubber dispersion is finer. The increase in torque after peroxide addition is, of course, attributed to crosslinking of the rubber phase by the usual free radical process. Buildup of large aggregates is prevented by the mixing action, the final result being a dispersion of crosslinked polybutadiene segments. Adhesion to the polystyrene continuous phase is provided by the attached polystyrene blocks.

Storage and loss moduli as a function of temperature are shown in Figure 2 for block copolymer C blend, with and without peroxide. Peaks in loss modulus occurred at temperatures characteristic of pure polybutadiene and polystyrene, indicating a two-phase system with no compatibility between the phases. Peroxide caused a general reduction in loss modulus with no shift in peak positions. Storage modulus was raised by peroxide at temperatures above T_g of polybutadiene and reduced at lower temperatures.

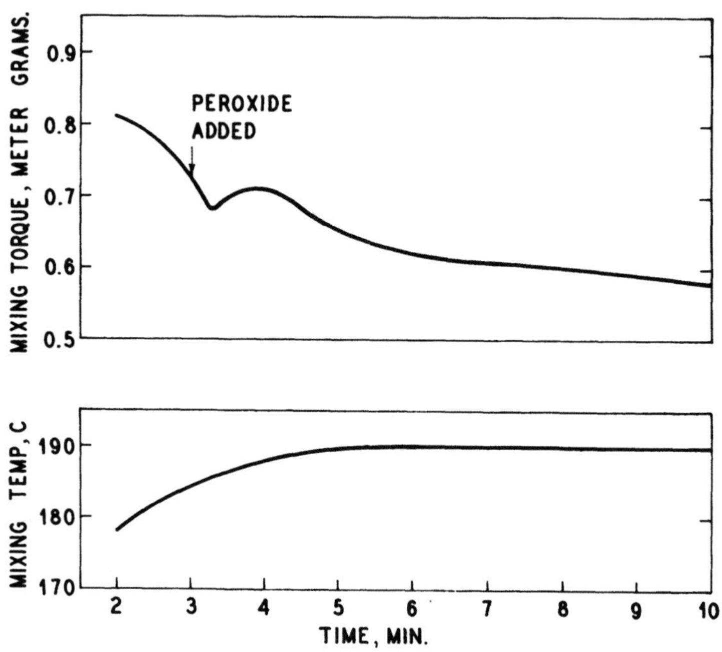

Fig. 1. Mixing torque and temperature of polystyrene blend with elastomer C and peroxide.

Rubber particles in the last two blends in Table II were easily seen with a phase contrast microscope. Each blend was hot mixed with additional polystyrene to a concentration of 1.0 percent or 0.25 percent based on rubber content. Since the rubber particles have been crosslinked, it is assumed that no change in their size distribution results from additional mixing. These dilute blends were dissolved in toluene from which films were cast. Figures 3 and 4 are phase contrast photomicrographs of random and block copolymer blends, respectively. Particles of the random copolymer range in size from about 2 up to 10 microns while those of the block copolymer were in the range of 2-3 microns.

Properties of peroxide treated blends in which rubber content was varied are listed in Table IV and plotted in Figure 5. It is apparent that increased levels of rubber raise impact strength while reducing melt flow and strength properties. In order to investigate separately the influence of rubber composition, molecular weight, and branching, as well as polystyrene molecular weight, comparisons were made at a constant rubber level of 25 percent.

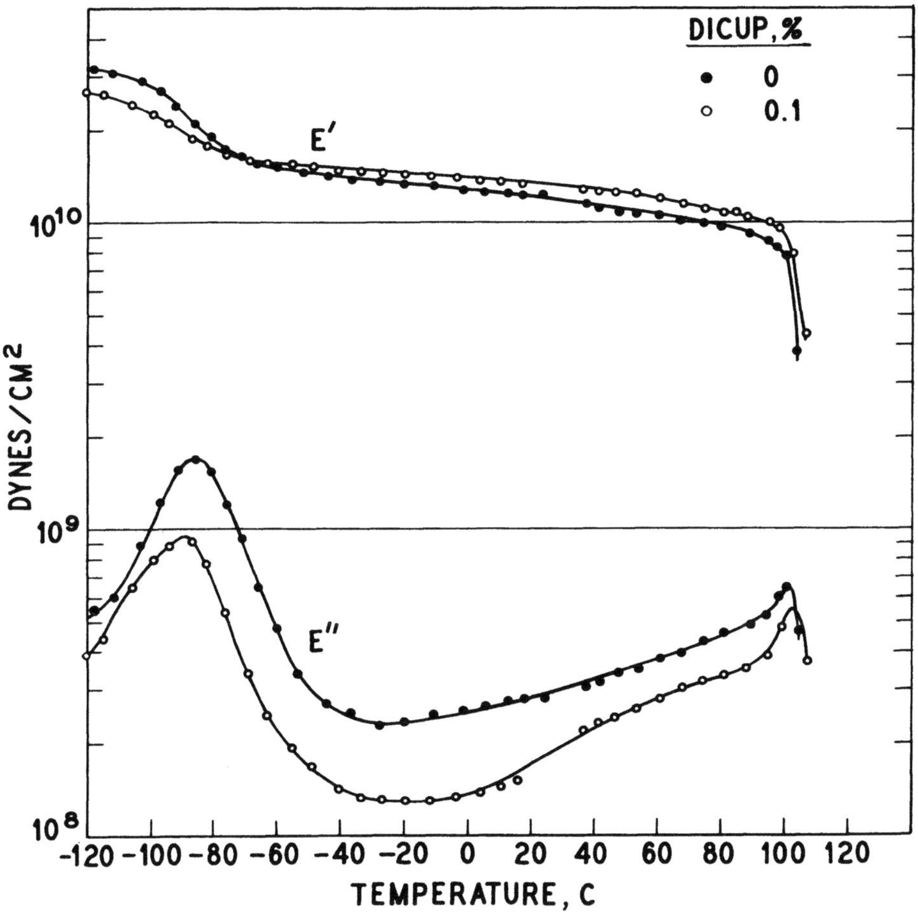

Fig. 2. Influence of peroxide on storage (E') and loss moduli (E") of polystyrene elastomer C blend.

TABLE IV

INFLUENCE OF ELASTOMER CONTENT ON PROPERTIES OF POLYSTYRENE BLENDS WITH ELASTOMER K AND PEROXIDE

Percent Elastomer	Melt Flow, 5 Kg, 200°C	Impact Strength, Ft Lb/In	Flexural Modulus, PSI x 10^{-3}	Tensile, PSI	Elong. %
10	1.4	0.3	360	5440	3
12.5	1.5	1.1	320	5030	6
15	1.4	2.7	280	4360	9
20	1.3	3.6	271	4080	16
25	0.9	4.2	235	3200	10

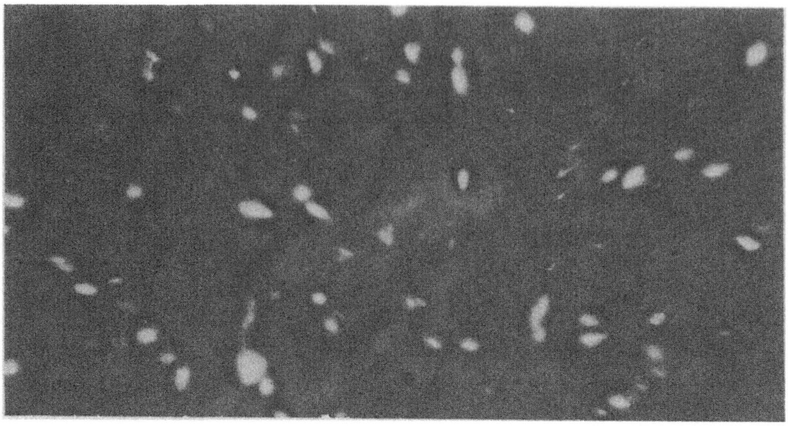

Fig. 3. Polystyrene/elastomer E blend with peroxide (phase contrast photomicrograph, 500X).

Polystyrene block length was first varied by maintaining elastomer composition constant and varying total chain length. Blends of polystyrene and peroxide with three block copolymers differing in molecular weight are shown in Table V. Impact strength is raised and melt flow reduced by increasing polystyrene block length. Hardness and modulus are unchanged, while tensile is raised slightly. These changes are attributed for the most part to improved adhesion between the phases. Rubber particle size observed under the phase contrast microscope was similar for the three blends. Dynamic properties of these blends shown in Figure 6 were identical, again illustrating the similarity of their morphology.

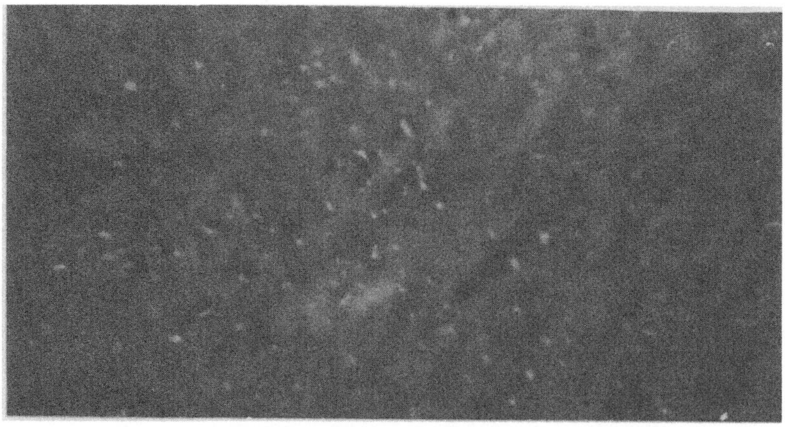

Fig. 4. Polystyrene/elastomer C blend with peroxide (phase contrast photomicrograph, 500X).

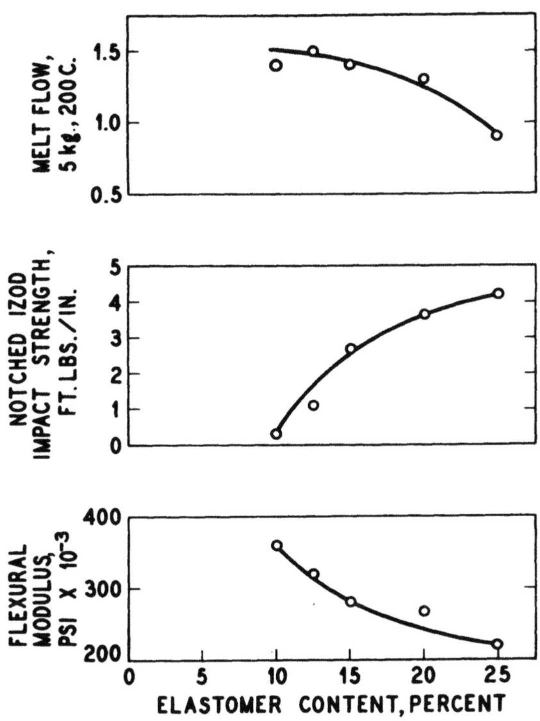

Fig. 5. Influence of elastomer content on properties of polystyrene blends with elastomer K and peroxide.

Polystyrene block length was next varied by maintaining composition and molecular weight of the elastomer constant while varying random and block copolymer content. Elastomer E was polymerized under conditions to produce constant composition along the chain. Another, F, varied in composition along the chain with a short polystyrene block at one end. The third had a pure polystyrene block and a copolymer block of low styrene content. Table VI illustrates that impact strength, modulus, tensile and hardness are all raised by increasing styrene block length.

Polystyrene block length was also varied in pure block copolymer elastomers with polystyrene content ranging up to 25 percent. Properties of blends, with peroxide and polystyrene block length ranging from 8,000 up to 36,000 molecular weight are listed in Table VII. In this set, impact strength increases with polystyrene block length up to about 23,000 molecular weight. Lower impact strength is found with a polystyrene block length of 36,000. Melt flow is maximum at M=16,000. Based on these results, adhesion between the two phases appears to be optimum for M=23,000.

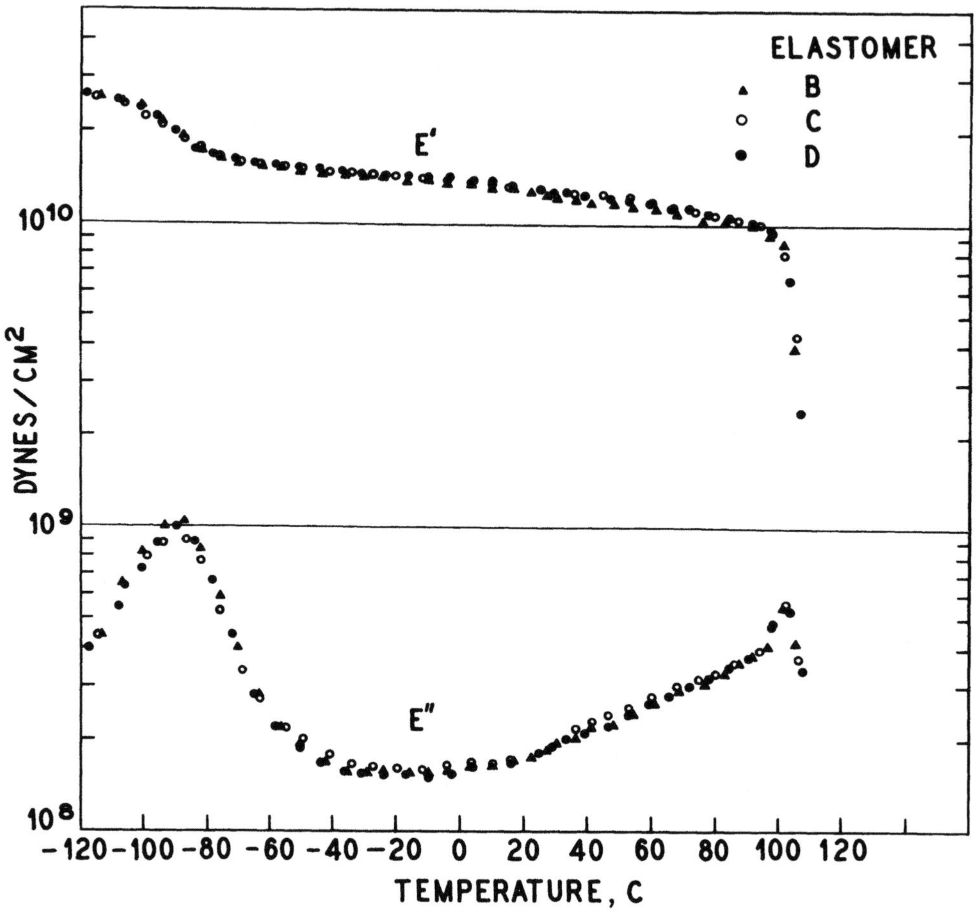

Fig. 6. Storage and loss moduli of polystyrene blends with peroxide and block copolymer elastomers of varying molecular weight.

TABLE V

INFLUENCE OF BLOCK COPOLYMER MOLECULAR WEIGHT ON BLENDS WITH POLYSTYRENE AND PEROXIDE

Elastomer	$M_w x10^{-3}$	Poly-styrene Block $M x10^{-3}$	Melt Flow, 5 Kg, 200°C	Impact Strength, Ft Lb/In	Flexural Modulus, PSIx10^{-3}	Tensile, PSI	Elong. %
B	148	30	0.5	4.6	234	3230	25
C	98	18	1.1	4.3	232	3180	47
D	50	9	1.5	2.5	235	2830	9

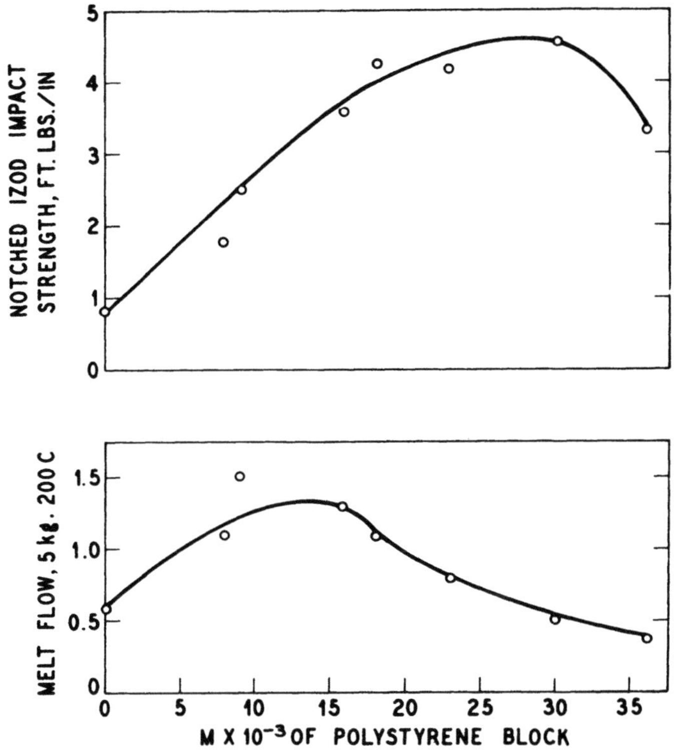

Fig. 7. Impact strength and melt flow of polystyrene/elastomer/peroxide blends as a function of polystyrene block length in elastomer (data from tables V and VII).

Plots of impact strength and melt flow as a function of polystyrene block length are shown in Figure 7. It is apparent that these properties are relatively insensitive to the original molecular weight and composition of the rubber phase.

Four polystyrenes of different molecular weights were prepared by free radical polymerization in an aqueous suspension. These were blended with peroxide and elastomer K which has an approximate polystyrene block molecular weight of 15,000. As shown in Table VIII, melt flow is reduced and all other properties are raised by increased polystyrene molecular weight. Figure 8 shows that melt flow and impact strength are smooth functions of polystyrene molecular weight at constant polystyrene block length of the elastomer.

Impact strengths of polystyrene blends with polybutadiene are also raised by peroxide and relatively small quantities of

TABLE VI

POLYSTYRENE BLENDS WITH PEROXIDE AND CONSTANT COMPOSITION ELASTOMERS

Elastomer	Polystyrene Block, $M \times 10^{-3}$	Melt Flow, 5 Kg, 200°C	Impact Strength, Ft Lb/In	Flexural Modulus, PSIx10^{-3}	Tensile, PSI	Elong. %
E	0	0.8	1.2	224	2500	21
F	0.8	0.9	2.0	229	2620	17
C	18	1.1	4.3	232	3180	47

TABLE VII

POLYSTYRENE BLENDS WITH PEROXIDE AND VARIABLE COMPOSITION ELASTOMERS

Elastomer	Polystyrene Block, $M \times 10^{-3}$	Melt Flow, 5 Kg, 200°C	Impact Strength, Ft Lb/In	Flexural Modulus, PSIx10^{-3}	Tensile, PSI	Elong. %
A	0	0.6	0.8	213	2310	6
G	8	1.1	1.8	239	2960	9
H	16	1.3	3.6	233	3240	8
I	23	0.8	4.2	243	3620	54
J	36	0.4	3.3	243	3870	30

TABLE VIII

INFLUENCE OF POLYSTYRENE MOLECULAR WEIGHT ON BLENDS WITH PEROXIDE AND ELASTOMER K

Polystyrene $M_v \times 10^{-5}$	Melt Flow, 5 Kg, 200°C	Impact Strength, Ft Lb/In	Flexural Modulus, PSIx10^{-3}	Tensile, PSI	Elong. %
3.7	0.31	5.4	246	3160	47
2.9	1.2	4.6	234	3270	40
1.5	5.3	2.7	227	3020	2.5
0.87	10.9	0.2	211	1790	1

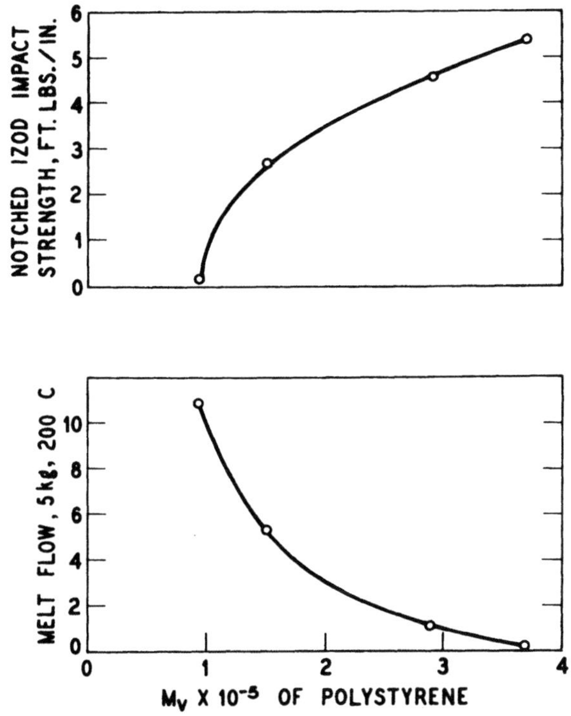

Fig. 8. Influence of polystyrene molecular weight on melt flow and impact strength of blends with elastomer K and peroxide.

block copolymers rich in styrene. In Table IX, polystyrene, polybutadiene and block copolymers containing between 47.3 and 95.6 percent styrene were blended with peroxide. Relative amounts of polybutadiene and block copolymer were adjusted to maintain total polystyrene and polybutadiene in the blend constant. Impact strength was raised by the block copolymers. Apparently, butadiene ends become attached to polybutadiene regions and the attached polystyrene blocks raise adhesion to the polystyrene phase.

Multiple block copolymer elastomers also respond to peroxide in polystyrene blends as shown in Table X although the changes are smaller than for simple block copolymers.

TABLE IX

INFLUENCE OF HIGH STYRENE BLOCK COPOLYMERS ON POLYSTYRENE BLENDS WITH POLYBUTADIENE AND PEROXIDE

Blend Components, %							
Poly-styrene	Elastomer A	Bd/S Block Polymer	Melt Flow, 5 Kg, 200°C	Impact Strength, Ft Lb/In	Flexural Modulus, PSIx10^{-3}	Tensile, PSI	Elong. %
75	25	-	0.6	0.8	213	2310	6
70.5	20.5	9(47.3)[a]	0.2	4.6	180	2930	41
70.5	23.5	6(72.8)[a]	0.2	5.8	191	2010	38
70.5	24.5	5(91.0)[a]	0.3	4.2	207	2790	17
70.5	24.8	4.7(95.6)[a]	0.4	1.7	210	2720	5

a - Styrene content of block copolymer

TABLE X

POLYSTYRENE BLENDS WITH ELASTOMERS CONTAINING MULTIPLE BLOCKS

Elastomer Type	Peroxide, %	Melt Flow, 5 Kg, 200°C	Impact Strength, Ft Lb/In	Flexural Modulus, $PSI \times 10^{-3}$	Tensile, PSI	Elong. %
SBS[a]	0	0.8	3.6	246	2960	15
Blocks	0.1	1.1	4.0	250	3610	32
Multi-[b]	0	2.3	2.9	249	2350	12
Chain	0.1	1.7	4.6	274	3460	7

a - S-B-S 12.5 - 75 - 12.5 Terblock elastomer.

b - B-S 75 - 25 Block copolymer with butadiene ends coupled.

References

1. Merz, E. H., G. C. Claver and M. Bare, J. Poly. Sci., 22, 325 (1956).

2. Turley, S. G., J. Poly. Sci., Pt. C, 1, 101 (1963).

3. Molau, G. E. and H. Keskkula, J. Poly. Sci., A1, 4, 1595 (1966).

4. Bucknall, C. B., Trans. Inst. Rub. Ind., 39, T221 (1963).

5. Newman, S. and S. Strella, J. Appl. Poly. Sci., 9, 2297 (1965).

6. Fletcher, K., R. N. Haward and J. Mann, Chem. Ind. (Lond.), 1854 (1965).

7. Bucknall, C. B. and R. R. Smith, Polymer, 6, 437 (1965).

8. Angier, D. J. and E. M. Fettes, Rubber Chem. & Tech., 38, 1164 (1965).

9. Childers, C. W., (to Phillips Petroleum Co.) U.S. 3,429,951 (Feb. 25, 1969).

10. Zelinski, R. P., (to Phillips Petroleum Co.) U.S. 2,975,160 (March 4, 1961) and U.S. 3,251,905 (May 17, 1966).

11. Kolthoff, I. M., T. S. Lee and C. W. Carr, J. Poly Sci., 1, 429 (1946).

12. Silas, R. S., J. Yates and V. Thornton, Anal. Chem., 31, 529 (1959).

13. Kurata, M. and W. H. Stockmayer, Fortschr. Hochpolym-Forsch., 3, 196 (1963).

MORPHOLOGY AND DEFORMATION BEHAVIOR OF CROSSLINKED BUTYLACRYLATE COPOLYMER GELS, PREPARED BY GRAFT REACTIONS

B. Vollmert, H. Stutz

Polymer-Institut, Universität Karlsruhe, Germany

THE PROBLEM AND THE METHOD OF FINDING THE ANSWER

The reaction of a solution of a OH-groups-containing polymer with a diisocyanate or a dicarboxylic acid dichloride leads to a gel, crosslinked by covalent bonds. The crosslink density depends at a given polymer concentration on the concentration of OH-groups in the polymer chains and on the conversion of the reaction.

We have carried out this reaction with a OH-groups containing n-butylacrylate-butanediol-monoacrylate (BMA) copolymer with 5 mol % BMA ($\bar{M}_w$ 500 000) and such amounts of hexamethylene diisocyanate that the molar ratio of functional groups OH/NCO was 1/1. The solvent was acetone. The same reaction was repeated with adipic acid dichloride as crosslinking agent in benzene as solvent at a reaction temperature of 70°C. The resulting reaction products were, as to be expected, polybutylacrylate gels.

The same reaction product should be obtained by reaction of the 5 mole % OH-groups-containing poly-n-butylacrylate with a 5 mole % NCO-groups-containing poly-n-butylacrylate, provided that the functional groups are available to each other. This will be true if the polymer coils can penetrate one another completely as predicted by the thread-model of polymer solutions or gels (Fig.1).

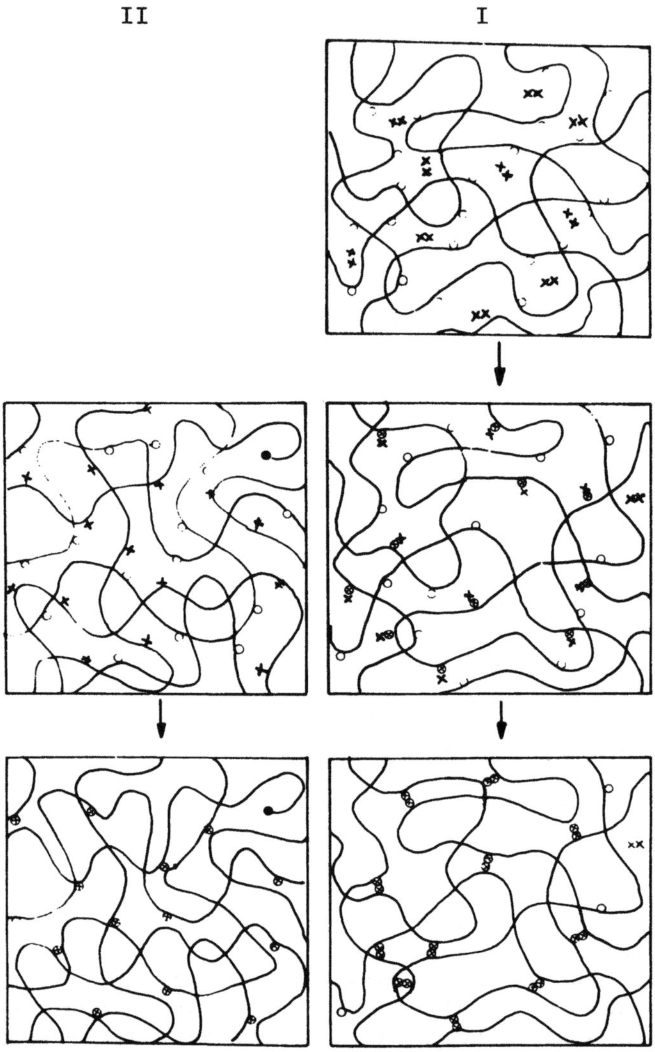

Fig. 1 Scheme of crosslinking of a butylacrylate-BMA-copolymer:

I Crosslinking by a conventional crosslinking agent such as hexamethylene diisocyanate or adipic acid dichloride

II Crosslinking <u>via</u> graft reaction of two butylacrylate copolymers by functional groups

o = OH groups; + = NCO groups (COCl, resp.);

⊕ = Urethane groups (ester groups, resp.)

++ = Hexamethylene diisocyanate (adipic acid dichloride, resp.)

BMA = Butanediolmonoacrylate

Fig. 1 shows the two crosslinking reactions in concentrated polymer solutions, depicted in the usual way. The intermediate stage of I after the first reaction step (reaction of diisocyanate with an OH group of a polymer chain) is an analogous example of the beginning stage of II, except for the fact that in system I macromolecules contain both types of functional groups in the same chain. If the thread model as shown in Fig. 1 correctly describes the state of concentrated solutions and gels, the maximum conversion available should be approximately the same in both reactions I and II, viz. close to 100 %, since in both experiments the functional groups are distributed statistically and homogeneously in the available volume.

There is another possibility to represent reactions I and II as shown in Fig. 2 (In Fig. 2 only the reaction II is formulated. Obviously, reaction I could be represented in an analogous way). Inasmuch as the relatively small molecules of the crosslinking agents hexamethylene diisocyanate or adipic acid dichloride can move without hindrance in the interior of the random coils, in reaction I by the new formulation according to Fig. 2 nothing would be changed with regard to the maximum conversion, quite contrary to reaction II. If the formulation in Fig. 2 is correct, and if the random coils of the two copolymers can not or not completely penetrate one another by mutual diffusion of chain segments, the highest realizable conversion should be only a relatively small portion of the theoretical conversion. Therefore, we should be able to use reaction II to measure the possibility and the degree of mutual penetration of macromolecules by analytical determination of the maximum conversion.

The following objections can be made with regard to this conclusion:

(1) It may be that the mutual penetration of coils, even if it is possible, is prevented by the crosslinks formed by the reaction II itself. The degree of this hindrance can not be calculated.

(2) It may be that the two butylacrylate copolymers become incompatible by the presence of different functional comonomer units even if the amount of functional comonomers M_X and M_Y being present in the copolymer chains is unimportant.

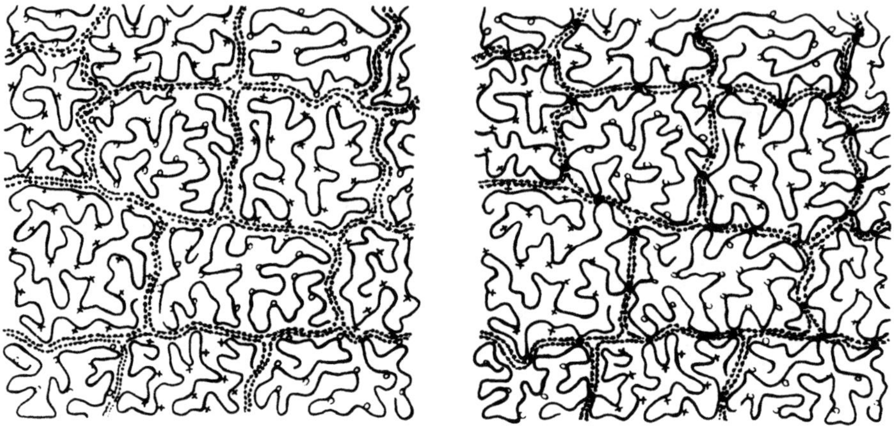

—o— butylacrylate-butanediolemonoacrylate-copolymer
—+— butylacrylate-isocyanatoethylacrylate-copolymer
::::: contact area between the coils

o OH-group
+ NCO-group
• urethane-group

Fig. 2 Scheme of crosslinking via graft reaction of two butylacrylate copolymers by functional groups (reaction II)

Both effects, the hindrance of coil penetration by crosslinking and by incompatibility, must become all the smaller, the smaller we make the number of functional groups X and Y, and both influences will disappear (become infinitely small) as the concentrations of X and Y go to zero. Therefore, by determination of the maximum conversion of reaction II at different concentrations of X and extrapolation of $X \longrightarrow 0$, we will obtain a conversion U_o at infinitely small X-values. This conversion U_o is a correct measure of the degree of the mutual coil penetration.

THE EXPERIMENTAL WORK

Maximum Conversion With Reaction I (see Fig. 1)

First the reaction was carried out with diisocyanate: A 20 % solution of butylacrylate-BMA copolymer with 10 mole-% BMA ($\bar{M}_W$ 500 000) in acetone was mixed with an equivalent amount of hexamethylene diisocyanate.

After a reaction time of 20 h at 25° C with rigorous exclusion of humidity, the UR spectrum showed a complete disappearance of the NCO absorption band, that is, the conversion of the reaction was nearly quantitative (within the scope of the sensitivity of the spectrometer).

This reaction was repeated with adipic acid dichloride at 70° C using benzene as the solvent. The conversion was estimated by analytical determination of the chlorine content after removal of HCl and solvent by vacuum evaporation. Five experiments showed a constant conversion between 95% and 97 %. This indicates a nearly quantitative conversion of reaction I according to Fig. 1 or analogous to Fig. 2.

Highest Obtainable Conversion of Reaction II (see Fig. 1)

A solution of a copolymer of butylacrylate and butanediolmonoacrylate (BMA) in acetone was mixed with a second copolymer of butylacrylate and isocyanatoethylacrylate.After addition of 0,03 % of diazabicyclo-2,2,2-octane (DABCO) the reaction was carried out at 25° C for 100 h. All monomers and solvents were dried rigorously and all operations were carried out under strict exclusion of humidity. As the experiments of BMA-copolymer with hexamethylene diisocyanate have shown, the reaction was complete within 20 h under the given conditions. The parameter r_1 and r_2 of the copolymerisation of the employed monomer pairs are close to 1 so that we could be sure to get a statistical distribution of the functional groups in all macromolecules present in the reaction mixture. The concentration of the functional monomers M_X, M_Y resp., was varied between 20 and 1 mole % and the polymer concentration in the reaction mixture was varied between 5 and 30 weight %. The reactions were carried out with such amounts of the reactive copolymer solutions that the molar ratio of $[M_X]$ / $[M_Y]$ was 1/1.

The maximum attainable conversions were determinated by measurements of the degree of swelling of the crosslinked reaction products according to reaction II. The relation between conversion (= experimental degree of crosslinking/ theoretical degree of crosslinking) and degree of swelling (theoretically described by the FLORY-REHNER equation) was experimentally determined by the reaction of butylacrylate-BMA copolymers with butylacrylate-acrylic acid chloride copolymers and by analytical determination of the remaining chlorine content (Fig. 3).

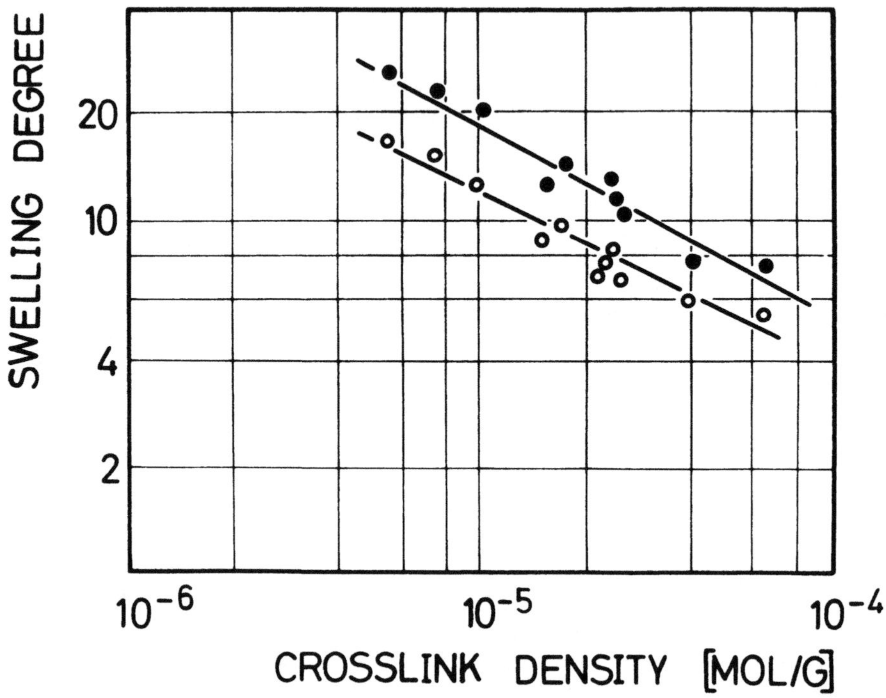

Fig. 3 Calibration curve between degree of swelling and degree of crosslinking of gels prepared by reaction II
o acetone ● benzene

With the aid of this calibration curve all reactions described above between the OH containing butylacrylate copolymers and the N=C=O containing butylacrylate copolymers could be evaluated. Fig. 4 shows the results.

DISCUSSION OF THE RESULTS

The curves of Fig. 4 are almost straight lines especially at the segments of low values of $[M_X]$ (in accordance with the work of J.C.H.HWA[1]) so that an extrapolation of $[M_X] \longrightarrow 0$ is allowed. Therefore, it is justified to consider the conversion U_0 as a measure for the degree of mutual penetration of the polybutylacrylate coils in a 30 % solution. Fig. 5a shows that the depth of the penetration zones depends on the polymer concentration. This is not very surprising. But it is remarkable that the overlapping contact zones of the macromolecules in concentrated solutions and gels are so small (Fig. 5b).

The decision between the two formulations of reaction II (Fig. 1 and 2) is clear now: The formulation in Fig. 2 is

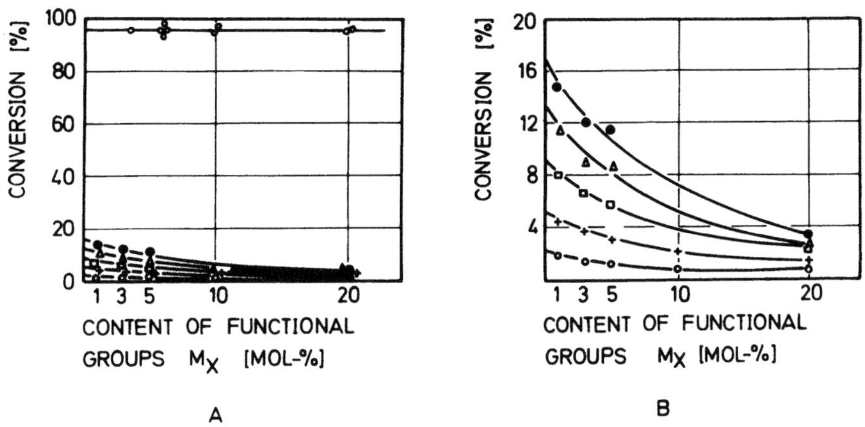

Fig. 4 Highest attainable conversion of two reactive polybutylacrylates, one of them containing OH groups and the other one containing N=C=O groups as a function of $[M_X]$ at different polymer concentrations
$[M_X]$ = concentration of the functional groups (mole %) in the reacting copolymers
At Fig. 4b the scale of the ordinate is extended in order to allow a better distinction between the individual curves.
Molecular weight of the copolymers: $\bar{M}_W \approx 500\,000$
Copolymer concentration during the reaction: curve 1: 30 % by weight; curve 2: 25 % ; curve 3: 20 %; curve 4: 15 % ; curve 5: 10 %

correct, and polymer gels of random coils forming polymers have a cell like tertiary structure as represented in Fig. 6.

With the described method the arrangement of the macromolecules is fixed at the state they have just at the moment of the crosslinking reaction. The received "snapshot" is "developed" by quantitative analysis of the non-reacted functional groups.

The result which was found, is in accordance with earlier works of SH. MARON et al[3], J.C.H. HWA[1], S.ONOGI et al[4], H. WESSLAU[2] and J. KRUSE and R. MEYER[5], who came to quite similar conclusions by means of different and more indirect ways.

THE DEFORMATION BEHAVIOR OF THE REACTION PRODUCTS

With regard to the cell structure and the degree of overlapping at the border zones of the coils there will be no difference between the gels prepared by reaction I and reac-

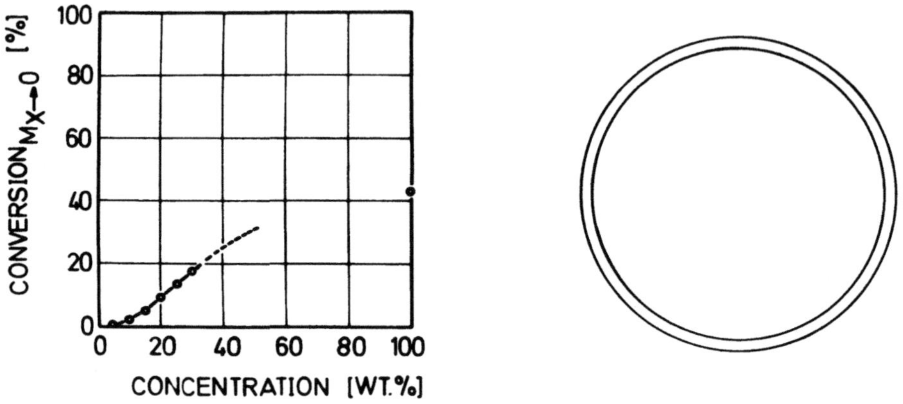

Fig. 5

a) Increase of the degree of overlapping contact zones (in the sense of Fig. 6) with increasing polymer concentration in solutions or gels
⊙ = experimental value of HWA [1] and WESSLAU[2]

b) Meridian section through a sphere with a shell having a volume of 20 % of the total sphere for demonstration of the breadth of contact zones in 30 % polymer gels.

tion II. The difference is in the different number and arrangement of the crosslinks in the two gels, schematically illustrated in Fig. 7.

Reaction I leads (analogous to the reaction scheme in Fig. 2) to a gel with homogeneous crosslink distribution of inter- and intramolecular crosslinks, whereas the gel prepared by reaction II (see Fig. 2 and 6) has exclusively intermolecular and no intramolecular crosslinks. The crosslinks of gel II (Fig. 7b) can only be in the contact zones of the polymer coils.

This different gel morphology should cause a different behavior of mechanical deformation.

We have studied the change of the mechanical moduli (measured by gel compression [6]) with increase of the polymer concentration during the crosslinking reactions (I) and (II). By this way of varying the degree of crosslinking the content of functional groups in all samples is the same. Fig. 8 shows the results.

The curves represented in Fig. 8 contain the following facts:
1. The modulus of gel (II) increases with the increase of polymer

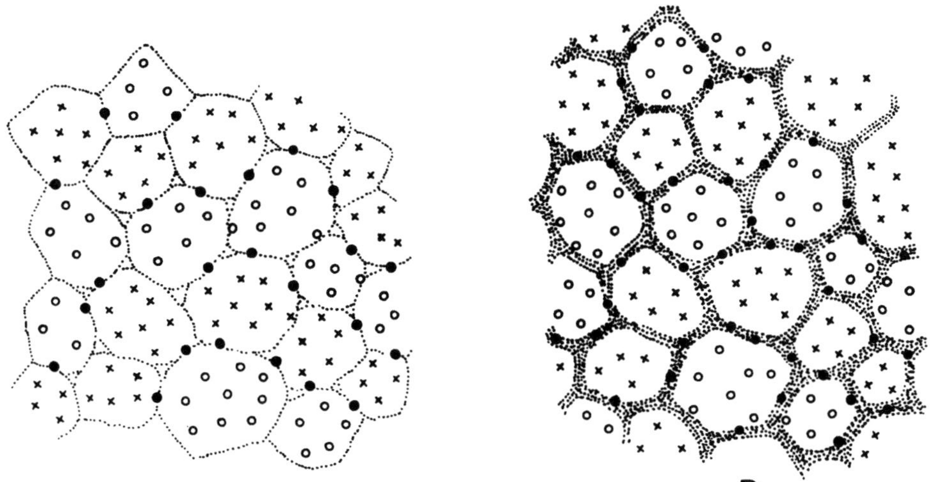

Fig. 6 Cell structure model of concentrated polymer solutions or gels of random coil forming polymers
a) Close to the critical polymer concentration (4 - 8 %, depending on molecular weight, type of solvent and chain structure)
b) Polymer concentration 30 - 50 %

concentration during the crosslinking reaction (Fig. 8a).

2. The modulus of gel (I) increases with the increase of polymer concentration during the crosslinking reaction (Fig. 8a).
3. The moduli of the (I)-gels are clearly higher than those of the (II)-gels (Fig. 8a).
4. In the case of the homogeneously crosslinked gels (I) the physically effective crosslink densities - which were calculated by means of the FLORY-equation and the experimentally determined moduli - are higher than the analytically determined intermolecular crosslink densities. In the case of the exclusively by intermolecular bonds crosslinked gels (II), however, the physically effective crosslink density is lower than the chemically determined crosslink density (see Fig. 8b).

Statement 1 is a clear confirmation of the results shown in Fig. 4 and 5a and of the cell structure model of Fig. 6, obtained by an independent way. With the common used thread model of polymer gels with complete mutual coil penetration, the maximum number of crosslinks formed by reaction (II) should be independent of the polymer concentration during the crosslinking reaction, i. e., the moduli should be constant.

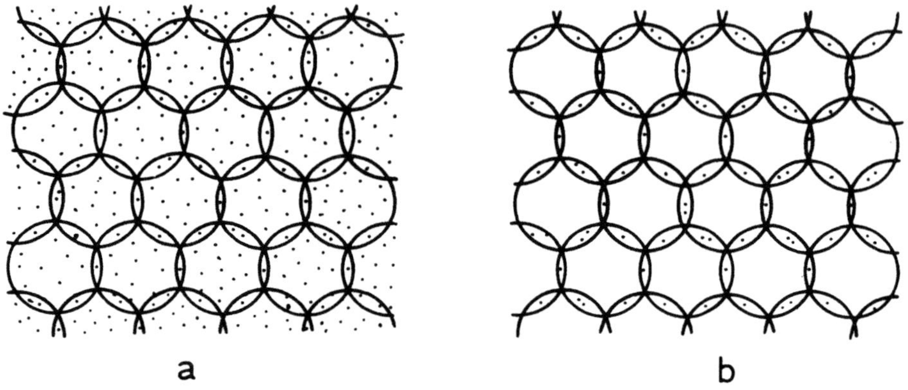

Fig. 7 Scheme of structure and crosslink distribution
a) of the gel prepared by reaction I (see Fig. 1 and 2)
b) of the gel prepared by reaction II (see Fig. 1 and 2)

According to the cell structure model, on the other hand, we have to expect an increase in the degree of penetration in the border zones of the coils with increasing polymer concentration and therefore an increase of the number of crosslinks, demonstrated by the higher moduli.

Statement 2 means that here is a homogeneously crosslinked system, whose moduli increase although the number of crosslinks is quite the same in all samples of curve I in Fig. 8a. What changes is only the ratio of intermolecular to intramolecular crosslinks. With increasing polymer concentration the portion of intermolecular crosslinks increases. To say it figuratively, the cell walls of the structure in Fig. 6 or 7 are reinforced or stiffened. This effect cannot be understood with the classical theory of networks, especially if we regard it together with statement 3.

According to the classical conception of networks, based on FLORY'S fundamental work[7], there are two types of crosslinks: elastically effective intermolecular and ineffective intramolecular crosslinks in the sense of loops. According to statement 3 intramolecular crosslinks are very well elastically effective even if there are only very few intermolecular cross-links present. We have therefore, to distinguish between two types of effective crosslinks, viz. inter- and intramolecular cross-links. Surely there will be many different types of intramolecular crosslinks (Fig. 9) with different degrees of elastical effectiveness, without any possibility to distinguish between them. What becomes clear from the results of Fig. 8 is the fact that it is meaningless, to

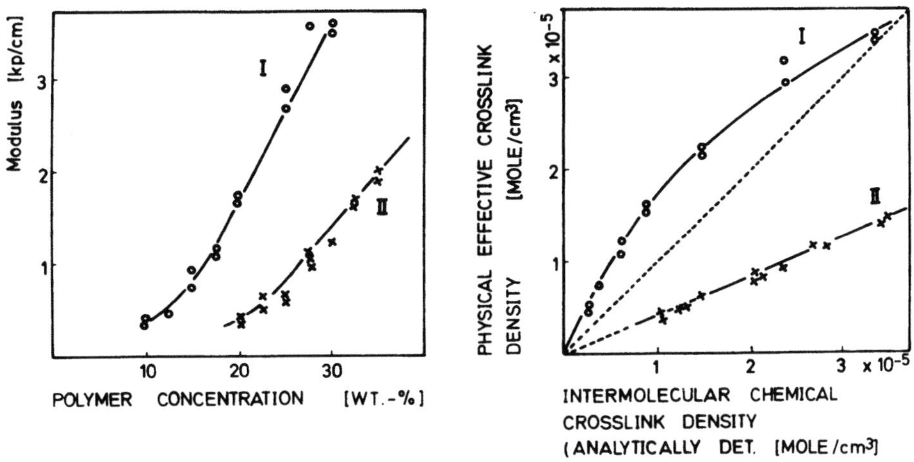

Fig. 8 Mechanical moduli of polybutylacrylate gels swollen in benzene (equilibrium state) $[M_X] = [M_{OH}] = 3{,}5$ mole % $= [M_Y]$ $= [M_{COCl}]$ = constant in all samples.

I Curves of homogeneously crosslinked gels illustrated in Fig. 7a

II Curves of exclusively intermolecularly crosslinked gels, illustrated in Fig. 7b

describe the elastical behavior of gels and rubbers by the number of cross-links.

Statement 4 is in conflict with another conventional idea: many authors have found in cross-linked polymer systems more physical determined than chemical determined intermolecular crosslinks. Curve II shows that in a polymer, which was prepared via a clear cut synthesis route and which contains exclusively intermolecular crosslinks, no additional cross-links of a physical nature are apparent. Therefore, we see no reason to assume entanglements in polymer solutions or gels.

General Scope

Since the fundamental work of W.KUHN and E.GUTH and H.MARK everybody is convinced that in diluted polymer solutions the long chains of macromolecules generally have the shape of random coils. The sedimentation and viscosity behavior of diluted polymer solutions have shown that the coils have the character of small gel particles. Our experiments have confirmed that even in concentrated solutions and gels the coils exist as individual particles as concluded already by S.H.MARON et al[3] from rheological data. Therefore in the

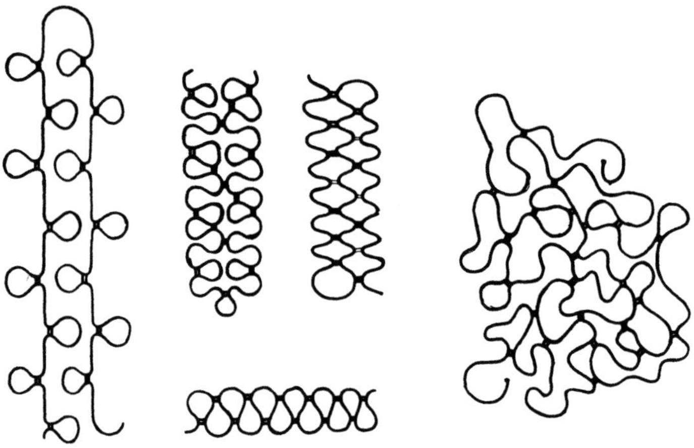

Fig. 9 Some possible types of intramolecular cross-links

case of coil forming polymers the specially favoured thread-model (spaghetti-model) of the gel-, rubbery- and glassy-state according to Fig. 1 should be tested and if necessary replaced by the cell structure model of Fig. 7.

The gels prepared by reaction I and II gave us the possibility to investigate separately the influence of inter- and intramolecular crosslinks on the mechanical moduli. As a result of this experiments we have found that there are two (or more) elastical effective types of crosslinks, which in general can not be determined separately. Therefore, the well known parameter ν_e, used in the equations of the statistical theory of rubber elasticity, is a general structural parameter, which should not be simply identified with the number of crosslinks. ν_e depends on the special way of preparation and on the structure of gels and cannot give an information about a distinct point of the gel structure as it is the number of cross-links.

Gels having exclusively intermolecular crosslinks does not give an intercept, if the mechanical determined crosslink density according to FLORY's equation is plotted against the analytical determined crosslink density. Therefore, the intercept found in gels prepared by the usual crosslinking methods, can not be explained by entanglements.

The cell structure model of gels, as proposed in Fig. 6 and 7, allows a good understanding of the properties and the deformation behavior of gels.

New aspects on some further phenomena will be made available by this I.C.S. model of the gel state.

1) The inside structure of coils (the mechanism of crystallisation of polymer chains).
2) The coil contraction on evaporation of the polymer solution.
3) The depencdence of the reaction of polymer molecules with reactive chain ends on the molecularweight (polycondensation reactions).
4) The phenomenon of incompatibility of the polymers
5) The phenomenon of high impact strength of polymers with stiffening structure units in the polymer chain (cell structure and partially oriented threadmodel).
6) The phenomenon of popcornpolymers.
7) The polymerisation of monomers in the presence of polymers, especially in the presence of crosslinked polymers.

Literature

1) J.C.H.HWA, J.Pol.Sc.58/715 (1962
2) H.WESSLAU, Makromol.Chem. 93/55 (1966)
3) S.H.MARON, N.NAKAJIMA, J.M.KRIEGER, J.Pol.Sc. 37/1 (1959)
S.H.MARON, J.Pol.Sc. 38/329 (1959)
4) S.ONOGI, T.MASUDA, N.MIYANAGA, Y.KIMURA, J.Pol.Sc.A-2 5/899 (1967)
S.ONOGI, T.KOBAYASHI, YKOJIMA, Y.TANIGUCHI, J.Appl.Pol. Sc. 7/847 (1963)
5) J.KRUSE, R.MEYER, Kautschuk u. Gummi, Kunststoffe 22/6, S. 299 (1969)
6) E.F.CLUFF, E.K.GLADDING, R.PARISER, J.Pol.Sc. 45/341 (1960)
H.WESSLAU, Makromol.Chem. 93/55 (1966)
7) J.P.FLORY, Principles of Polymer Chemistry, Cornell University Press, New York (1953)
8) A.M.BUECHE, J.Pol.Sc. 19/225 (1956)
H.WESSLAU, Makromol.Chem. 93/55 (1966)
J.JANACEK, J.HASA, Coll.Czechoslov.Chem.Comm. 31/2186(1966)
J.HASA, M.ILAVSKY, Preprints of Papers, Czechoslov. Acad. of Sc. Inst. Macromol.Chem. 40/68

MICROPHASE SEPARATION IN MIXTURES OF BLOCK COPOLYMERS WITH THE CORRESPONDING HOMOPOLYMERS: ZEROTH APPROXIMATION

Sonja Krause

Department of Chemistry

Rensselaer Polytechnic Institute, Troy, N.Y. 12181

ABSTRACT

The conditions necessary for microphase separation in mixtures of block copolymers with one of the corresponding homopolymers are developed as a function of block copolymer composition, molecular weight, number of blocks per molecule, interaction parameter between the corresponding homopolymers, and molecular weight of the added homopolymer. These conditions for microphase separation are compared with the conditions necessary for ordinary phase separation in mixtures of random copolymers with one of the corresponding homopolymers. Both the amount and the molecular weight of added homopolymer have a larger effect on the conditions for phase separation in random-copolymer-homopolymer mixtures than they have on the conditions for microphase separation in block copolymer-homopolymer mixtures.

INTRODUCTION

In a previous work,[1] the conditions necessary for microphase separation in block copolymers containing an arbitrary number of blocks of two monomers, A and B, were developed as a function of block copolymer composition, molecular weight, number of blocks per molecule, and interaction parameter between the corresponding homopolymers. Since block copolymer samples, as synthesized, often contain one or both of the corresponding homopolymers, it becomes of interest to investigate the effects of the presence of such homopolymer on the phase separation of the block copolymer. The main question studied in this work is the following: does the presence of one of the corresponding homopolymers in the block copolymer sample enhance the (thermodynamic) likelihood of microphase separation or does it inhibit the separation, i.e., does the homopolymer compatibilize the blocks in the copolymer molecules? In the cases investigated here, the presence of homopolymer almost always enhances the probability of microphase separation, although this enhancement is negligible in some cases.

It must be emphasized here that the present paper deals only with the conditions under which a homopolymer-block copolymer mixture changes from a true solution to a system in which the block copolymer has undergone microphase separation with all of the homopolymer dissolved in the microphase made up of the corresponding blocks. The present work does not deal with the situation studied experimentally by Inoue,[2] in which the conditions under which added homopolymer forms a microphase of its own instead of entering the microphases are of interest. The emulsifying effect of block copolymer on the corresponding homopolymers (also being studied by Molau[3]), is therefore only of secondary interest in the present work.

ASSUMPTIONS

The block copolymers considered here are monodisperse samples in which every molecule has the same molecular weight, the same average composition in terms of the two monomers A and B, and the same sequence distribution in terms of the number of blocks of each type, A and B. The blocks are fairly long, so that one may assume that there is no change in molecular disorder within each block <u>except</u> directly at the very sharp boundary between microphases. The homopolymer, HA, to be considered here is also monodisperse in molecular weight, and dissolves only in microphase A after the microphase separation occurs. This assumption appears reasonable if only the work of Inoue[2], Molau[3], and McIntyre[18] is considered; electron micrographs of homopolymer block copolymer mixtures show complete microphase separation with sharp boundaries between microphases just like electron micrographs of the block copolymer alone. However, there is evidence that some mixing of different repeat units occurs even in some pure block copolymers,[4] and there is no *a priori* reason to assume that homopolymer A will never mix with microphase B after microphase separation occurs. A more complete treatment of microphase separation requires consideration of diffuse boundaries between microphases with a concommitant minimization of the free energy in order to determine the amount of mixing; such a treatment has been reported by Leary and Williams[5] for pure A-B-A block copolymers. The simple approximation, involving sharp boundaries between microphases and no mixing of *any* A and B units after microphase separation will nevertheless be used in this work and it is expected to yield qualitative trends correctly.

ENTHALPY

Using these assumptions and the lattice theory of solutions, the enthalpy change on microphase separation is

$$\Delta H = -\frac{kTV}{V_r} v_A^M v_B^M \chi_{AB} \left(1-\frac{2}{z}\right) \quad (1)$$

where V is the total volume of the system, T is the absolute temperature, k is the Boltzmann constant, V_r is the volume of a lattice site, z is the coordination number of the lattice, v_A^M and v_B^M are volume fractions of monomer repeat units A and B in the total mixture, respectively, and χ_{AB} is the interaction parameter between A units and B units. Equation 1 is the Hildebrand[6]-Scatchard[7]-Van Laar[8] heat of mixing that is lost on microphase separation if we consider the fixed composition of each polymer chain. For each monomer unit, excepting end units, in each polymer chain, only z-2 contacts are random before microphase separation, since two contacts are fixed by the composition of the chain. It is also assumed that the only contacts between A and B after microphase separation are at the surfaces between microphases and consist only of A-B junctions within block copolymer molecules. This assumption restricts the discussion to small and moderate volume fraction of added homopolymer.

ENTROPY

The total entropy change when microphase occurs is

$$\frac{\Delta S}{k} = N_c \ell n (v_A^M)^{v_A^C} (v_B^M)^{v_B^C} + N_{HA} \ell n v_A^M$$

$$-2N_c(m-1)\left(\frac{\Delta S_{dis}}{R}\right) + N_c \ell n(m-1) \quad (2)$$

where N_c and N_{HA} are the numbers of block copolymer molecules and homopolymer molecules in the system, respectively, v_A^C and v_B^C are the volume fraction of monomer A and B in the copolymer molecules, respectively, and m is the number of blocks in each block copolymer molecule. The first term on the right-hand side of eq. 2 is concerned with the decrease in available volume for each block in the copolymer molecules after microphase separation assuming that there is no volume change on mixing. The second term on the right-hand side of eq. 2 is concerned with the decrease in available volume for the homopolymer molecules after microphase separation. Terms of this type were discussed in a previous paper.[9] The third term on the right-hand side of eq. 2 is the additional entropy decrease caused by immobilization at the microphase surface of the A and B units that link the blocks in each copolymer molecule. The fourth term arises for block copolymers containing three or more blocks because of the large number of sites available for the blocklinking segments on the surfaces between microphases. These last two terms have also been discussed previously,[1] especially the values that could be used for $(\Delta S_{dis}/R)$, the disorientation entropy gain on fusion per segment of a polymer. In agreement with previous calculation, a value of $(\Delta S_{dis}/R)=1.0$ will be used in this paper; actual values of $(\Delta S_{dis}/R)$ range from 0.85 to 4.3 for various polymers.[10]

FREE ENERGY

The free energy change on microphase separation for the whole system of N_c copolymer molecules and N_{HA} homopolymer molecules occupying volume V is $\Delta G=\Delta H-T\Delta S$ where ΔH and ΔS are the expressions in eqs. 1 and 2, respectively. When $\Delta G=0$, the separated system is in equilibrium with the completely mixed system; solution of this equation allows calculation of critical values of χ_{AB}, $(\chi_{AB})_{cr}$, for various ratios of N_{HA} to N_c, for various values v_A^C and m, and for various degrees of polymerization of the two polymers. Since

$$V = N_c(V_A n_A^C + V_B n_B^C) + N_{HA} V_A n_{HA} \quad (3)$$

and

$$v_A^M = \frac{N_{HA} V_A n_{HA} + N_C V_A n_A^C}{V} \quad (4)$$

Where V_A and V_B are the volumes of A and B repeat units, respectively, n_A^C and n_B^C are the number of A and B units in each copolymer molecule, respectively, and n_{HA} is the number of A units in each homopolymer molecule; we get

$$(\chi_{AB})_{cr} = \frac{zV_r}{(z-2)V_B v_A^M n_B^C}\Bigg[-\ell n(v_A^M)^{v_A^C}(v_B^M)^{v_B^C}$$

$$-\frac{N_{HA}}{N_c}\ell n\, v_A^M + 2(m-1)(\frac{\Delta S_{dis}}{R}) - \ell n(m-1)\Bigg] \quad (5)$$

When eq. 5 is used to make predictions, it is convenient to let $V_r = V_A = V_B$,[1] and to set a convenient value for z, possibly z=8, and for $(\Delta S_{dis}/R)$, possibly the value 1.0 mentioned above.

Eq. 5 was derived using essentially an approach suitable for a one-component system, although a mixture of homopolymer with block copolymer is a two-component system. This approach was possible because it was assumed that the homopolymer HA did not mix with microphase B after microphase separation, i.e., the system was not allowed to act like an ordinary two-component system.

RANDOM COPOLYMER-HOMOPOLYMER MIXTURES

The enthalpy of mixing of a random copolymer AB, made up of A monomers and B monomers, with homopolymer HA is[11]

$$\Delta H = \frac{kTV}{V_r} v_{HA}^M v_C^M \chi_{AC}(1-\frac{2}{z}) \quad (6)$$

where χ_{AC} is the interaction parameter between homopolymer HA and random copolymer AB. Eq. 6 comes from eq. 1 of ref. 11, with the added factor (1-2 /z), which should be included whenever mixing of polymer molecules with each other, but not with solvent, is discussed. From eq. 2 of ref. 11,

$$\chi_{AC} = (v_B^C)^2\chi_{AB} \quad (7)$$

so that the free energy change on mixing can be written in terms of χ_{AB}. Using the Flory[12,13]-Huggins[14,15] entropy of mixing with eqs. 6 and 7 in the free energy of mixing, we get

$$\frac{\Delta G_{mix}}{kT} = \frac{V}{V_r}\Bigg[(v_B^C)^2 v_{HA}^M v_C^M \chi_{AB}(\frac{z-2}{z})$$

$$+ \frac{v_{HA}^M}{n_{HA}}\ell n\, v_{HA}^M + \frac{v_C^M}{n_A^C + n_B^C}\ell n\, v_C^M\Bigg] \quad (8)$$

from which phase relations for the mixtures can be derived. Although the conditions for equilibrium phase separation (the binodal) would be the most interesting to compare with the conditions for microphase separation in the block copolymer-homopolymer system, the absolute stability limits (the spinodal) of this system are easier to calculate:

$$(\chi_{AB})_{sp} = \frac{z}{2(z-2)(v_B^{\ C})^2}\left[\frac{1}{n_{HA}v_{HA}^{\ M}} + \frac{1}{(n_A^{\ C} + n_B^{\ C})v_C^{\ M}}\right] \quad (9)$$

where $(\chi_{AB})_{sp}$ refers to the value of χ_{AB} along the spinodal for the random copolymer-homopolymer system.

CALCULATED RESULTS

Tables I and II show values of $(\chi_{AB})_{cr}$ for block copolymers whose degrees of polymerization, $n_A^{\ C} + n_B^{\ C}$, equal 400 and 800, respectively. The value chosen for the coordination number of the lattice, z, was 8, and $(\Delta S_{dis}/R = 1)$. Values are shown for five different degrees of polymerization of homopolymer, n_{NA} = 100, 200, 400, 800, and 1600, for four different volume fractions of homopolymer in the mixture, $v_{HA}^{\ M}$ = 0.10, 0.25, 0.50, and 0.75, for three different composition of the block copolymer, $v_A^{\ C}$ = 0.25, 0.50, and 0.75, and for a diblock copolymer, m = 2, a triblock copolymer, m = 3, and a tetrablock copolymer, m = 4. The Tables also show values of $(\chi_{AB})_{cr}$ for the pure block copolymers, with no added homopolymer in the system.

Table III shows values of $(\chi_{AB})_{sp}$, the value of χ_{AB} above which a particular mixture of random copolymer and homopolymer must separate into two phases. Values are shown for mixtures containing random copolymer of degree of polymerization, $n_A^{\ C} + n_B^{\ C}$, 400 or 800, plus homopolymer HA, of degree of polymerization 100, 200, 400, 800, or 1600, at volume fractions 0.10, 0.25, 0.50, or 0.75 in the mixture. The random copolymer contains volume fraction of A, $v_A^{\ C}$, of 0.25, 0.50, or 0.75.

DISCUSSION

Tables I and II show that addition of homopolymer A to a block copolymer of A and B lowers $(\chi_{AB})_{cr}$, i.e., increases the likelihood of microphase separation, in practically all cases.

Table I. $(\chi_{AB})_{cr}$ for Mixtures of Homopolymers with Block Copolymers having $n_A^C + n_B^C = 400$, $z = 8$, $(\Delta S_{dis}/R) = 1$

		$(\chi_{AB})_{cr}$ m = 2			$(\chi_{AB})_{cr}$ m = 3			$(\chi_{AB})_{cr}$ m = 4		
n_{HA}	v_{HA}^M	v_A^C=0.25	v_A^C=0.50	v_A^C=0.75	v_A^C=0.25	v_A^C=0.50	v_A^C=0.75	v_A^C=0.25	v_A^C=0.50	v_A^C=0.75
100	0.10	0.042	0.036	0.046	0.060	0.052	0.069	0.082	0.071	0.096
200	"	0.039	0.034	0.045	0.057	0.050	0.068	0.078	0.069	0.095
400	"	0.037	0.034	0.045	0.055	0.049	0.067	0.077	0.069	0.095
800	"	0.036	0.033	0.044	0.054	0.049	0.067	0.076	0.068	0.094
1600	"	0.036	0.033	0.044	0.054	0.049	0.067	0.075	0.068	0.094
100	0.25	0.038	0.036	0.047	0.051	0.050	0.068	0.067	0.067	0.094
200	"	0.033	0.032	0.045	0.046	0.046	0.066	0.062	0.063	0.092
400	"	0.030	0.031	0.043	0.043	0.045	0.065	0.059	0.062	0.091
800	"	0.028	0.030	0.043	0.041	0.044	0.064	0.058	0.061	0.090
1600	"	0.028	0.029	0.043	0.041	0.043	0.064	0.057	0.060	0.090
100	0.50	0.034	0.035	0.048	0.043	0.047	0.068	0.054	0.061	0.092
200	"	0.027	0.030	0.044	0.036	0.042	0.064	0.048	0.056	0.088
400	"	0.024	0.028	0.042	0.033	0.039	0.062	0.044	0.054	0.086
800	"	0.022	0.026	0.041	0.031	0.038	0.061	0.043	0.052	0.085
1600	"	0.021	0.026	0.040	0.030	0.037	0.060	0.042	0.052	0.085
100	0.75	0.032	0.036	0.050	0.039	0.045	0.069	0.048	0.058	0.091
200	"	0.025	0.030	0.044	0.032	0.040	0.063	0.041	0.052	0.086
400	"	0.021	0.027	0.042	0.029	0.037	0.060	0.037	0.049	0.083
800	"	0.020	0.025	0.040	0.027	0.035	0.059	0.036	0.047	0.082
1600	"	0.019	0.024	0.040	0.026	0.034	0.058	0.035	0.047	0.081
-	0	0.047	0.036	0.047	0.071	0.053	0.071	0.100	0.075	0.100

Table II. $(\chi_{AB})_{cr}$ for Mixture of Homopolymers with Block Copolymers having $n_A^C + n_B^C = 800$, $z = 8$, $(\Delta S_{dis}/R) = 1$

		$(\chi_{AB})_{cr}$ m = 2			$(\chi_{AB})_{cr}$ m = 3			$(\chi_{AB})_{cr}$ m = 4		
n_{HA}	v_{HA}^M	v_A^C=0.25	v_A^C=0.50	v_A^C=0.75	v_A^C=0.25	v_A^C=0.50	v_A^C=0.75	v_A^C=0.25	v_A^C=0.50	v_A^C=0.75
100	0.10	0.024	0.020	0.024	0.033	0.027	0.035	0.044	0.037	0.049
200	"	0.021	0.018	0.023	0.030	0.026	0.034	0.041	0.036	0.048
400	"	0.019	0.017	0.023	0.028	0.025	0.034	0.039	0.035	0.048
800	"	0.018	0.017	0.022	0.027	0.025	0.034	0.038	0.034	0.047
1600	"	0.018	0.017	0.022	0.027	0.024	0.033	0.038	0.034	0.047
100	0.25	0.025	0.021	0.026	0.031	0.028	0.036	0.039	0.037	0.049
200	"	0.019	0.018	0.023	0.026	0.025	0.034	0.034	0.033	0.047
400	"	0.016	0.016	0.022	0.023	0.023	0.033	0.031	0.032	0.046
800	"	0.015	0.015	0.022	0.021	0.022	0.032	0.030	0.031	0.045
1600	"	0.014	0.015	0.021	0.021	0.022	0.032	0.029	0.030	0.045
100	0.50	0.024	0.023	0.028	0.028	0.029	0.038	0.034	0.036	0.050
200	"	0.017	0.018	0.024	0.021	0.024	0.034	0.027	0.031	0.046
400	"	0.013	0.015	0.022	0.018	0.021	0.032	0.024	0.028	0.044
800	"	0.012	0.014	0.021	0.016	0.020	0.031	0.022	0.027	0.043
1600	"	0.011	0.013	0.020	0.016	0.019	0.030	0.021	0.026	0.043
100	0.75	0.023	0.024	0.031	0.026	0.029	0.040	0.031	0.035	0.051
200	"	0.016	0.018	0.025	0.019	0.023	0.034	0.024	0.029	0.046
400	"	0.012	0.015	0.022	0.016	0.020	0.032	0.020	0.026	0.043
800	"	0.011	0.013	0.021	0.014	0.018	0.030	0.019	0.024	0.042
1600	"	0.010	0.013	0.020	0.013	0.018	0.029	0.018	0.024	0.041
-	0	0.023	0.018	0.023	0.034	0.027	0.034	0.048	0.037	0.048

There are a few cases in which the addition of a homopolymer to the block copolymer increases $(\chi_{AB})_{cr}$ very slightly, i.e., makes the system slightly more compatible; in the case of diblock copolymers, this happens only when the degree of polymerization of the homopolymer is equal to or less than one-quarter the degree of polymerization of the block copolymer; in the case of triblock or tetrablock copolymer, this happens only when the degree of polymerization of the homopolymer is equal to or less than one-eighth the degree of polymerization of the block copolymer. Addition of high molecular weight homopolymer, on the other hand, can decrease $(\chi_{AB})_{cr}$ appreciably, i.e., greatly increase the likelihood of microphase separation.

The degree of polymerization of added homopolymer has a much greater effect on the phase separation in random copolymer-homopolymer mixtures, as shown in Table III. In general, this effect is so great that low molecular weight homopolymers are more compatible with random copolymers than with block copolymers, while high molecular weight homopolymers are less compatible with random copolymers than with block copolymers. These comparisons must be left in very general terms, because the exact values of $(\chi_{AB})_{cr}$ in Tables I and II are sensitive to the value chosen for $(\Delta S_{dis}/R)$, and should not be taken too seriously, except insofar as they indicate trends. In spite of this, it is tempting to compare the expectations for actual copolymers with these calculations. For example, if we take $\chi_{AB} = 0.01$[16] for styrene = A and methyl methacrylate = B, we would expect that none of the mixtures shown in Table I would exhibit microphase separation, but that a few of the mixtures containing diblock copolymers and high molecular weight homopolymers in Table II would be on the verge of microphase separation. A number of the random copolymer-homopolymer mixtures from Table III, those containing the higher molecular weight polymers, would be expected to exhibit phase separation. If, on the other hand, we take $\chi_{AB} = 0.07$, a value that might correspond to styrene = A, butadiene = B if appropriate values[17] of the solubility parameter are chosen for these polymers, then most of the mixtures in Tables I, II, and III, should exhibit phase separation except for a few in Table I containing tetrablock copolymers mixed with very low molecular weight homopolymers, and a few of the mixtures containing low molecular weight homopolymers and random copolymers in Table III.

ACKNOWLEDGEMENT

This work has been supported in part by the National Aeronautics and Space Administration under Grant NGL-33-018-003.

Table III. $(\chi_{AB})_{sp}$ for Mixtures of Homopolymers with Random Copolymers, z = 8

		$(\chi_{AB})_{sp}$ when $n_A^C+n_B^C$ = 400			$(\chi_{AB})_{sp}$ when $n_A^C+n_B^C$ = 800		
n_{HA}	v_{HA}^M	v_A^C=0.25	v_A^C=0.50	v_A^C=0.75	v_A^C=0.25	v_A^C=0.50	v_A^C=0.75
100	0.10	0.122	0.28	1.09	0.12	0.27	1.08
200	"	0.063	0.14	0.56	0.060	0.14	0.55
400	"	0.033	0.075	0.29	0.031	0.071	0.28
800	"	0.019	0.041	0.16	0.016	0.037	0.15
1600	"	0.011	0.024	0.096	0.009	0.020	0.081
100	0.25	0.052	0.116	0.47	0.049	0.11	0.44
200	"	0.028	0.063	0.25	0.025	0.057	0.23
400	"	0.016	0.036	0.14	0.013	0.031	0.12
800	"	0.009	0.023	0.089	0.008	0.017	0.070
1600	"	0.007	0.016	0.063	0.005	0.011	0.044
100	0.50	0.029	0.067	0.27	0.027	0.060	0.24
200	"	0.017	0.040	0.16	0.015	0.033	0.13
400	"	0.012	0.027	0.107	0.009	0.020	0.080
800	"	0.009	0.020	0.080	0.005	0.013	0.053
1600	"	0.007	0.016	0.067	0.004	0.011	0.040
100	0.75	0.028	0.063	0.25	0.021	0.049	0.20
200	"	0.020	0.044	0.17	0.013	0.031	0.12
400	"	0.016	0.036	0.14	0.009	0.023	0.089
800	"	0.013	0.031	0.12	0.008	0.017	0.070
1600	"	0.013	0.029	0.12	0.007	0.016	0.063

REFERENCES

1. S. Krause, Macromolecules, 3, 84 (1970).
2. T. Inoue, T. Soen, T. Hashimoto, and H. Kawai, Macromolecules, 3, 87 (1970).
3. G.E. Molau and W.M. Wittbrodt, Macromolecules, 1, 260 (1968).
4. J.F. Beecher, L. Marker, R.D. Bradford, and S.L. Aggarwal, Polymer Preprints, 8, 1532 (1967).
5. D.F. Leary and M.C. Williams, submitted to Polymer Letters.
6. J.H. Hildebrand, J. Amer. Chem. Soc., 57, 866 (1935).
7. G. Scatchard, Chem. Rev., 7, 321 (1931).
8. J.J. Van Laar, Z. Physik. Chem., A137, 421 (1928).
9. S. Krause, J. Polym. Sci., Part A-2, 7, 249 (1969).
10. A. Bondi, "Physical Properties of Molecular Crystals, Liquids, and Glasses," John Wiley & Sons, Inc., New York, N.Y., 1968, Chapter 6.
11. S. Krause, A.L. Smith, and M.G. Duden, J. Chem. Phys., 43, 2144 (1965).
12. P.J. Flory, J. Chem. Phys., 9, 660 (1941).
13. P.J. Flory, J. Chem. Phys., 10, 51 (1942).
14. M.L. Huggins, J. Chem. Phys., 9, 440 (1941).
15. M.L. Huggins, Am. N.Y. Acad. Sci., 43, 1 (1942).
16. W.H. Stockmayer, L.D. Moore, Jr., M. Fixman, and B.N. Epstein, J. Polym. Sci., 16, 517 (1955).
17. J. Brandrup and E.H. Immergut, eds., Polymer Handbook, Interscience, New York, N.Y., 1966.
18. D. McIntyre and E. Campos-Lopez, Macromolecules, 3, 322 (1970).

DILUTE SOLUTION BEHAVIOUR OF COPOLYMERS

LECHOSLAW A. UTRACKI

GULF OIL CANADA LTD., SHAWINIGAN RESEARCH
STE. ANNE DE BELLEVUE, QUEBEC
CANADA

Recently a two parameter dilute solution theory of homopolymers was successfully applied[1] to statistical copolymer-solvent and block polymer-solvent systems. The following basic assumptions were made:

I. Under a hypothetical Θ-condition all individual long range interaction parameters β_{ij} vanish identically and the size of the copolymer coil is solely defined by short range interaction parameters a_{ij} (i, j = 1, 2,...n, in a n-component system):

$$a_c^2 = \Sigma_{ij} (a_{ii}^2 X_{ii} + a_{ij}^2 X_{ij})$$
$$a_i^2 = M_{oi} (\langle r_\Theta^2 \rangle_i / M_i) \qquad (1)$$

where subscripts c and i refer to the copolymer and its component i respectively, X_{ij}'s are the appropriate probabilities of i-j contact along the backbone chain, M_{oi} and M_i are segmental molecular weight calculated per one backbone carbon atom and total molecular weight of component i respectively, $\langle r_\Theta^2 \rangle$ is the unperturbed mean square end-to-end distance.

II. Under normal experimental conditions at least one $\beta_{ij} \neq 0$. Provided the solvent is not much poorer than Θ-solvent for any of the macromolecular components of the system we can write:

$$\beta_c = \Sigma_{ij} \beta_{ij} Z_{ij} , \qquad (2)$$

where β_{ij} is the binary cluster integral and Z_{ij} probability of i-j contact within a coil volume.

These two assumptions with a proper evaluation of X_{ij} and Z_{ij} are sufficient for a total description of a multicomponent copolymer - solvent system.

BLOCK POLYMERS

Here the probability of heterocontacts along the backbone chain as expressed by X_{ij} is relatively small:

$$X_{ij} \ll X_{ii} \quad \text{and} \quad X_{ii} = x_i \tag{3}$$

where x_i is the mole fraction of species i. The probability of long range interaction contacts, Z_{ij}, can be arbitrarily expressed as:

$$Z_{ij} = x_i x_j \tag{4}$$

This assumption is strictly valid for a random distribution of disconnected i and j segments within a coil volume.

Equations (1) - (4) have been used to analyze the intrinsic viscosity ($[\eta]$) and/or second virial coefficient (A_2) solution data of the following four block polymers: I Styrene (S) - Butadiene (B), SB and SBS blocks in toluene, dioxane and cyclohexane[1]. II S - Methylmethacrylate (M), SM and MSM blocks in benzene and toluene[2]. III Recently Kotaka et al. [3] analyzed their data of MSM in six solvents. IV MSM in toluene and butanone[4]. Using the given[4] values of $[\eta]$, M_w and M_n, as well as the short and long range interaction parameters calculated before[2], β_{12} was computed by means of the Stockmayer-Fixman relation[5]. In Fig. 1 two sets of β_{12} calculated for IV are plotted vs. M_w/M_n - one calculated using the weight average molecular weight, the second using the number average molecular weight of the copolymer samples. It can be seen that extrapolation to $M_w/M_n = 0$ leads to a common intercept: $\beta_{12} = 2.0 Å^3$ for toluene and $\beta_{12} = 1.5 Å^3$ for butanone. These values are in good agreement with the previous computations[2].

In all four cases the method was found to be adequate. For the system I the values of M_c calculated from $[\eta]$ and/or A_2 agreed within 10% with the values measured. For systems II and III the stability of β_{12} calculated from $[\eta]$'s was within $\pm$ 0.4Å^3.

STATISTICAL AND ALTERNATE COPOLYMERS

In statistical copolymers the X_{ii} and X_{ij} must be calculated from the monomers reactivity ratios[2,3], available either from the literature[6] or from kinetic measurements[2]. The parameters Z_{ij} can either be expressed by eq. (4) as for block polymers[2,7], or computed as dyadic segmental interactions with a solvent[3]. The following three systems were analyzed: I S-M statistical

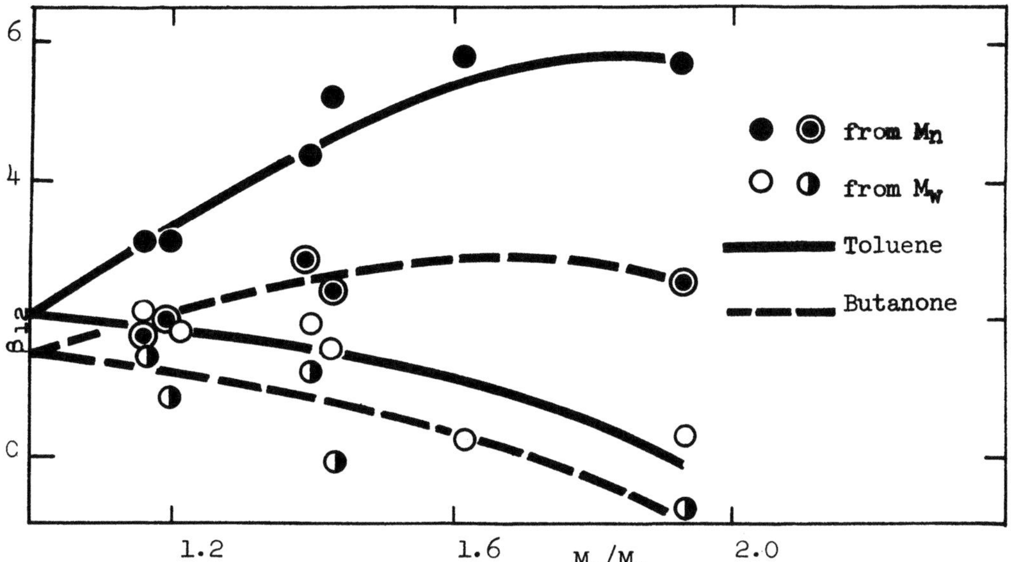

Fig. 1. β_{12} as a function of polydispersity.

copolymer in toluene, butanone, diethylmalonate, 1-chloro-n-butane[2]. II S-Methylmethacrylate statistical copolymer in benzene, toluene and butanone[7]. III S-M statistical and alternating copolymers in six solvents[3]. As before, the method was found to be adequate, judged by the stability of a_{12} and β_{12} within each of the systems.

It is worth mentioning that for the three systems a_{12}^{2}'s were found to have an arithmetic mean value of a_{ii}^{2}. The same observation[12] was also made for M-n-butylacrylate statistical copolymer. However, 10% and 35% variance from the mean values was computed for S-acrylonitrile and methyl styrene-ethylene statistical copolymers, respectively[7]. This phenomenon is probably related to the difference in both sterical hindrance of the constitutive monomers and the magnitude of their mutual interaction potential, which influences coisotacticity of the molecules, as suggested recently[3].

BRANCHED MOLECULES

The solution properties of graft copolymers can be treated as those of a comb-like homopolymer superimposed on the properties of a copolymer. The conformation of a branched homopolymer molecule is usually characterized by a ratio of a mean square radius of gyration of a branched molecule to that of a linear one of the same molecular weight, i.e.:

$$g = \langle s^2 \rangle_b / \langle s^2 \rangle \qquad (5)$$

As the theoretical calculations are performed on an unperturbed model, eq. (5) should be valid under Θ-conditions only. According to Zimm and Kilb[8]:

$$([\eta]_b/[\eta])_\Theta = g^{1/2} \qquad (6)$$

for several classes of nonlinear molecules in a Θ-solvent. This relation was found to be valid in numerous systems[9]. At $T > \Theta$ one should expect the influence of long range interactions to be less pronounced for branched than for linear molecules. In branched systems the coil volume being smaller, there is less possibility of long range interactions, as expected from the eq. (2). This effect should be proportional to the decrease of the pervaded volume, i.e. as a first approximation to $g^{1/2}$. For a branched copolymer molecule we can rewrite the Stockmayer-Fixman relation[5] in the form:

$$[\eta]_{cb}/(M_c g)^{1/2} = 2.87 \cdot 10^{21}[(a_c^2/M_{oc})^{3/2} + 0.51(\beta_c/M_{oc}^2)(M_c g)^{1/2}] \qquad (7)$$

where c and cb subscripts refer to linear and branched copolymer molecules respectively. In the limit eq. (7) should also be applicable to branched homopolymer-solvent systems. For small values of β, g, and M, eq. (7) takes the form of the Zimm-Kilb relation[8] while for g=1 the original Stockmayer-Fixman relation reappears. Defining the effective molecular weight of a branched molecule as:

$$M_{eff} = M_b g \quad , \qquad (8)$$

the data for linear, randomly branched and comb-like poly(vinyl acetate) (PVAc) samples in benzene at 35°C were plotted in Fig. 2 as $[\eta]/M_{eff}^{1/2}$ vs. $M_{eff}^{1/2}$.

For the calculation, values for $[\eta]_b$, M_{wb} and g were taken from ref. 10. In Fig. 2 the solid line was computed by a least squares method for the linear PVAc samples. The standard deviations from this line for the linear, randomly branched and comb-like structures are respectively: ± 0.124, ± 0.224 and ± 0.224. (The range of values of $K_\Theta = 2.87 \cdot 10^{21}(a^2/M_o)^{3/2}$ quoted in literature[11] gives a standard deviation of ± 0.216 for linear PVAc in various solvents at about 25°C). No systematic trend of the deviation plotted as a function of g or M_{wb} was observed.

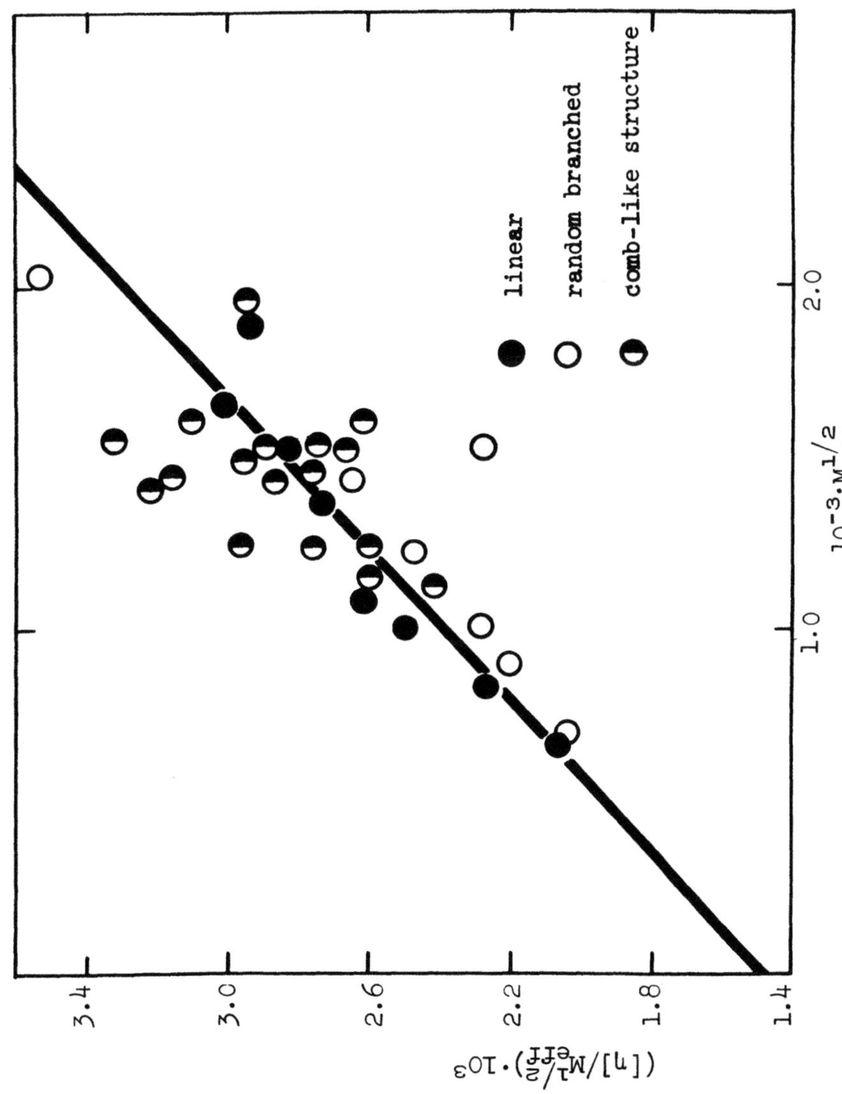

Fig. 2. Stockmayer-Fixman dependence for linear and branched PVAc.

Equation (7) predicts that the solvent power dependence of the ratio $[\eta]_b/[\eta]$ is determined by a product: $(\beta_c/M_{oc}^2)(1 - g^{1/2})M_c^{1/2}$ i.e. as g approaches unity, no solvent power dependence should be observed[9].

Equations (7) - (8) suggest an easy method for the determination of g. Plotting $[\eta]$ vs. M for a linear polymer one can construct a calibration plot, from which for a given $[\eta]_b$ one can determine M_{eff}. Knowing the values of M_b from independent measurements one can calculate g from eq. (8). The method will not be sensitive for polymers with low branching densities, with low molecular weights and/or in a poor solvent[12].

GRAFT COPOLYMERS

Anionically grafted styrene on fractions of poly(methyl methacrylate) is the easiest graft copolymer to prepare in a controllable way. Its solution properties in bromoform at 20°C were published recently by Eskin and Baranovskaya[13], and in benzene, butanone and chloroform at 25°C by Polowinski[14]. The short and long range interaction parameters calculated for styrene-methylmethacrylate-solvent systems are given in Table I. The values for bromoform solutions were computed from the solubility parameters[15].

Table I

Interaction Parameters for Styrene - Methylmethacrylate Polymers

Solvent	t°C	a_1^2	a_2^2	β_1	β_2	β_{12}
Benzene	25°C	25.5	20.5	2.13	2.08	2.93
Chloroform	25°C	"	"	2.31	3.51	3.69
Butanone	25°C	"	"	0.46	1.20	1.98[a]
Bromoform	20°C	"	"	1.0[b]	-0.3[b]	3.6[c]

Subscripts 1 and 2 refer to styrene and methylmethacrylate segments, respectively.
a - value for fractions, Fig. 2, ref. 2.
b - from solubility parameters.
c - value for polydisperse samples, Fig. 2, ref. 2.

The parameter g was calculated from the following two relations, derived for comb-like structures with branches of equal length:

$$g = [p(3p-2)r^3 + p(p+1)(p+2)r^2 + p(2p+1)(p+1)r + (p+1)^3][p(r+1)+1]^{-3} \quad (9)$$

for branches spaced evenly along the backbone chain[16], and

$$g_r = [(1+r)^2 p + 2r^2(1 + r/p) + 3r^3]/[p(1+r)^3] \quad (10)$$

for branches placed randomly along the backbone chain[17]. Here p is the number of branches per macromolecule and r is the ratio of the number of statistical segments in a branch to the number of statistical segments in the backbone chain between two neighboring branching points. In the Tables II and III the branching frequency (f), defined as an average number of segments of a backbone chain between two branching points, is also given.

Using the values of p and r quoted in the original paper[13] $g_r^{1/2}$ and $g^{1/2}$ were calculated from the eqs. (9) and (10). The results presented in the columns 7 and 8 of Table II indicate that the two models lead to virtually the same value of the parameter g. A similar conclusion was published by Casassa and Tagami[18].

Table II

Graft Copolymer Data in Bromoform at 20°C.[13]

No.	$M_w \cdot 10^{-6}$	x	f	p	r	$g_r^{1/2}$	$g^{1/2}$	[η] (obs)	[η] (calc)
1	31.6	0.7426	83	720	2.8846	0.5094	0.5096	2.40	(6.70)
2	18	0.8039	59	580	4.1007	0.4463	0.4465	1.67	(3.70)
3	11	0.7219	93	324	2.5962	0.5315	0.5319	4.00	4.00
4	7.3	0.9069	25	235	9.7400	0.3213	0.3218	0.97	0.92
5	5	0.8348	135	111	5.0519	0.4277	0.4288	2.00	2.01
6	4	0.8655	3	1941	6.4357	0.3682	0.3683	0.24	(0.95)
7	2.7	0.8449	22	589	5.4456	0.3982	0.3985	2.62	(1.68)
8	1.9	0.8966	12	205	8.6731	0.3387	0.3393	0.82	0.59
9	1	0.1256	134	43.3	0.1437	0.9355	0.9356	1.38	1.12
10	0.5	0.8964	2	315	8.6538	0.3331	0.3335	0.45	0.29
11	0.4	0.9688	0.6	1132	31.1000	0.1833	0.1834	0.25	0.27

Table III

Graft Copolymer Data in Benzene, Butanone and Chloroform at 25°C.[14]

No.	$M_w \cdot 10^{-6}$	x	f	p	r	$g^{1/2}$
1	0.529	0.885	10.5	56.1	7.70	0.395
2	0.828	0.865	10.6	101	6.44	0.395
3	1.19	0.901	10.8	98.9	9.06	0.352
4	1.21	0.861	12.9	125	6.22	0.395
5	1.19	0.856	12.6	130	5.96	0.401
6	2.85	0.906	8.33	312	9.60	0.320
7	10.70	0.984	1.63	1003	61.8	0.137
8	3.65	0.954	3.65	450	20.6	0.229
9	2.48	0.931	6.08	270	13.5	0.281
10	0.467	0.640	42.2	38.9	1.78	0.625
11	0.348	0.893	2.00	180	8.37	0.346
12	1.03	0.892	2.63	407	8.28	0.337
13	0.303	0.552	13.5	98.3	1.23	0.676
14	0.307	0.589	36.3	33.9	1.43	0.664

	$[\eta]_c$(obs)			$[\eta]_c$(calc)		
No.	Benzene	Butanone	Chloroform	Benzene	Butanone	Chloroform
1	0.440	0.294	-	0.417	0.310	-
2	0.540	0.340	0.54	0.509	0.400	0.533
3	0.545	0.362	-	0.552	0.426	-
4	0.710	0.431	-	0.654	0.500	-
5	0.664	-	-	0.661	-	-
6	0.945	0.511	-	0.881	0.632	-
7	0.340	-	-	(0.674)	-	-
8	0.528	-	-	0.654	-	-
9	0.634	-	-	0.667	-	-
10	0.652	-	-	0.638	-	-
11	0.117	-	-	(0.248)	-	-
12	0.482	-	-	0.475	-	-
13	0.580	-	-	0.523	-	-
14	0.533	-	-	0.516	-	-

Tables II and III show the characteristics of the graft polymers and $[\eta]_c$'s observed by the authors[13,14] and computed from the interaction parameters of Table I by means of eqs. (2)-(4), 7 and (9). The $[\eta]_c$'s calculated for the highest molecular weight samples of the two series of data show a distinct deviation from the measured quantities. Apparently for $M_c > 10^7$ the coil volume is smaller than calculated. This may be due to an incomplete penetration of these large molecules by a solvent, inadequacy in the theories of branched molecules, non applicability of the Stockmayer-Fixman equation[5] or experimental errors. Also the values of $[\eta]_c$ calculated for samples Nos. 6 and 7, Table II, and No. 11, Table III, do not coincide with the observed ones. It is impossible to correlate this variance with any of the independent variables. The standard deviation for the graft copolymer data in bromoform is 0.27, in benzene 0.067 and in butanone 0.17, not taking into account the values in parentheses in Tables II and III.

CONCLUSIONS

The method of calculation of copolymer solution data proposed earlier[1,2,7] for statistical, alternate, and block polymer solution data, has been successfully extended at the present time to include graft copolymers as well. The treatment is not a general one and cannot be applied to a system in which $\beta \ll 0$. The deviations are large but random, reflecting the experimental difficulties for the systems. Recently Dondos[19] reported that $[\eta]$'s of some graft copolymers plotted vs. temperature pass through local extrema. The reported fluctuations amount to ca. ± 4% i.e. outside the range of sensitivity of the data discussed here.

The use of eqs. (7) - (9) for computation of g-parameters from solution data of branched homopolymers was found to be easy and consistent.

SUMMARY

Recently we proposed[1] a new method of evaluation of dilute solution data for block random copolymers. The method is based on separation of the short and long range interactions, as expressed by a and β parameters respectively. For a complete description of a system one needs to know a_i's and β_i's which characterize the polymer-i-solvent couple and a_{ij}'s and β_{ij}'s characteristic of the polymer-i - polymer-j short and long range interaction forces.

We analyzed a number of polymer-1 - polymer-2 - solvent systems where the solvent was a good solvent for at least one parent polymer and not much poorer than Θ-solvent for the other. For a random copolymer a_{12}, was calculated from kinetics of polymerization; for a block polymer $a_{12} \ll (a_1 + a_2)$ and can be neglected. We found β_{12} to be independent of molecular weight and composition of the copolymer macromolecule and of number and size of blocks in it. However β_{12} does vary with solvent power. The β_{12}'s are positive and numerically large, comparable to β_i for a polymer in a good solvent[12]. The constancy of β_{12} for each of the investigated systems was accepted as a proof of validity of the procedure.

In the present paper the method has been used to analyze dilute solution data of graft copolymers as well.

REFERENCES

1. L. A. Utracki, and R. Simha and L. J. Fetters, J. Polymer Sci., Part A-2, 6, 2051 (1968).

2. L. A. Utracki and R. Simha, Macromolecules, 1, 505 (1968).

3. T. Kotaka, T. Tanaka, H. Ohnuma, Y. Murakami and H. Inagaki, Polymer J. (Japan), 1, 245 (1970).

4. H. Ohnuma, T. Kotaka and H. Inagaki, Polymer, 10, 501 (1969).

5. W. H. Stockmayer and M. Fixman, J. Polymer Sci., Part C, 1, 137 (1963).

6. e.g. "Copolymerization", G. E. Ham, Ed., Interscience Publ. New York (1964).

7. R. Simha and L. A. Utracki, Polymer Preprints, 10, 522 (1969).

8. B. H. Zimm and R. W. Kilb, J. Polymer Sci., 37, 10 (1959).

9. See for example: N. Nagasabramanian, O. Saito and W. W. Graessley, J. Polymer Sci., Part A-2, 7, 1955 (1969); W. W. Graessley in "Characterization of Macromolecular Structure", D. McIntyre, Ed., Ntl., Acad. Sci. Publ. Washington, D. C. 1968, pp. 371 - 388.

10. L. M. Hobbs and V. C. Long, Polymer, 4, 479 (1963); G. C. Berry and R. A. Cragg, ibid. 5, 19 (1964), G. C. Berry, L. M. Hobbs and V. C. Long, ibid. 5, 31 (1964).

11. "Polymer Handbook", I. Brandrup and E. H. Immergut Eds., Interscience Publ., New York (1966).

12. M. Abe, M. Iwama and T. Homma, Kogyo Kagaku Zasshi, 72, 1572, (1969).

13. E. Eskin and I. A. Baranovskaya, Vysokomol. Soedin., All, 764, (1969).

14. S. Polowinski, "Badania nad Lepkoscia Roztworow Kopolimerow Szczepionych", Habilitation Thesis, Lodz 1966; J. Polymer Sci., Part A-2, 5, 891 (1967).

15. G. M. Bristow and W. F. Watson, Trans. Faraday Soc., 54, 1731 (1958).

16. G. C. Berry and T. A. Orofino, J. Chem. Phys., 40, 1614 (1964).

17. G. C. Berry, J. Polymer Sci., Part A-2, 6, 1551 (1968).

18. E. F. Casassa and Y. Tagami, J. Polymer Sci., Part A-2, 6, 63 (1968).

19. A. Dondos, Europ. Polym. J., 5, 767 (1969).

THE ADSORPTION BEHAVIOR OF POLYSTYRENE-POLY (METHYL METHACRYLATE) MIXTURES CONTAINING BLOCK COPOLYMERS

R. A. Botham, C. P. Shank, C. Thies

Polymer Research Section, Materials Research

The National Cash Register Company, Dayton, Ohio 45409

INTRODUCTION

The preparation and characterization of block copolymers has been actively pursued in recent years (1). Although such copolymers have been recognized as unique interface active species, their behavior at solid/liquid interfaces has received relatively little attention. Accordingly, a study has been made of the adsorption behavior of samples containing S/MMA block copolymers of the AB type. The styrene/methyl methacrylate (S/MMA) system was chosen because the adsorption behavior of S and MMA homopolymers has been studied, both as individual polymers and as mixtures (2). Since the S/MMA block copolymer synthesis procedure consistently yielded a mixture of S and MMA homopolymer plus S/MMA block copolymer, all measurements thus far have been confined to such mixtures.

EXPERIMENTAL

S/MMA block copolymerizations were carried out in toluene at -78°C after the method of Cunningham and Treiber (3) and with n-butyl lithium as catalyst. All transfer operations were carried out in a glove box (argon atmosphere). Anionic S (PS-1) and MMA (PMMA-1) homopolymers were similarly prepared with n-butyl lithium as catalyst. The anionic PMMA homopolymer had a J value of 56 which indicates it has a large degree of isotacticity (4). Free radical PMMA (PMMA-2) was prepared in bulk at 60°C with n-dodecyl mercaptan as chain transfer agent to control molecular weight and azobisisobutyronitrile as initiator. The homopolymers and S/MMA block copolymer samples were isolated by precipitation into methanol and freeze dried from benzene. After isolation, a portion of sample C was reprecipitated into n-heptane from benzene and then extracted briefly with acetonitrile, a solvent for MMA homopolymer and nonsolvent for S homopolymer. This treatment yielded Sample C-1. Brief extraction of sample C-1 with methylcyclohexane ($C_6H_{12}CH_3$), a solvent for S and nonsolvent for MMA homopolymer, yielded sample C-2. Sample A was extracted with $C_6H_{12}CH_3$ to give two fractions, A-1 ($C_6H_{12}CH_3$ soluble) and A-2 ($C_6H_{12}CH_3$ insoluble). Fraction A-1 was isolated by precipitation into n-heptane. Fraction A-2 was extracted with dimethylsulfoxide (DMSO) to yield two fractions, A-3 (DMSO soluble) and A-4 (DMSO insoluble). Fraction A-3 was isolated by precipitation into isopropanol. All fractionated samples were freeze dried from benzene. In all cases, the effectiveness of the extractions was limited due to the formation of stable colloidal dispersions.

Table I contains conversions of the various polymerization reactions, catalyst concentrations, and number

TABLE I

Polymer Characterization Data

Sample	Conversion,%	Catalyst Conc.,* Moles x 10^3	M_n
A	58.2	0.8	27,400
A-1	-	-	39,200
A-3	-	-	36,900
A-4	-	-	39,500
B	61.8	1.4	22,200
C	75.4	2.0	22,700
C-1	-	-	31,700
C-2	-	-	70,200
PS-1	97	0.6	38,100
PMMA-1	35	1.0	80,000
PMMA-2	25	(Free Radical)	55,000

* per 15 g. monomer used

average molecular weights (M_n) for all samples. Values of M_n were determined in toluene (37°C) with a Hewlett-Packard Model 502 high speed membrane osmometer (Hewlett-Packard Corporation, North Hollywood, California). Gel permeation chromatography analyses indicate all block samples have comparatively narrow molecular weight distributions.

Procedures for construction of the polymer adsorption isotherms have been reported (2). Isotherms for the S and MMA homopolymers are plotted as polymer adsorbance (x/m, mg/g silica) versus concentration of polymer remaining in solution (C_s,g/100 ml). Isotherm data for the block copolymer samples have been plotted as values of C_s or x/m versus weight of silica (w,g). Infrared bound fraction values, p, were determined as described previously (2). The adsorbent, Cab-O-Sil M-5 (Cabot Corporation, Boston, Massachusetts), was the same as that used previously (2) and was dried at 105°C. The solvent was Chromatoquality trichloroethylene (Matheson, Coleman and Bell, Norwood, Ohio). All infrared measurements were made with a Perkin-Elmer 457 recording spectrophotometer (Perkin-Elmer Corporation, Norwalk, Connecticut). Infrared absorptivities of the block copolymer samples (700 cm^{-1} and 1734 cm^{-1}) were determined by the method of Harwood and coworkers (5). The measured absorptivities agreed with values obtained from the normal Beer's Law relation of each of the individual homopolymers.

RESULTS AND DISCUSSION

Figure 1 contains adsorption isotherms for anionic S and MMA homopolymers prepared under conditions similar to those used in the S/MMA block copolymer syntheses. Included are data for an

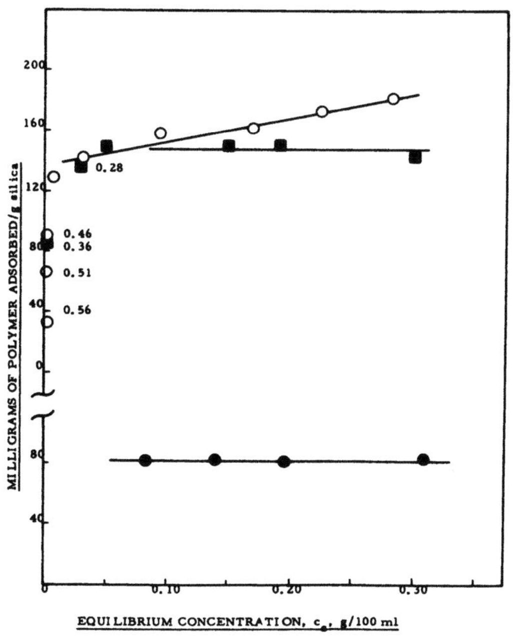

Figure 1. Adsorption isotherms on silica from trichloroethylene at 25°C. The numbers on the graph are the values of p, the bound fraction, at that point: ●, PS-1; ○, PMMA-1; ■, PMMA-2.

atactic sample (PMMA-2) with a M_n value similar to that of PMMA-1. The PS-1 and PMMA-2 isotherms have a typical steep initial rise in adsorbance followed by a flat plateau region. For PMMA-1, adsorbance increases linearly with increasing values of C_s. Numbers on the isotherms represent values of p at that point. For both PMMA samples, values of p decrease with increasing surface coverage. This contrasts with the essentially constant values of p found at all surface coverages for atactic PMMA of higher M_n (2). The high values of p for PMMA-1 relative to PMMA-2 are attributed to the high degree of isotacticity of PMMA-1. This also presumably accounts for the difference in shapes of their isotherms.

Figure 2 contains an isotherm representative of the isotherms found for S/MMA block samples A, B, and C. All of the isotherms are similar and consist of three distinct regions. In the first region (silica wts. $<$ 0.1 - 0.2 g), the MMA concentration decreases significantly, but there is little change in S concentration. The second region is characterized by a simultaneous decrease in S and MMA concentration. In the third region, no MMA is left in solution, but a significant amount of S still is present. S adsorption often appears to taper off in this region.

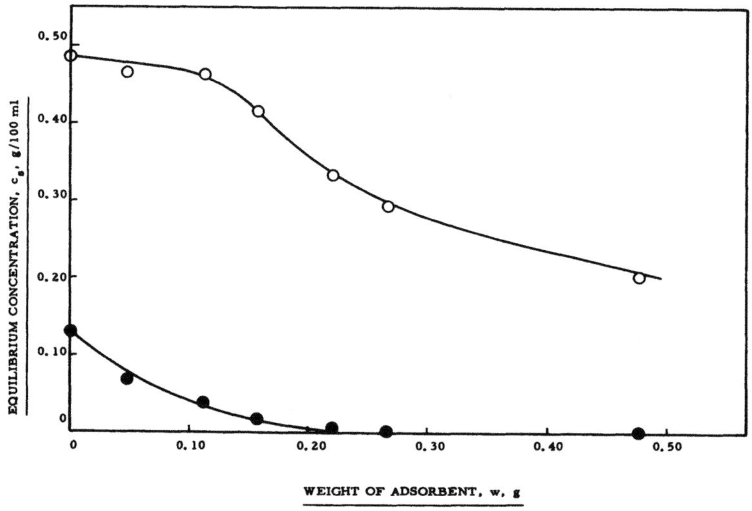

Figure 2. Adsorption isotherm for unfractionated S/MMA block sample B on silica from trichloroethylene at 25°C: ○, S concentration; ●, MMA concentration.

Although isotherms like that in Figure 2 are complex, they establish that the samples being studied contain S/MMA block copolymer as well as S and MMA homopolymer. Physical mixtures of homopolymers free of block copolymer yield isotherms like that shown in Figure 3.

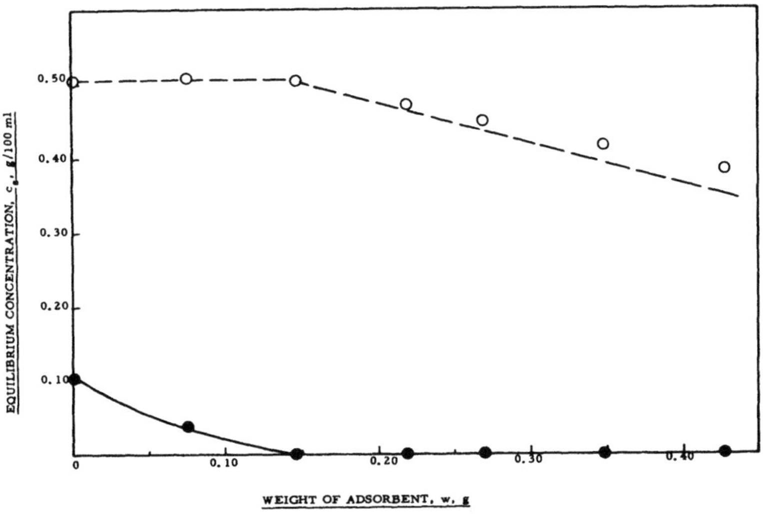

Figure 3. Adsorption isotherm for physical mixture of PS-1 and PMMA-1 on silica from trichloroethylene at 25°C: ○, PS-1 concentration; ●, PMMA-1 concentration.

No S is adsorbed until MMA adsorption is complete and once S adsorption occurs, the S x/m value is slightly less than that of S adsorbed when no MMA is present. These observations were first made with free radical S and MMA homopolymers (2), but have now been found to be true for the anionic S and MMA homopolymers made in the same way as the block samples. Thus, Figure 2 shows that low weights of silica cause a marked decrease in MMA concentration with little parallel change in S concentration. This indicates preferential adsorption of MMA homopolymer. Any slight decrease in S concentration at this point must be due to simultaneous adsorption of a small amount of S/MMA block copolymer, perhaps block molecules with a long MMA sequence. The S left in the supernatant solution once MMA adsorption is complete must be homopolymer. The simultaneous decrease in S and MMA concentration observed for all block samples when $0.1 - 0.2$ g $< w < 0.25 - 0.4$ g is direct evidence for the presence of S/MMA block copolymer. If the S is to be adsorbed simultaneously with MMA when unadsorbed MMA is present, the S must be chemically bound to the MMA.

Because the S is chemically tied to the MMA, one can achieve S x/m values which exceed those found for S homopolymer of similar or higher M_n. This is shown in Figure 4 where x/m values of S for

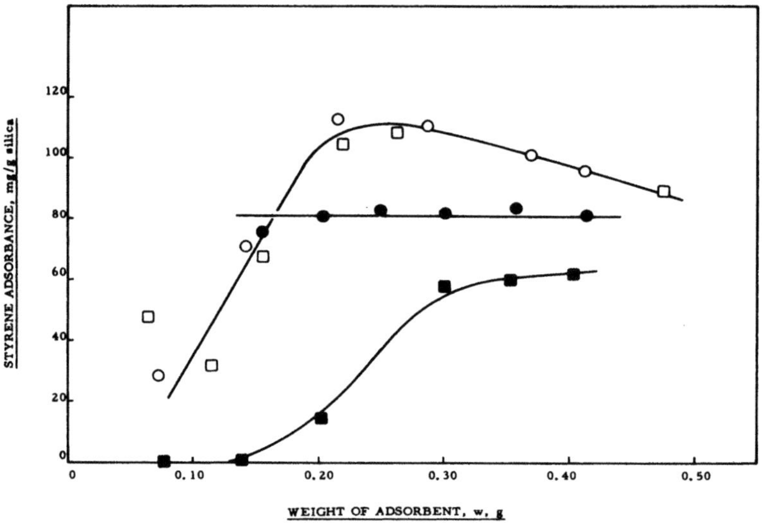

Figure 4. Styrene adsorption isotherms on silica from trichloroethylene at 25°C: ●, PS-1; ■, unfractionated block sample C; ○, unfractionated block sample A; □, unfractionated block sample B. All block samples had an initial solution concentration of 0.62 g/100 ml.

samples A, B, and C are plotted versus w. Included for comparison purposes are data for a S homopolymer similar to the ones used to prepare the block samples. In two cases (Samples A and B), S x/m values for the block samples exceed those found for S homopolymer over a substantial portion of the adsorption isotherm. Because these two samples behave similarly, one curve has been used to represent their behavior. The low S x/m values for all samples at w < 0.1 - 0.2 g are due to preferential adsorption of MMA homopolymer. In all cases, the S adsorbances do not include the MMA also adsorbed on the silica. Figure 5 contains a plot of total polymer adsorbance, S plus MMA, versus w. For samples A and B, the total adsorbance passes through a maximum which exceeds that found for higher molecular weight MMA or S homopolymer (2). The adsorbance for sample C declines sharply at first, but then undergoes only a slight further decrease over a wide range of silica weights. The different behavior of sample C in Figures 4 and 5 relative to samples A and B is believed to be due to the higher MMA homopolymer content of the latter.

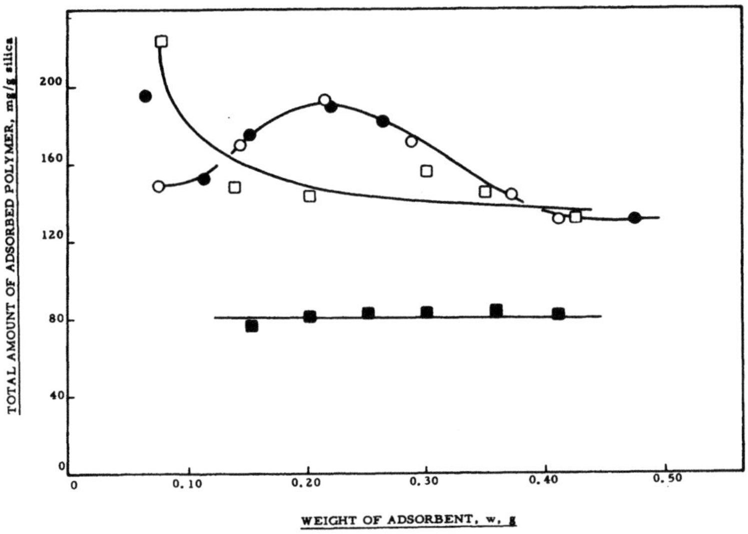

Figure 5. Isotherms showing total weight of polymer on silica from trichloroethylene at 25°C: ■, PS-1; ●, unfractionated block sample B; ○, unfractionated block sample A; □, unfractionated block sample C. All block samples had an initial solution concentration of 0.62 g/100 ml.

The data in Figures 4 and 5 illustrate how S/MMA block copolymers can greatly increase not only the S adsorbance, but also the total weight of polymer adsorbed. The MMA segments presumably are preferentially adsorbed and the S segments are not attached directly to the surface until the MMA segments are completely exhausted from solution. Figure 2 indicates C_s for S changes slope once MMA adsorption is complete. This may indicate that once MMA adsorption is complete and free surface sites become available, S segments of adsorbed S/MMA block molecules adsorb preferentially to the S of S homopolymer.

In an effort to characterize the block samples more fully and isolate fractions with higher block contents, samples A and C were submitted to a variety of extraction procedures. Sample C was reprecipitated from benzene into n-heptane and then extracted briefly with acetonitrile, a solvent for MMA homopolymer and nonsolvent for S homopolymer. The goal was to remove much of the MMA homopolymer in the original sample. However, the isotherm for the reprecipitated sample (sample C-1) is very similar to that of the unextracted sample which indicates little change in sample composition has occurred. Further extraction with $C_6H_{12}CH_3$, a nonsolvent for MMA but solvent for S homopolymer, yielded a sample (C-2) which was largely MMA homopolymer as determined by infrared spectroscopy. This confirmed the presence of MMA homopolymer as a product of the block copolymerization reaction. Better separations were achieved when Sample A was extracted first with $C_6H_{12}CH_3$ to yield a soluble fraction (A-1) that was 97% styrene. A second fraction (A-2) insoluble in $C_6H_{12}CH_3$ was extracted with DMSO to yield two more fractions, A-3 (DMSO soluble) and A-4 (DMSO insoluble). Both A-3 and A-4 have S/MMA contents substantially greater than unfractionated sample A. Figure 6 shows that the overall shape of the isotherm for A-3 still resembles that of unfractionated A, thereby indicating a complete separation of the mixture was not achieved. However, it also shows a major increase

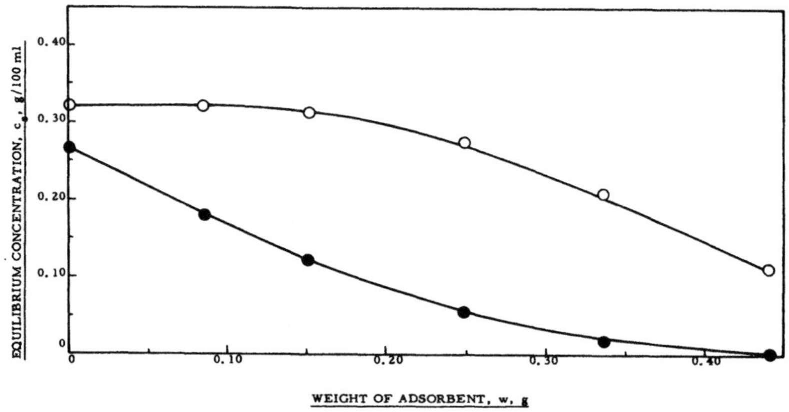

Figure 6. Adsorption isotherm for fractionated S/MMA block sample (A) on silica from trichloroethylene at 25°C: ○, S concentration; ●, MMA concentration.

in that fraction of the sample which is S/MMA block, since simultaneous adsorption of S and MMA occurs over a much broader range of the isotherm. The M_n data in Table I show that all fractions of the block samples have M_n values greater than the unfractionated samples, but in most cases the increase is small. Only C-2 (essentially MMA homopolymer) has a M_n several times greater than the starting material. Difficulty in achieving a better separation of the MMA homopolymer from the S/MMA block copolymers was due to the formation of stable emulsions in both acetonitrile and DMSO.

From isotherm data like that in Figures 2 and 6, one can

estimate the amount of S and MMA homopolymer plus S/MMA block present in each sample. For a solution containing these three components, the total polymer concentration (C_t,g/100 ml) is given by

$$C_t = C_S + C_{MMA} + C_{S/MMA}$$

where C_S = concentration of styrene homopolymer (g/100 ml), C_{MMA} = concentration of MMA homopolymer (g/100 ml), and $C_{S/MMA}$ = concentration of S/MMA block (g/100 ml). The concentration of S remaining in solution after complete exhaustion of MMA by adsorption on the silica equals C_S. An estimate of C_{MMA} is obtained from the concentration of MMA adsorbed in that region of the isotherm where the concentration of S is essentially unchanged due to preferential adsorption of MMA homopolymer. Values of $C_{S/MMA}$ are obtained by difference. The isotherm data also enable one to estimate compositions of the S/MMA block copolymer molecules present in each sample. The concentration of S and MMA associated with block S/MMA is calculated from the S and MMA concentration changes in that region of the isotherms where both are adsorbed simultaneously in significant amounts.

Both of the above calculations must be considered approximations, since one must usually estimate the exact points at which MMA adsorption is complete and when S adsorption begins. Nevertheless, the data obtained illustrate how adsorption isotherms can be used to help characterize a complex block copolymer sample or mixture. Results are given in Table II.

TABLE II

Composition of S/MMA Block Copolymer Samples Calculated from Adsorption Data

	Overall Sample Composition			S/MMA Block Composition	
Sample	S,%	MMA,%	S/MMA,%	S,%	MMA,%
A	65	11	24	77	23
A-1 (66% of A)	97*	3*	-	-	-
A-2 (34% of A)	57*	43*	-	-	-
A-3 (71% of A-2)	21	22	57	58	42
A-4 (29% of A-2)	30	6	64	59	41
B	47	15	38	81	19
C	38	28	34	76	24
C-1 (80% of C)	28	28	44	76	24

* Compositions determined by infrared analysis; no adsorption data obtained.

They show that S/MMA block is present in all unfractionated samples, but it is not the major component. Selective solvent extractions increase the block content, although in no case did this achieve complete separation of the S/MMA blocks from the S and MMA homopolymers. In fact, it would appear from the isotherm data that selective polymer adsorption onto a substrate such as silica provides an efficient means of separating block copolymers from their corresponding homopolymers. The block composition calculations indicate that the S/MMA blocks contain more S than MMA which is reasonable. Control experiments consistently showed that the anionic S polymerizations were carried to 80-95% conversion before the MMA was added. Block molecules with an overall 1:1 S/MMA composition ratio would require an overall conversion of 80-95% if no homopolymer formed. The lower conversions actually achieved are due largely to premature termination of active S chain ends which explains why all unfractionated samples contained much S homopolymer. If this is taken into account and one assumes the extreme case that all the MMA formed was associated with active S chains, there still is not enough MMA present in any of the unfractionated samples to yield blocks with overall S/MMA composition ratios less than 1:1. Since there is some MMA homopolymer in all samples, only blocks with average S contents $>$ 50% could be formed, which is consistent with the data in Table II. Some individual block molecules could contain more MMA than S, but most blocks must be rich in S.

Table III contains infrared bound fraction data for the various S/MMA block samples. These data represent the average fraction of MMA segments attached directly to the silica. The fraction of bound S segments was not determined although S always was simultaneously adsorbed with the MMA on the silica. Values of p for all samples examined fall between 0.48 and 0.59. Many of these values are similar to those found for equivalent amounts of anionic MMA homopolymer adsorbed on silica, but several samples have values of p that differ somewhat from those observed for MMA homopolymer. Unfractionated A and B had lower values of p; several fractions of samples A and C gave high values of p. The latter may indicate that more MMA segments of adsorbed S/MMA blocks are bound to the silica than in the case of MMA homopolymer. This trend, if it actually occurred, would be expected for such samples since they contain larger amounts of block. However, the data obtained are inconclusive. The two-step adsorption experiments were specifically designed to resolve this point, but did not. In the two-step experiments a small amount of silica was added to a solution of the block sample. This adsorbed mostly MMA homopolymer. The silica plus adsorbed MMA was then removed from the system before additional silica was added and values of p for the MMA segments adsorbed in the second step were determined. Presumably these adsorbed MMA segments belong largely to S/MMA block copolymer molecules. Although these experiments yielded the highest MMA p values found in this study, the differences were not great. Further experiments with samples high in block content are needed to firmly establish whether or not adsorbed MMA segments of adsorbed S/MMA block copolymers have a conformation similar to that of anionic MMA homopolymer. The data obtained thus far indicate that insofar as infrared bound fraction measurements are concerned, any differences are small.

TABLE III

Block Sample	Concn. of MMA,g/100 ml*	x/m,MMA,mg/g	p
A	0.1069	38.9	0.50
A-3	0.2610	88.0	0.49
A-3	0.2635	89.5	0.50
A-4	0.1570	52.7	0.53
B	0.0886	31.2	0.48
B	0.1323	41.7	0.54
B**	0.0314	10.7	0.59±0.03
C	0.2065	72.1	0.48
C-1	0.2470	78.3	0.56
C-1**	0.1328	49.7	0.59

*Concn. of MMA in mixture before adsorption. In all cases, equilibrium supernatant concn. of MMA = 0.

**Two step experiment.

REFERENCES

1. "Graft Copolymers", H. A. J. Battaerd and G. W. Tregear, Interscience Publishers, New York (1967), Ch IV.
2. C. Thies, J. Phys. Chem. 70, 3783 (1966).
3. R. E. Cunningham and M. R. Treiber, J. Appl. Pol. Sci. 12,23 (1968).
4. M. Maruhashi and H. Takida, Makromol. Chem., 124, 172 (1969).
5. H. J. Harwood, T. S. Ang. R. G. Bauer, N. W. Johnston, and K. Shimizu, Polymer Preprints 7, 980 (1966).

DILUTE SOLUTION BEHAVIOR OF BLOCK COPOLYMERS: STYRENE-METHYL METHACRYLATE AND α-METHYL STYRENE-METHYL METHACRYLATE SYSTEMS

Tadao Kotaka, Hiroshi Ohnuma,* and Hiroshi Inagaki

Institute for Chemical Research, Kyoto University

Uji, Kyoto-fu 611, Japan

The structural feature of block copolymers is that each polymer chain consists of chemically different subchains, each being a long stretch of monomers of a single type. Block copolymers in solution exhibit characteristic behavior resulting from thermodynamic interactions of subchains with solvent molecules and with each other. The nature of such interactions should resemble, in some respects, to the interactions found in a ternary system involving the parent homopolymers and the solvent. The situation is in contrast to the solution behavior of statistical and alternating copolymers, in which chemically different monomers are distributed along each chain in highly alternating sequences (1,2).

It has been known that the incompatibility of two different polymers in the presence as well as in the absence of diluent is a general rule rather than an exception (3). The fact had led to an idea that block copolymers composed of an incompatible homopolymer pair would exhibit a phenomenon of <u>intrachain phase separation</u> even in a solvent that dissolves both parent homopolymers: Many authors had attempted to interpret anomalies found in the behavior of block copolymer solutions as the result of this phenomenon (4,5).

The phenomenon of intrachain phase separation becomes apparent for certain block copolymers in bulk phase (6). However in view of compatibilizing nature of diluent in ternary systems (3), we suspect whether or not the phenomenon would take place in block copolymer chains under infinite dilution. Rather we expect that the phenomenon

---

* Present address: Department of Physics, Faculty of Science Nagoya University, Chikusa-ku, Nagoya 464, Japan.

would be influenced by the extent of dilution and, to some extent, by the nature of solvent, besides many other factors such as molecular weight, composition, chain-type etc., of the block copolymer chains.

To make these points clear, we have attempted to examine the solution properties of block copolymers, particularly in relation with the incompatibility of corresponding ternary systems (7). We have chosen pairs of styrene (ST) with methyl methacrylate (MMA) and of α-methyl styrene (MST) with MMA as the model systems. The choice is made because polystyrene (PST) and poly(methyl methacrylate) (PMMA) are known to be an incompatible pair, whereas poly(α-methyl styrene) (PMST) and PMMA are exceptionally a compatible pair (3,8).

EXPERIMENTAL

Polymer Samples. Polymer samples used here were prepared by an anionic polymerization technique (9). Block copolymer samples of BAB-type, in which A was either PST or PMST and B was PMMA, were prepared in tetrahydrofuran (THF) at -78° with sodium biphenyl-THF solution as initiator (7b). A few PST-PMMA diblock samples were prepared by the same procedures except that n-butyl lithium-n-hexane solution was used as initiator (7d). The butyl lithium solution was prepared by employing the method reported by Fujimoto, Ozaki, and Nagasawa (9d).

As reported previously (7b), the preparation of each block copolymer sample was carried out through the following steps. First, initiator was allowed to react with ST (or MST) monomer, yielding living PST (or PMST). An aliquot of the living polymer solution was taken and the polymer was recovered as a reference homopolymer sample. An equivalent amount (with respect to living ends) of 1,1-diphenyl ethylene was added to the rest of the living polymer solution. The step was employed, as suggested by Freyss, Rempp, and Benoit (9c), to suppress hazadous reaction of living PST (or PMST) with ester groups of MMA units when added later. Finally MMA monomer was added, and the living polymer initiated the polymerization of MMA, yielding a block copolymer sample. A few homo-PMMA samples were prepared by the similar procedures, in which sodium biphenyl was allowed to react directly with MMA.

Crude samples of BAB-type may be contaminated by homo-PST (or PMST) and diblock chains; and likewise crude samples of AB-type by homo-PST. Although rather unlikely, but still they may possibly be contaminated by homo-PMMA. Each crude sample of triblock type was subjected to boiling-point extraction with cyclohexane (CHX) and then with acetonitrile (4c). Later it turned out that contaminating diblock chains might have been removed by this procedure.

On the contrary, the procedure posed a problem when applied to treat diblock samples. As shown later, PST-PMMA diblock copolymers often become soluble in either cyclohexane or acetonitrile (7e). The diblock samples were washed simply by cold cyclohexane.

As judged from the above steps, PST (or PMST) subchains of each block copolymer sample must have the same characteristics, i. e., molecular weight, polydispersity, microtacticity, etc., with each precursor PST (or PMST). Homo-PMMA samples were found to have typically atactic configurations, to which the microtacticity of PMMA subchains of block copolymer samples must be identical. As the result of characterization experiments, each block copolymer sample was found to be reasonably homogeneous and to be free from any contaminating branched chains (7b).

Measurements. Comonomer composition was determined by combustion analysis. Osmotic-pressure measurement was made on a Mechrolab high speed membrane osmometer model 502 with ultracellafilter membranes (7b). Light-scattering measurement was made on a Shimadzu photometer of modified Brice type (7b,10a). Sedimentation velocity (10b) and equilibrium (10c) experiments were made on a Hitachi analytical ultracentrifuge model UCA-1A equipped with a schlieren and Rayleigh interference optics. Determination of intrinsic viscosities [η] was made on Ubbelohde dilution viscometers (10a). The details of these procedures were described elsewhere (7,10).

Besides these experiments, we have made determination of critical miscibility temperature T_c (2) and 'Θ' temperature (7a,c) of block copolymer solutions, and also determination of critical miscibility concentration c_m of ternary systems (7d). Here the 'Θ' temperatures were determined by employing the same procedures established for homopolymer solutions (11), namely as the temperature at which osmotic second virial coefficient A_2 is zero or the T_c in the limit of infinitely large molecular weight.

RESULTS AND DISCUSSION

Solubility and 'Θ' Temperatures

A simple measure of the solubility of nonelectrolytes may be provided by the Hildebrand-Scott solubility parameter δ (12a). For PST and PMMA pair, the difference in δ is about 0.5 units (12b): The pair is incompatible, at least, in the absence of diluent. For PMST the value of δ is not yet known, but it is reported that PMST and PMMA are compatible and yield uniform mixture even without diluent (8).

The results of solubility test are summarized in Tables I and II, which list either T_c or 'Θ'. Here we should specify the meaning

Table I. The T_c and 'Θ' (in ^{o}C) for di- and triblock copolymers of ST and MMA in various solvents.*

	PMMA	PST	PST-PMMA		PMMA-PST-PMMA			
code	14M	16H	39B	46B	20B	28B	26B	16B
$10^{-4}M_n$ §	7.2	20.6	5.5	26	16	7	26	39
ST-mole fr.	0.0	1.00	0.48	0.45	0.36	0.49	0.45	0.72
cyclohexane	N	Θ=35	20	N	----- not soluble -----			
1-chloro n-hexane	N	S	(S)	((S))	----- not soluble -----			
cyclohexanol	Θ=79	Θ=81	Θ = 82		Θ=81	Θ = 81		Θ=84
2-methyl cyclohexanol	28	55	35	-	37	35	40	54
2-ethoxy ethanol	Θ=39	N	(S)	(S)	Θ=70	Θ= 81		N
acetonitrile	Θ=35	N	((S))	((S))	((40))	((30))	((35))	N

* Values of 'Θ' were estimated from A2 versus T data (11). N = not soluble at 100°C; S = soluble at -15°C; ( ) = soluble but slightly turbid; (( )) = soluble but heavily turbid.
§ M_n = number average molecular weight.

of the term 'soluble' used in these tables. It has been known that block copolymers in solution, especially dissolved in selective solvents, are not necessarily in the state of molecular dispersion: They often yield visibly turbid solutions but do not undergo further precipitation, indicating formation of micelles, i. e., stable aggregates (4b,6a,13). In many examples listed in the tables, we encountered such visibly turbid solutions.

From Tables I and II, the following crude generalizations might be made. (i) PST-PMMA diblock copolymers and presumably any other diblock copolymers become soluble in selective solvents that dissolve either one of the parent homopolymers. Quantitatively the solubility depends on composition and molecular weight of the sample. On the other hand, (ii) PMMA-PST-PMMA triblock copolymers appear to be more largely influenced by the solubility of side PMMA subchains. They are usually soluble in selective solvents that dissolve PMMA but not PST. In such cases, presumably soluble PMMA subchains may wrap around insoluble PST subchains and protect against precipitate formation. However the opposite would be more unlikely. They are unstable in solvents of the opposite selectivity (7d).

Table II. The T_c and 'Θ' (in ^oC) for triblock copolymers of MST and MMA in various solvents.*

	PMMA	PMST	PMMA-PMST-PMMA	
code	38M	36H	32B	36B
$10^{-4}M_n$	34.6	46.3	53.3	117
MST-mole fr.	0.0	1.00	0.41	0.38
2-octanone	34.0	98.5	87.2	88.5
	Θ = 43 (1.10)	Θ =130 (1.90)	Θ = 94 (0.85)	
3-heptanone†	18.3	67.5	73.5	74.5
	Θ = 25 (0.44)	Θ = 83 (0.35)	Θ = 84 (0.43)	
2-methyl cyclohexanol	32.5	67.3	57.7	60.0
	Θ = 41 (0.41)	Θ = 74 (0.90)	Θ = 62 (1.40)	
1-chloro n-hexane	92.5	11.7	84.8	88.0
	Θ =101 (0.80)	Θ = 33 (0.26)	Θ = 93 (0.57)	

* Values of 'Θ' were estimated from T_c versus $M_n^{-1/2}$ plots (11). Values in parentheses are the entropy of mixing parameter.
† Values of 'Θ' were also estimated from A_2 versus T data as 25° for PMMA, 79° for PMST, and 78° for triblock samples.

Nevertheless, PMMA-PMST-PMMA triblock copolymers appear to be insoluble in solvents that dissolve PMMA but not PMST and also that dissolve PMST but not PMMA. Namely (iii) in this system, less soluble subchains appear to govern the solubility of the whole chains, irrespective of their location in the chains. This might be due to the better compatibility of PMST and PMMA pair. The protective action of soluble subchains toward insoluble partners might be better achieved in block copolymers of an incompatible pair rather than those of a compatible pair.

Behavior in 'Θ' Solvents

We should emphasize that the 'Θ' condition of block copolymer solutions determined as mentioned above must be distinguished from the Θ condition of homopolymer solutions. The latter is the state wherein all types of excluded-volume interactions between intra and intermolecular segment pairs vanish simulteneously, thereby polymer molecules behave as random-flight chains (11). The former is merely the state wherein overall intermolecular interactions do vanish, but all types of excluded-volume interactions do not necessarily vanish simulteneously.

Intrinsic viscosity $[\eta]_\Theta$ of random-flight chains may be related to the unperturbed mean-square molecular radius $\langle s^2 \rangle_0$ as

$$[\eta]_\Theta = K_0 M^{1/2}; \quad K_0 = \Phi_0[6(\langle s^2 \rangle_0/M)]^{2/3} \qquad \text{(1a,b)}$$

where Φ_0 is the Flory-Fox viscosity constant (11,14), and M is molecular weight of the polymer chain. Experimentally weight average value M_w is to be used instead of M. As pointed out as above, block copolymer chains would not always behave as random-flight chains under the 'Θ' condition: Then eq. 1 may not necessarily hold for such systems.

First we assume a hypothetical unperturbed state for block copolymer chains. Then $\langle s^2 \rangle_0$ of the block copolymer may be readily correlated with the corresponding quantities of the parent homopolymers (2,15).

$$(\langle s^2 \rangle_0/M) = b^2 = xb_A^2 + (1 - x)b_B^2 + \Delta \qquad (2)$$

$$\text{AB-type:} \quad \Delta = x(1 - x)(2x - 1)(b_A^2 - b_B^2) \qquad \text{(3a)}$$

$$\text{BAB-type:} \quad \Delta = (x/2)(1 - x)(1 + x)(b_A^2 - b_B^2) \qquad \text{(3b)}$$

Here x is composition in weight fraction of A. The subscripts A and B denote the quantities characteristic of the respective parent homopolymers: The quantity b_i^2 (i = A or B) reflects primarily the short-range interactions of the chain and is independent of M.

With seven samples of PMMA-PST-PMMA triblock copolymers, of which $10^{-4}M_w$ ranged from 3.4 to 147 and x from 0.42 to 0.62, the $[\eta]_\Theta$ versus M_w relations were examined in two 'Θ' solvent systems, i. e., in 2-ethoxy ethanol (2-EE) at 81° and in cyclohexanol (CHL) at 81° (7a,c). The results are shown in Figure 1a.

It is seen that $[\eta]_\Theta$ is not proportional to $M_w^{1/2}$ in 2-EE. The result is not surprising at all, since 2-EE is a nonsolvent for PST and a Θ solvent for PMMA. The PMMA-PST-PMMA chains would not assume random-flight conformations even when A_2 vanishes. The block copolymers might form micelles. Even if they are molecularly dispersed, the PST-subchains collapse and each molecule as a whole is barely dispersed by being protected by the soluble PMMA-subchains. Interestingly the block copolymers appear to behave as if they were homo-PMMA chains with M_w equal to the sum of the two PMMA-subchains.

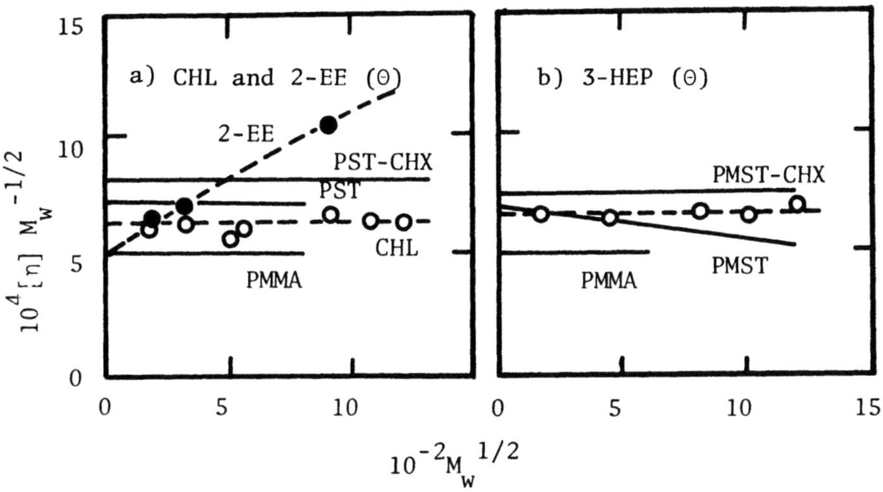

Figure 1. Plots of $[\eta]M_w^{-1/2}$ versus $M_w^{1/2}$ for nearly equimolar (a) PMMA-PST-PMMA and (b) PMMA-PMST-PMMA triblock copolymers in 'Θ' solvents. Circles for block copolymer data; and solid curves for homopolymer data in CHX, CHL, and 3-HEP at their respective Θ temperatures (see Tables I and II).

In CHL at 81°, the $[\eta]_\Theta$ data appear to be proportional to $M_w^{1/2}$. The solvent happens to have almost equal Θ temperatures for both PST and PMMA, to which values of 'Θ' of the block copolymers are also quite close (7a,c). The fact implies that both the PST : PST and PMMA : PMMA interactions practically vanish at this condition. Interestingly a phase separation test on ternary system PST : PMMA : CHL at 85° showed that the system is highly compatible: For example, an equimolar mixture of PST-26H and PMMA-14M did not undergo a phase separation even at the concentration as high as 20% total polymer (7d). This implies that the PST : PMMA interactions might as well vanish at this condition. Consequently all types of the excluded-volume interactions might be regarded as substantially zero, and the block copolymer chains might assume random-flight conformations under this particular 'Θ' condition.

With five samples of PMMA-PMST-PMMA triblock copolymers, of which $10^{-4}M_w$ ranged from 3.5 to 151 and x from 0.38 to 0.50, the $[\eta]_\Theta$ versus M_w relations were examined in 3-heptanone (3-HEP) at 81° (7f). The results are shown in Figure 1b. The $[\eta]_\Theta$ data appear to follow eqs. 1-3, in spite of the considerable difference in Θ of PMST and PMMA in this solvent (cf., Table II). The conformational anomalies resulted from this difference might be small so as not to induce appreciable anomalies in the $[\eta]_\Theta$ versus M_w relation.

Intrinsic Viscosity of Block Copolymers and Incompatibility of Ternary Systems

The [η] versus M_w relations of the nearly equimolar triblock copolymer samples and the phase equilibrium behavior of ternary systems involving their parent homopolymers were examined in several solvents. For the PMMA-PST-PMMA triblock samples, solvents chosen were THF, toluene (TOL), 2-butanone (MEK), diethyl malonate (DEM), 1-chloro-n-butane (1-CB), and p-xylene (p-XY). The solvents, THF, TOL, and MEK, are all common solvents for both PST and PMMA; DEM is a Θ solvent for PST and a moderate solvent for PMMA; while 1-CB and p-XY are Θ solvents for PMMA and moderate or good solvents for PST. For the PMMA-PMST-PMMA samples, solvents chosen were TOL and p-XY. The former is a common solvent for both PMST and PMMA, while the latter is a Θ solvent for PMMA and a moderate solvent for PMST.

A phase separation test on the parent homopolymer pairs, PST : PMMA and PMST : PMMA, was carried out in the same solvents as those used in the viscometric studies. As anticipated, the PMST : PMMA pair was confirmed to be compatible: In either solvents (of course in p-XY at the temperature above the T_c for PMMA), phase separation was not observed in mixtures of all proportions examined.

As mentioned above, the PST : PMMA pair was found to be fairly compatible in CHL (85°): Phase separation did not take place even at the highest concentrations investigated (7d), and we could not locate the critical concentration c_m for this system. However in other solvents the PST : PMMA pair was found to be more or less incompatible.

On the incompatibility of ternary systems involving polymer A : polymer B : solvent, it has been found that the critical concentration c_m decreases with increasing molecular weight according to the following empirical relation (16):

$$c_m = C'\tilde{M}^{-2/3} + C''; \quad \tilde{M} = (M_A M_B)^{1/2} \tag{4a,b}$$

Here C' and C'' are constants characteristic of a given ternary system and $\tilde{M}$ is the geometric mean-molecular weight of the two polymers.

First we examined the dependence of c_m (g-total polymer/ml-total solution) on the mixing ratio of the two polymers for a few different pairs of PST and PMMA. The critical concentration c_m was found for equimolar mixtures. Further test was conducted on several equimolar mixtures (7d).

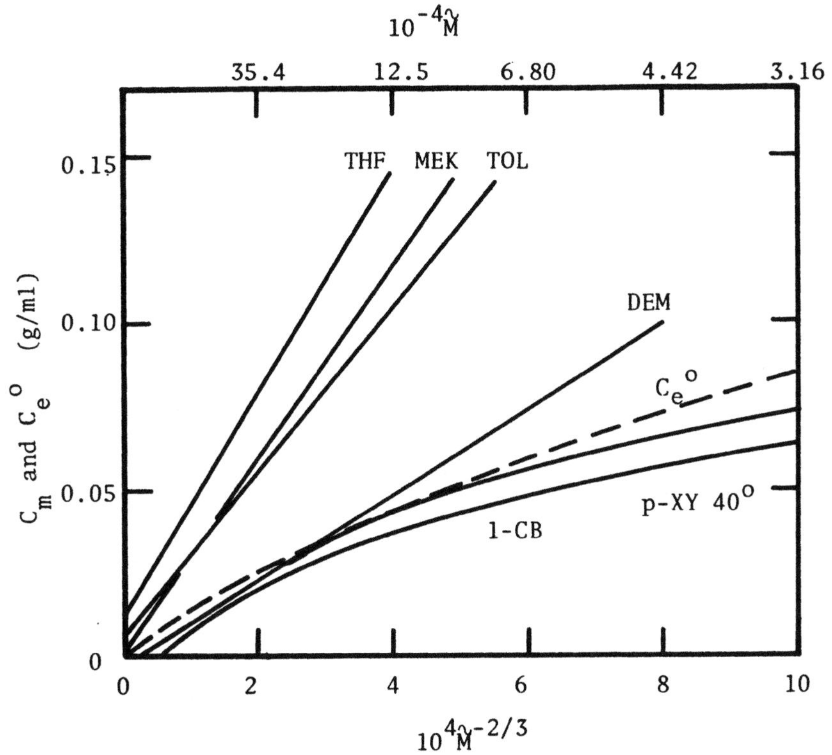

Figure 2. Plots of c_m versus $\tilde{M}^{-2/3}$ for PST : PMMA : solvent systems (for codes see text). Dashed curve shows c_e^o for nearly equimolar PMMA-PST-PMMA triblock copolymers (cf., eqs. 10a and b).

The critical concentration c_m data for the PST : PMMA : solvent systems are plotted according to eqs. 4a and b. The results are shown in Figure 2. The data for THF, TOL, MEK, and DEM systems appear to be well described by eqs. 4a and b, whereas those for 1-CB and p-XY do not. The reason for this difference is not clear at the moment. Apparently the fact that the latter two solvents are Θ solvents for one of the constituent homopolymers could not be the sole reason for the difference, since it has not been observed in DEM that also is a Θ solvent for one of the constituent, PST. Anyway the PST : PMMA pair becomes more and more incompatible in the order CHL (85°) < THF < TOL < MEK < DEM < 1-CB $\cong$ p-XY. The pair is more strongly incompatible in selective solvents rather than in common solvents. The tendency would be true for any incompatible homopolymer pairs.

Now we turn our attention to the problem of estimating the excluded-volume (long-range) interaction parameter from $[\eta]$ data of block copolymer solutions. For describing $[\eta]$ versus M_w relations of homopolymer solutions, several viscosity equations have been proposed so far (11,17,18). Among them an equation proposed by Fujita et al. (18a,b) reads as follows

$$[\eta]M_w^{-1/2} = 1.05K_0 + 0.83\ \Phi_0 BM_w^{1/2} \qquad (5)$$

$$(1.2 < \alpha_\eta^3 = [\eta]/[\eta]_\Theta < 2.3)$$

The parameter B represents the excluded-volume interactions for intrachain segment pairs. The equation 5 was so constructed that the values of B are made to agree with those from mean-square radius $< s^2>$ data and from A_2 data (18). In that sense the equation is consistent within the framework of the two parameter theories that demand the identity of the interaction parameters B for intrachain segment pairs (those relevant to $< s^2>$ and $[\eta]$ data) and for interchain segment pairs (those relevant to A_2 data) (17,18,19).

Application of eq. 5 for treating block copolymer data requires further consideration. The problem is concerned with conformational anomalies of block copolymer chains wich might prevail in certain particular solvents. For block copolymer solutions the parameter B may be correlated with those of the parent homopolymer solutions, B_{AA} and B_{BB}, as follows:

$$B = p_{AA}B_{AA} + p_{BB}B_{BB} + 2p_{AB}B_{AB} \qquad (6a)$$

Here p_{ij} (i, j = A, B) is the probability of i-j contact which is characterized by the parameter B_{ij}: The p_{ij}'s depend on the composition and the chain architecture such as AB-type (5) or BAB-type (7f) of the block copolymer chain.

The parameter B_{AB} may be correlated with a similar quantity χ_{AB}, which characterizes the incompatible interactions in ternary system:

$$\chi_{AB} \propto \Delta B_{AB} = B_{AB} - (B_{AA} + B_{BB})/2 \qquad (6b)$$

Phase separation of a ternary system should take place, when χ_{AB} exceeds certain critical value which is, as a zeroth approxiamtion, inversely related to polymer concentration and molecular weight (3b).

On literal interpretation of the theory (1,3b), χ_{AB} and hence ΔB_{AB} should be determined primarily by the nature of the homopolymer pair, and the solvent has only secondary effect through its molal volume. However this is not necessarily true: Namely both the viscometric ΔB_{AB} and the χ_{AB} for ternary systems are significantly dependent on the nature of solvent (3c,7d). It is difficult to achieve a quantitative agreement between these two quantities.

On the other hand, a calculation of p_{ij}'s requires a knowledge of the segment distribution of block copolymer chains under the perturbed state. Instead of carrying out such calculations, we assume that the distribution may be approximated by a random distribution of disconnencted segments. The probability becomes $p_{ij} = m_i m_j$ with m_i being the mole fraction of i-segment, i. e., $m_A = m$ and $m_B = 1 - m$ (1,20). Then the parameter B may be given by either one of the following two forms:

$$B = m^2 B_{AA} + (1 - m)^2 B_{BB} + 2m(1 - m) B_{AB} \quad (7a)$$

$$B = m B_{AA} + (1 - m) B_{BB} + 2m(1 - m) \Delta B_{AB} \quad (7b)$$

In some special cases the excluded-volume interactions between chemically unlike segments might induce conformational anomalies of the block copolymer chain: Namely the chain might assume a particular conformation in which the domains of the unlike subchains are completely segregated, i. e., a segregated chain model (5). In such a case, eq. 5 would become invalid, since it is primarily based on a random coil model with uniform expansion approximation.

On the other hand, the assumption involved in eq. 7 also is equivalent to the assumption that block copolymer chains would not assume such anomalous conformations. The two equations, eqs. 5 and 7, involve similar approximations. Therefore, when $[\eta]M_w^{-1/2}$ versus $M_w^{1/2}$ relation for a given block copolymer solution obeys eq. 5, yielding a reasonable estimate of K_o, then it would be legitimate to assume that its slope would allow a reasonable estimate of B, which can be approximated by eq. 7.

The criterion mentioned above is apparently too crude, but may be employed as a working hypothesis. On this bases we have examined the $[\eta]$ versus M_w relations of the nearly equimolar triblock copolymer samples of PMMA-PST-PMMA and PMMA-PMST-PMMA in the solvents as mentioned above. As typical examples of the viscometric behavior of these triblock copolymers, Figures 3 and 4 show the $[\eta]M_w^{-1/2}$ versus $M_w^{1/2}$ plots obtained respectively in TOL and in p-XY solutions together with those of the parent homopolymers.

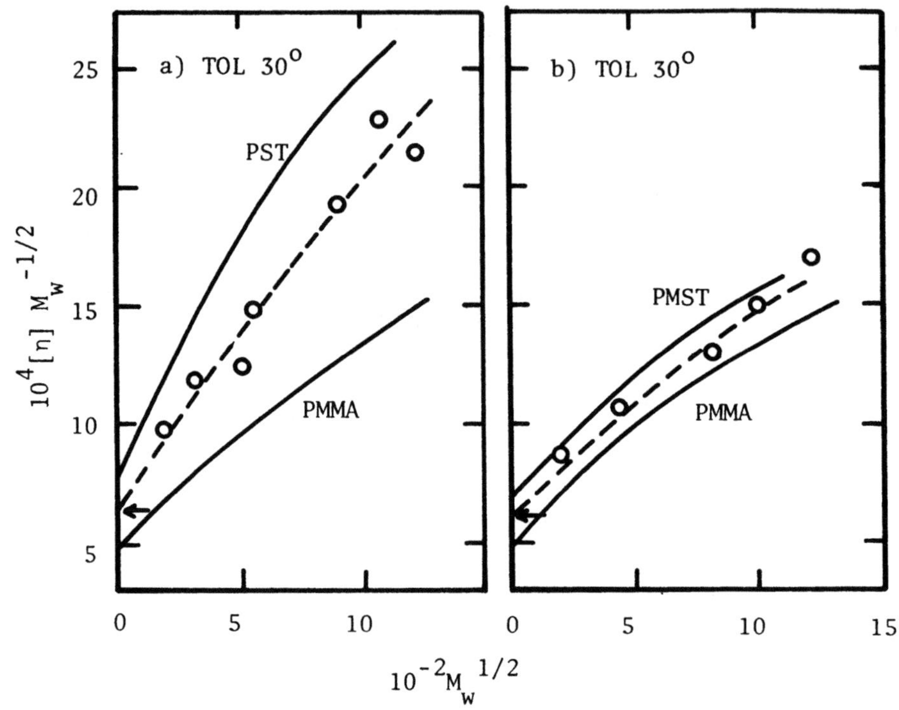

Figure 3. Plots of $[\eta]M_w^{-1/2}$ versus $M_w^{1/2}$ for nearly equimolar (a) PMMA-PST-PMMA and (b) PMMA-PMST-PMMA triblock copolymers in TOL at 30°C: Circles for block copolymer data; solid curves for parent homopolymer data. Arrows indicate the values of K_0 estimated by eqs. 1-3.

As seen in Figure 3, in TOL solutions no appreciable anomalies are observed for both samples, as opposed to some previously reported results (4). The plots appear to be well described by eq. 5, and allow reasonable estimates of K_0. The values of K_0 are in agreement with those computed by eqs. 1-3 by using appropriate values of b_i^2 for the parent homopolymers (21). Although the plots are not shown here, the $[\eta]$ versus M_w data of the PMMA-PST-PMMA samples in THF, MEK, and even in DEM, show no anomalies at all (2,7d). It should be noted that these solvents are all moderate or good solvents for PMMA, although their solubility toward PST differ considerably from one another.

On the contrary, the $[\eta]$ versus M_w data of these triblock copolymers exhibit significant anomalies in the solvents that have poor solubility toward PMMA. As is demonstrated in Figure 4, the

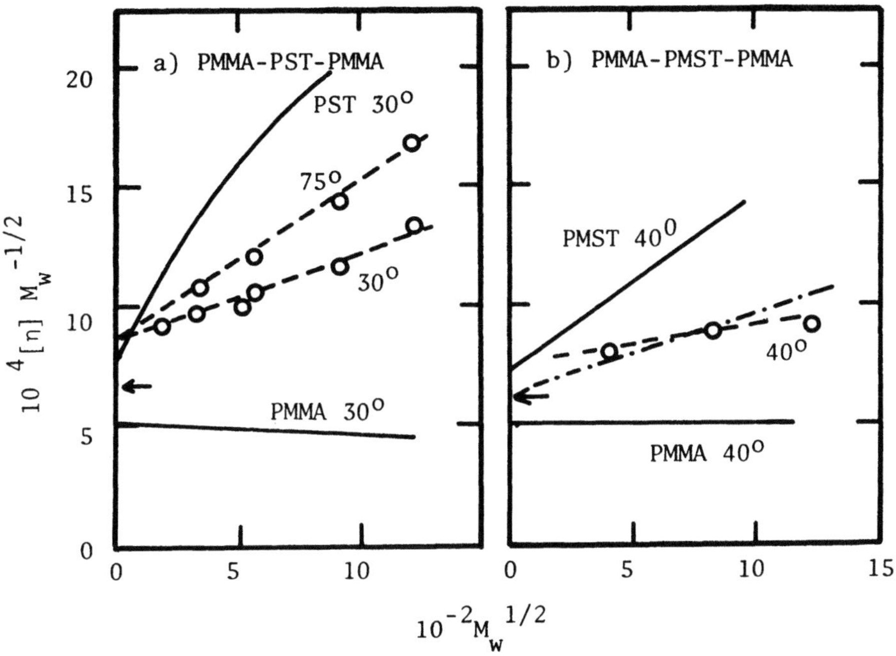

Figure 4. Plots of $[\eta]M_w^{-1/2}$ versus $M_w^{1/2}$ for nearly equimolar (a) PMMA-PST-PMMA and (b) PMMA-PMST-PMMA triblock copolymers in p-XY at the temperatures as indicated: Chain line in (b) is the relation calculated by eqs. 5 and 7b by arbitrarily assuming $\Delta B_{AB} = 0$. For other symbols, see Figure 3.

$[\eta]M_w^{-1/2}$ versus $M_w^{1/2}$ plots of the PMMA-PST-PMMA samples in p-XY yield unduly large value of K_0 as compared with the value computed by eqs. 1-3. The anomalies persist even at elevated temperatures. The similar anomalies are found also in 1-CB solutions (7d). Although less significant, but the $[\eta]$ versus M_w data of the PMMA-PMST-PMMA samples in p-XY also show anomalies of the similar trend, as demonstrated in Figure 4b. For such data, eq. 5 or any other viscosity equations of similar nature (17,18) would not be applicable to estimate the values of B.

These anomalies should be attributed to the fact that the solvents are Θ solvents for one of the constituent homopolymers. In addition for the PMMA-PST-PMMA solutions the poor compatibility of PST : PMMA pair may be respeonsible. However it should be reminded that DEM is also a selective Θ solvent, and the PST : PMMA pair has poor compatibility in this solvent. Nevertheless anomalies were not observed in this solvent (7d). Then the difference of the behavior in 1-CB and p-XY and in DEM must be responsible to

Table III. Values of excluded-volume interaction parameters B and ΔB_{AB} for nearly equimolar PMMA-PST-PMMA samples.

solvent (^{o}C)	PST $10^{-30}B_{AA}$	PMMA $10^{-30}B_{BB}$	PMMA-PST-PMMA $10^{-30}B$	$10^{-30}\Delta B_{AB}$
cyclohexanol (81)	0.0	1.0	0.0	- 1.0
tetrahydrofuran (30)	65.2	43.8	55.0	1.2
toluene (30)	62.5	23.8	45.2	4.0
2-butanone (30)	10.7	28.3	25.6	12.2
diethyl malonate (30)	- 6.5	14.9	18.5	38.6

their chain architecture. The solvents, 1-CB and p-XY, are Θ solvents toward PMMA, which constitute the two side-subchains of the triblock samples. While DEM is a Θ solvent toward the central PST subchains.

From the [η] versus M_w data of the PMMA-PST-PMMA samples in CHL, THF, TOL, MEK, and DEM, we have estimated the values of B and ΔB_{AB} by using eqs. 5 and 7b. The results are listed in Table III. Interestingly the ΔB_{AB} increases in the order CHL (81°) < THF < TOL < MEK < DEM. The ΔB_{AB} for the PMMA-PMST-PMMA samples in TOL was found to be nearly zero. The tendency of increasing ΔB_{AB} of the block copolymer solutions just corresponds to the tendency of increasing incompatibility of the corresponding ternary systems.

Possibility of Intrachain Phase Separation

From the results mentioned above, we may conclude that there is at least a qualitative correlation between the excluded-volume interactions within unlike subchains of a block copolymer chain and the incompatibility of its parent homopolymer pair. Now the question to be asked is whether or not the phenomenon of intrachain phase separation is actually taking place in a block copolymer chain at an infinite dilution. And if so, what conformation would the block copolymer chain assume?

To examine these points, we assume an optimistic point of view such as the following. Namely it would be legitimate to assume that the phenomenon would take place, if the effective concentration c_e of polymer segments within the domain of an isolated block copolymer chain exceeds the critical concentration c_m of the corresponding ternary system (7b,d). On the contrary the intrachain phase separation would not take place, if the c_e is lower than c_m. The latter part of the assumption might sound too crude and too optimistic. However we may at least expect this to be the case.

To estimate the value of c_e for a block copolymer chain, we replace it by a sphere with radius $<s^2>^{1/2}$, and obtain

$$c_e = (3/4\pi)(M/N_a < s^2>^{3/2}) = c_e^{0}/\alpha^3 \qquad (8)$$

where N_a is the Avogadro number, c_e^o is the effective segment concentration for the unperturbed chain; and $\alpha^2 = <s^2>/<s^2>_o$ is the expansion factor due to the excluded-volume interactions. Therefore in any good solvent c_e must be lower than c_e^o.

The geometric mean-molecular weight $\tilde{M}$ of the chemically different subchains may be written for a triblock copolymer of symmetric BAB-type as

$$\tilde{M} = [x(1 - x)/2]^{1/2}M \qquad (9)$$

Using the empirical value of $10^{18} <s^2>_o/M_w = 6.35$ obtained from the $[\eta]_\Theta$ data of the PMMA-PST-PMMA samples in CHL at 81°, and assuming the block copolymers being monodisperse, i. e., $M = M_w$, we obtain

$$c_e^{0} = 24.8M_w^{-1/2}; \quad \tilde{M} = M_w/2.828 \qquad (10a,b)$$

From these equations we can construct c_e^{0} versus $\tilde{M}^{-2/3}$ relation, which in turn may be compared with the c_m versus $\tilde{M}^{-2/3}$ relations for the ternary systems such as shown in Figure 2.

It is seen from Figure 2 that in a group of solvents such as CHL (85°), THF, TOL, and MEK, the c_m is much higher than the c_e^o and hence the c_e at all values of $\tilde{M}$. Whereas in DEM the two concentrations become comparable in the region of large $\tilde{M}$, and similarly in 1-CB and p-XY they become comparable in the almost entire region of $\tilde{M}$. Judging from these results we may say that PMMA-PST-PMMA block copolymer chains (and presumably any ST-MMA block copolymers) would behave just as ordinary random coils in the former group of solvents. The $[\eta]$ versus M_w data may be satisfactorily described by eq. 5 with the modification of the parameters Ko and B by eqs. 1-3 and eq. 7, respectively. The phenomenon of intrachain phase separation is quite unlikely to be taking place in such solvents (7d).

Table IV. Intrinsic viscosities of PMMA-PST-PMMA triblock and PST-PMMA diblock copolymers compared with those of their precursor PST in p-xylene at 30°C.

Precursor PST			ST-MMA block copolymers			
code	$10^{-4}M_w$	[η]	code	$10^{-4}M_w$	x	[η]
			PMMA-PST-PMMA triblock type			
27H	67.1	1.620	27B	147.4	0.42	1.590
22H	38.0	1.090	22B	80.7	0.45	1.033
26H	13.6	0.556	26B	30.8	0.46	0.582
28H	5.2	0.273	28B	9.2	0.47	0.289
			PST-PMMA diblock type			
46H	14.0	0.570	46B	30.0	0.45	0.660
47H	5.0	0.249	47B	11.0	0.42	0.295
39H	--	0.176	39B	--	0.48	0.190

In the latter group of solvents, i. e., DEM, 1-CB, and p-XY, the fact that they are Θ solvents for either one of the parent homopolymers appears to be important. The corresponding subchains might collapse in its Θ solvent, and this might induce the intrachain phase separation, resulting in conformational anomalies. However the [η] versus M_w data of the DEM solutions did not show any anomalies at all, or at least, those detectable by the [η] measurements. In this solvent, DEM, the conformational change takes palce at the central PST subchains, whereas the side PMMA subchains behave normally and protect the collapsed PST subchains. Such a change may not result in easily detectable anomalies of the [η] versus M_w relation.

On the other hand, in 1-CB and p-XY, both of which are Θ solvents toward PMMA, the behavior of the triblock copolymers must be entirely different from those in DEM. In fact significant anomalies were observed in the [η] versus M_w relations such as shown in Figure 4. The anomalies become more evident when the [η] data of the PMMA-PST-PMMA samples obtained in p-XY at 30° are compared with those of their precursor PST samples. Table IV lists the data. Surprisingly we notice that for high molecular weight samples the [η] data of the whole block copolymer chains are smaller than those of the precursor PST.

The anomalies mentioned above must have been arisen from the facts that the block copolymer samples are of BAB-type, in which B is PMMA, and that the solvent, p-XY, is a Θ solvent for PMMA and a very good solvent for PST. In this solvent homo-PST chains may assume expanded conformations and consequently have large $[\eta]$ value. On the contrary in PMMA-PST-PMMA triblock chains, the two PMMA side-chains would presumably undergo intrachain association, thereby severely restricting allowable conformations of the central PST subchain. In addition the poor compatibility of PST : PMMA pair in this solvent might enhance the anomalies. The similar trend is observed also in PMMA-PMST-PMMA triblock chains, but the anomalies are much less significant. This might be due to that p-XY is only moderately good toward PMST and that PMST : PMMA pair is compatible.

Anyway, if the anomalies could have been resulted from the phenomenon of intrachain association of the two B sidechains, then the behavior of PST-PMMA diblock chains in the Θ or nonsolvents for PMMA must be entirely different from those of the triblock chains. In fact our preliminary $[\eta]$ data on PST-PMMA diblock samples in p-XY at 30° (cf., Table IV) and in other solvents of similar selectivity show no anomalies of this nature (7f). Results of sedimentation equilibrium and other experiments suggest that the diblock copolymers undergo intermolecular association and stable micelles are formed in either type of selective solvents.

CONCLUSION

From all the results obtained here we might be able to draw the following conclusions: (i) Block copolymer chains (of any architecture) would behave just as ordinary random coils in dilute solution of common solvents, unless the parent homopolymer pair is highly incompatible. Of course if the homopolymer pair is highly incompatible, it should be difficult to find appropriate common solvents. (ii) Block copolymers of AB-type would become 'soluble' in dilute solution of any selective solvents, i. e., that dissolve either one of the constituent homopolymers, presumably by forming micelles. The insoluble subchains accumulate in the cores of such micelles, while the soluble subchains wrap around the cores and protect them from further aggregation. On the ohter hand, the behavior of BAB triblock copolymers may be more easily influenced by the selectivity of solvents. Namely (iii) in a nonsolvent of A subchains, a single block copolymer molecule can be a micelle particle, in which insoluble central A subchain is protected by soluble B side chains. The solvent becomes much poorer, they may form intermolecular aggregates. On the contrary (iv) in a solvent of the opposite selectivity, i. e., in a nonsolvent of B subchains, BAB triblock copolymers are usually insoluble. Even when they are in the state of molecular dispersion, the insoluble B subchains would undergo intrachain association within the same molecule, thereby restricting allowable conformations of the central A subchain. (v) The stability of the micelles including the unimolecular dispersion of BAB triblock chains appears to be better for block copolymers composed of incompatible homopolymer pair rather than those of compatible homopolymer pair. (vi) Of course the detailed features should be influenced by many other factors such as the molecular weight and composition of the block copolymers.

Acknoledgments. We wish to thank Messers T. Tanaka and M. Nakatani for their assistance in carrying out experiments. We are indebted also to Dr. Teruo Fujimoto of Nagoya University for his valuable advices on the anionic polymerization technique. A part of this work was supported by the Ministry of Education, Japan, under Grant #A4037, which is gratefully acknowledged.

REFERENCES

(1) W. H. Stockmayer, L. D. Moore, Jr., M. Fixman, and B. N. Epstein, J. Polymer Sci., 16, 517 (1955).
(2) T. Kotaka, T. Tanaka, H. Ohnuma, Y. Murakami, and H. Inagaki, Polymer J., 1, 245 (1970).
(3) (a) A. Dobry and F. Boyer-Kawenoki, J. Polymer Sci., 2, 90 (1947): (b) R. L. Scott, J. Chem. Phys., 17, 279 (1949): (c) R. J. Kern and R. J. Solocombe, J. Polymer Sci., 15, 183 (1955).
(4) (a) G. M. Burnett, D. Meares, and C. Paton, Trans. Faraday Soc., 58, 737 (1962): (b) S. Krause, J. Phys. Chem., 68, 1948 (1964): (c) J. R. Urwin and J. M. Stearne, Makromol. Chem., 78, 197, 204 (1964); Europ. Polymer J., 1, 227 (1965). (d) H. Inagaki, Makromol. Chem., 86, 289 (1965); H. Inagaki and T. Miyamoto, ibid., 87, 166 (1965).
(5) (a) D. Froelich and H. Benoit, ibid., 92, 224 (1966): (b) D. Froelich, J. Chim. Phys., 64, 1311 (1968).
(6) See, for example, (a) G. E. Molau and W. M. Wittbrodt, Macromolecules, 1, 260 (1968): (b) T. Inoue, T. Soen, H. Kawai, M. Fukatsu, and M. Kurata, J. Polymer Sci. B, 6, 75 (1968); T. Inoue, T. Soen, Y. Hashimoto, and H. Kawai, Macromolecules, 3, 87 (1970).
(7) (a) T. Kotaka, H. Ohnuma, and Y. Murakami, J. Phys. Chem., 70, 4099 (1966): (b) H. Ohnuma, T. Kotaka, and H. Inagaki, Polymer, 10, 501 (1969): (c) T. Kotaka, H. Ohnuma, and H. Inagaki, ibid., 10, 517 (1969): (d) H. Ohnuma, T. Kotaka, and H. Inagaki, Polymer J., in press: (e) T. Kotaka, T. Tanaka, and H. Inagaki, ibid., to be submitted: (f) T. Tanaka, M. Nakatani, and T. Kotaka, unpublished.
(8) Y. Sakurada, K. Imai, and S. Matsumoto, Preprint at the 16th symposium on polymer science, Japan, October 1967, Fukuoka.

(9) (a) M. Szwarc and A. Rembaum, J. Polymer Sci., 22, 189 (1956); M. Szwarc, Fortschr. Hochpolym. Forsch., 2, 275 (1960): (b) M. Baer, J. Polymer Sci., A, 2, 417 (1964): (c) D. Freyss, P. Rempp, and H. Benoit, ibid., B, 2, 217 (1964): (d) T. Fujimoto, N. Ozaki, and M. Nagasawa, ibid., A, 3, 2259 (1965).

(10) (a) T. Kotaka, Y. Murakami, and H. Inagaki, J. Phys. Chem., 72, 829 (1968): (b) T. Kotaka and N. Donkai, J. Polymer Sci., A2, 6, 1745 (1968): (c) T. Kotaka, N. Donkai, and H. Inagaki, ibid., submitted.

(11) P. J. Flory, "Principles of Polymer Chemistry," Cornell Univ. Press, Ithaca, N. Y., 1953.

(12) (a) J. H. Hildebrand and R. L. Scott, "The Solubility of Nonelectrolytes," Rheinhold, New York, 1950: (b) H. Burrell and B. Immergut, "Solubility Parameter Values" in J. Brandrup and E. H. Immergut, editors, "Polymer Handbook," Interscience Publishers, New York, 1965, Chapter IV-11.

(13) See, for example, (a) Y. Gallot, M. Leng, H. Benoit, and P. Rempp, J. Chim. Phys., 59, 1093 (1962): (b). S. Ye. Bresler, L. M. Pyrkov, S. Yu. Frenkel, L. A. Laius, and S. I. Klenin, Vysokomol. Soedin., 4, 250 (1962): (c) G. E. Molau, J. Polymer Sci., A, 3, 1267 (1965); ibid., A, 3, 4235 (1965).

(14) P. J. Flory and T. G Fox, J. Am. Chem. Soc., 73, 1904 (1951); see also ref. 11, chapter 14.

(15) Th. G. Scholte, J. Polymer Sci., A2, 5, 461 (1967).

(16) D. Berek, B. Bohmer, and D. Lath, Plaste u. Kautschuk, 14, 556 (1967).

(17) (a) M. Kurata and W. H. Stockmayer, Fortschr. Hochpolym. Forsch., 3, 196 (1963): (b) W. H. Stockmayer and M. Fixman, J. Polymer Sci., C, 1, 137 (1963).

(18) (a) T. Norisuye, K. Kawahara, A. Teramoto, and H. Fujita, J. Chem. Phys., 49, 4330 (1968): (b) K. Kawahara, T. Norisuye, and H. Fujita, ibid., 49, 4339: (c) G. Tanaka, S. Imai, and H. Yamakawa, ibid., 52, 2639 (1970).

(19) W. H. Stockmayer, Makromol. Chem., 35, 54 (1960).

(20) (a) L. A. Utracki, R. Simha, and L. J. Fetters, J. Polymer Sci., A2, 6, 2051 (1968): (b) L. A. Utracki and R. Simha, Macromolecules, 1, 505 (1968).

(21) See, for example, M. Kurata, M. Iwama, and K. Kamada, "Viscosity-Molecular weight Relationships and Unperturbed Dimensions of Linear Chain Molecules," in J. Brandrup and B. Immergut, editors, "Polymer Handbook," Interscience Publishers, New York, 1965, Chapter IV-1.

SURFACE MORPHOLOGIES OF STYRENE-ETHYLENE OXIDE BLOCK COPOLYMERS

Richard G. Crystal

Rochester Research Laboratories

Xerox Corporation

Rochester, New York 14603

INTRODUCTION

Control of block copolymer bulk morphologies via preferential solvent techniques has been the subject of several investigations (1,2,3). Effects of preferential solvents on the bulk morphologies of styrene-ethylene oxide AB block copolymers having both varied compositions and molecular weights have been recently reported from this laboratory. (3)

This system is an interesting model since both components are incompatible and have contrasting properties. Polystyrene (PS) is amorphous, non polar, and hydrophobic while poly (ethylene oxide) (PEO) is semicrystalline, polar, and hydrophilic. As a continuation of the above work, we have examined the effects of preferential solvent castings and crystallizations from the melt on the surface morphologies of these same materials.

EXPERIMENTAL

Synthesis and characterization of polymers used in this study have been previously described by O'Malley et. al.[4] Three copolymers with varying PS/PEO compositions were chosen for systematic study which offered variations in both relative composition and chain lengths. Table 1 summarizes the compositions and molecular weights of these samples.

All specimens described in this paper are A-B type block copolymers. For brevity, samples will be identified in this text in terms of their weight percent polystyrene; i.e. sample #54 (ref. 4) will be referred to as "70.0% PS."

The principal casting solvents used in this study were nitromethane (preferential to PEO), chloroform (mutual solvent for both species) and ethyl benzene (preferential to PS). Since surface evaporation rate could have marked effects on resultant surface morphologies, films were cast from a 5% solids solution and placed under a glass dish along with a dish of appropriate solvent to minimize evaporation rates. In addition, melt crystallized films were heated to the prescribed temperature and subsequently air cooled to room temperature.

TABLE 1 BLOCK COPOLYMER COMPOSITIONS AND MOLECULAR WEIGHTS

Wt % PS	Mole % PS	$\overline{MW}$ Total	$\overline{MW}$ PS	$\overline{MW}$ PEO
19.6	9.6	108,000	21,000	87,000
39.3	21.4	33,500	13,200	20,300
70.0	49.5	34,800	24,400	10,400

Specimens were examined by scanning electron microscopy (SEM) as well as conventional electron microscopy. The SEM proved valuable in studying phase separated morphologies because of easy sample preparation, its moderately high resolution capability, its great depth of field even at higher magnifications, and its unique wide magnification range (20X to 50,000X). The depth of field feature enabled examination of previously undetectable gross contours while the wide magnification range proved essential in relating the basic phase separated domain structures to the macroscopic structures associated with spherulitic textures. For SEM studies, films were cast directly onto polished SEM stubs which were then vacuum overcoated with a conductive thin gold layer to inhibit charge buildup. The specimens were observed in a Cambridge Stereoscan instrument. Details of this technique have been described by Short and Crystal (5). Conventional electron microscopy was accomplished by examining one step Pt-C shadowed carbon replicas. These replicas were prepared by vacuum evaporation onto block copolymer films followed by film dissolution in chloroform.

RESULTS AND DISCUSSION

Before examining block copolymer surface morphologies, it is useful to look at surfaces of homopolymer polyblends in which large scale phase separation occurs. Figure 1, a scanning electron micrograph, was taken from a 50/50 PS/PEO mixture cast from chloroform, a mutual solvent for both components. Notice the rough and rugged texture brought out by the vivid three dimensionality of scanning electron micrographs. Optical microscopy has shown that the spherical structures in this figure are single PEO spherulites while the cracked fragments and interspherulite material are amorphous PS. Gross phase separation has resulted in a non uniform, rough surface texture.

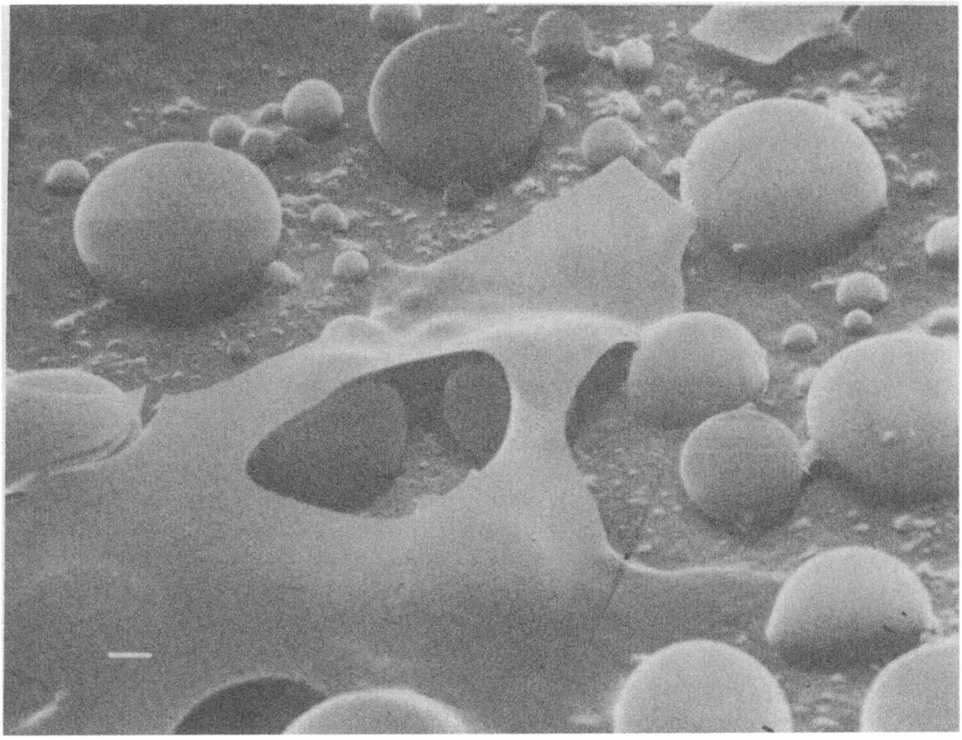

Figure 1 - Scanning Electron Micrograph of a 50/50 PS/PEO Homopolymer Blend Cast From Chloroform (Mutual Solvent) Bar is 10u

The rough surfaces observed with a polyblend would not be expected with block copolymers since phase separation is limited to rather small domains. We would expect, however, that surface morphology would change with both composition and solvent treatment in the PS-PEO block system primarily as a result of surface textures induced by PEO crystallization. The two anticipated extreme morphologies would be well developed spherulitic textures at high PEO loadings and smooth amorphous textures at high PS loadings. A complete description of morphological structures resulting from changes in composition, segmental molecular weight, and casting solvent would be far too lengthy. Discussion will center primarily around the description of two compositions, i.e. 19.6%, and 70.0% PS cast from the three previously mentioned solvent systems and also melt cast. As a first approximation, we would expect that casting from nitromethane (preferential to PEO) would result in more crystalline surface textures while casting from ethylbenzene (preferential to PS) would result in more amorphous textures. From observations of bulk PS-PEO block copolymer morphologies (3) in which no less than 16 different types of bulk spherulitic or pseudo spherulitic textures are realized, we would expect that a number of surface morphologies would be created.

SOLVENT CAST SAMPLES

Looking first at the 70.0% PS sample, figures 2, 3, and 4 are a series of one step replica electron micrographs of specimens cast from nitromethane, chloroform, and ethyl benzene respectively. From Figure 2, it is clear that the predominant PS phase precipitates from the PEO preferential solvent into spherical domains approximately 500Å in diameter which cover the major portion of the surface. Small fragments of crystalline PEO fibrils are also present to some degree and appear to splay out of the surface at several locations. The faint smooth textures near PEO fibrils are believed to be amorphous PEO. Thus, both amorphous PS clusters and a lesser amount of PEO crystallites and possibly amorphous PEO comprise the surface morphology.

Figure 3, the mutual solvent casting, illustrates an entirely different smooth, amorphous texture generated from the same material. In this case, the PS phase stays in solution before casting, resulting in a smooth PS surface morphology. Because

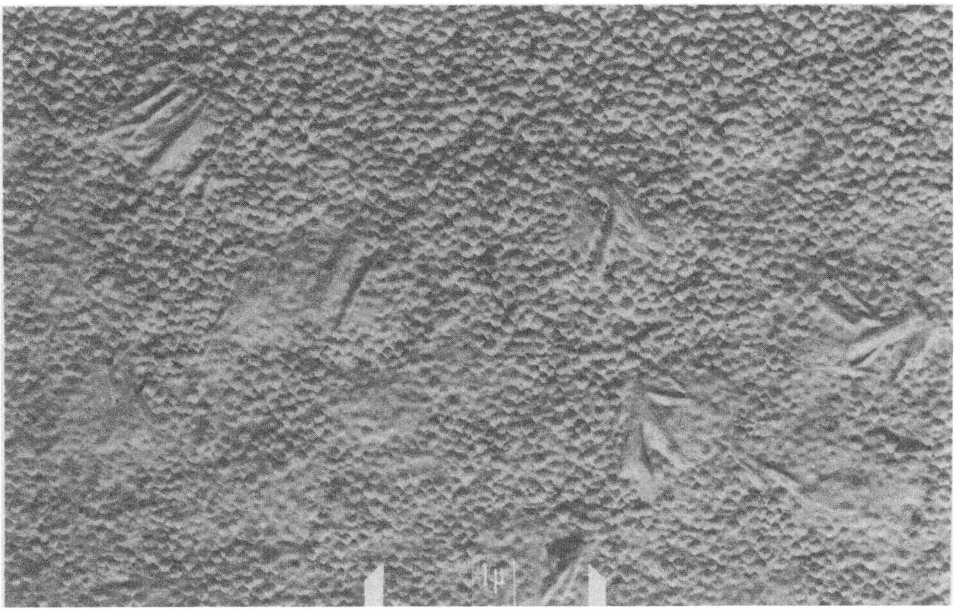

Figure 2 - Surface of a 70.0% PS Block Copolymer Cast From Nitromethane (Preferential to PEO) One Step Pt-C Replica Electron Micrograph

both components remain in solution, neither PEO crystallization nor PS clustering interfere with the formation of a smooth texture.

Figure 4, the PS preferential solvent casting, illustrates the effect of crystalline PEO precipitation before film casting. In comparison to figure 3, a distinct grainy texture is generated even though PS comprises 70% of the specimen. Unlike the clusters in figure 2, these grains are approximately cylindrical in shape, indicating that they are PEO crystallite fragments resulting from precipitation of the insoluble PEO phase. Unlike the radially oriented fibers realized with free spherulitic crystallization, these PEO crystallites show no obvious interconnection or preferred orientations. Based on previous bulk observations[3], these PEO crystalline fragments are isolated in a PS matrix. Thus, in spite of the rough texture, the principal component at the upper most surface is PS.

Figure 3 - Surface of a 70% PS Block Copolymer Cast From Chloroform (Mutual Solvent) One Step Pt-C Replica Electron Micrograph

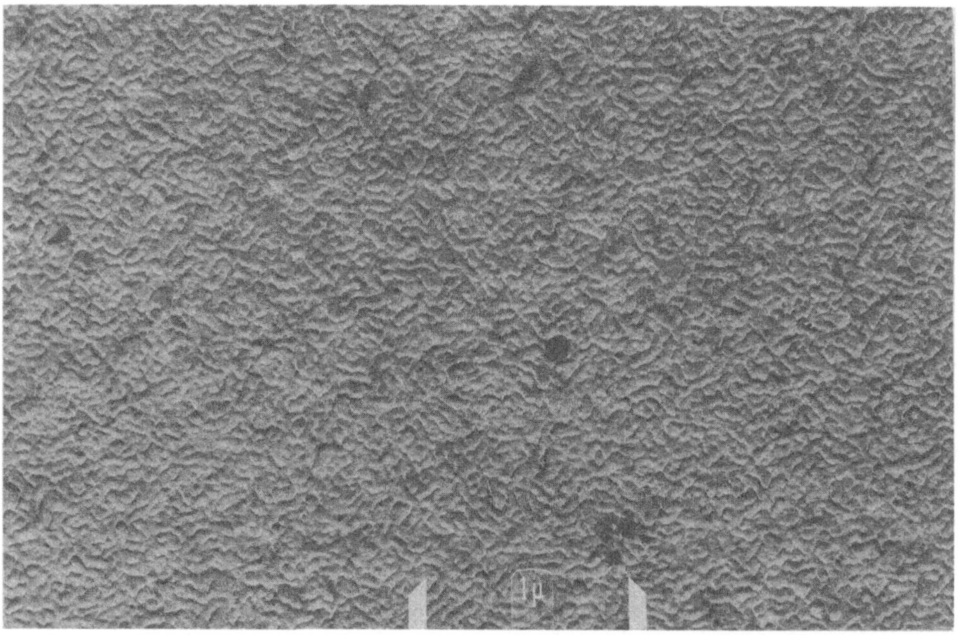

Figure 4 - Surface of a 70% PS Block Copolymer Cast From Ethyl Benzene (Preferential to PS). One Step Pt-C Replica Electron Micrograph

The next sequence of figures illustrate changes in surface morphology when a more crystallizable block copolymer containing 19.6% PS is cast from the same three solvent systems.

Figure 5 illustrates the type of surface texture generated by casting from a PEO preferential solvent. In comparison to the high PS content structure illustrated in figure 2, the PS insoluble clusters are still apparent, but the higher PEO content results in a fibrous spherulitic PEO texture interdispersed with a lesser amount of PS clusters. As mentioned in a previous paper,[3] the PS clustering strongly interferes with free PEO spherulitic crystallization. Figure 6 is a SEM photograph which illustrates the larger scale surface roughness caused by spherulitic crystallization. Notice that high and low points are generated in the surface due to PEO fibril twisting while interspherulitic borders result in low points.

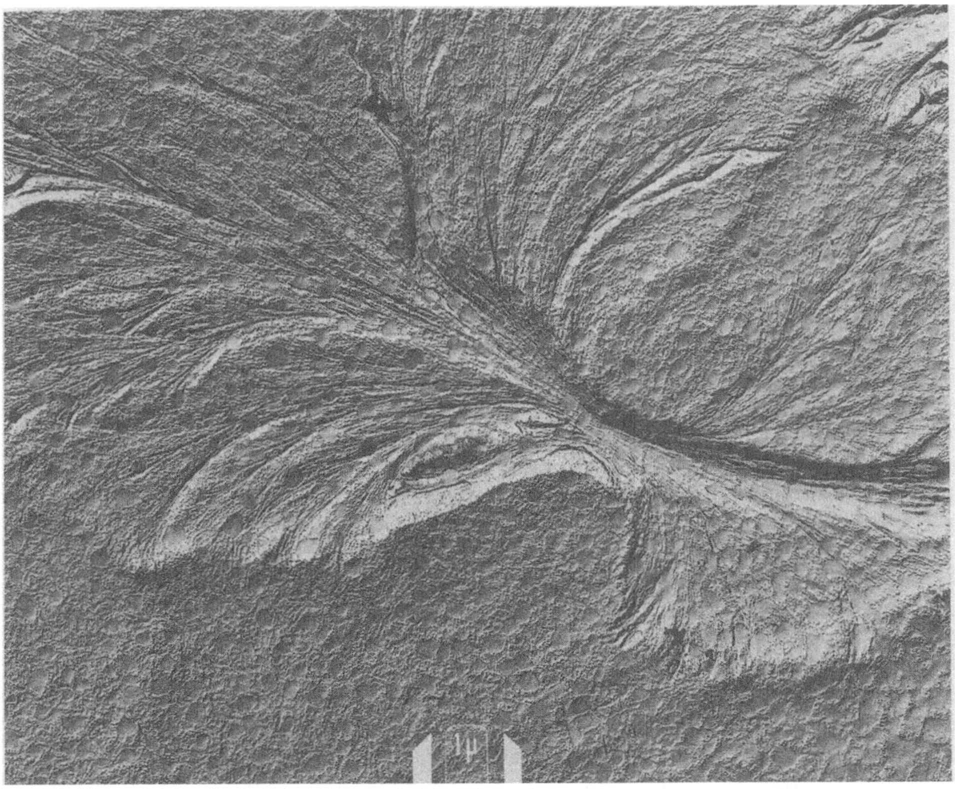

Figure 5 - Surface of a 19.6% PS Block Copolymer Cast From Nitromethane (Preferential to PEO). One Step Pt-C Replica Electron Micrograph

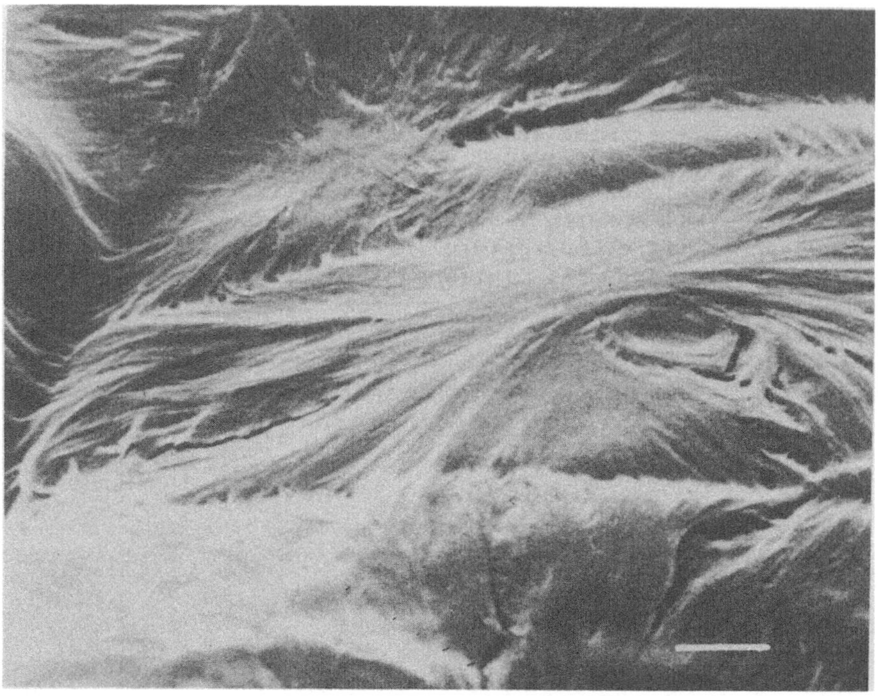

Figure 6 - Scanning Electron Micrograph of a 19.6% PS Block Copolymer Cast From Nitromethane. The Bar is 1u.

Figure 7 is an electron micrograph illustrating the surface texture generated by casting from a mutual solvent. In this case, PS clusters are absent while the fibrous PEO texture dominates the surface. Since both species are in solution, the PEO is free to crystallize without interference from the PS phase. Optical microscopy has shown that PEO crystallites form first,[3] resulting in a PS overcoating on the PEO fibrils. Thus when cast from a mutual solvent, the amorphous PS phase dominates the uppermost surface skin of both high and low PS concentration samples since the PS is the last component to come out of solution. Figure 8 illustrates the surface morphology generated with the same polymer by casting from a PS preferential solvent. In comparison to figure 4, the insoluble PEO fibrils formed are longer and appear to associate in aggregates aligned radially from the nucleation source. The PS phase forms a matrix between individual fibrils within these aggregates as well as between aggregates. The amorphous PS phase still dominates at the uppermost surface skin.

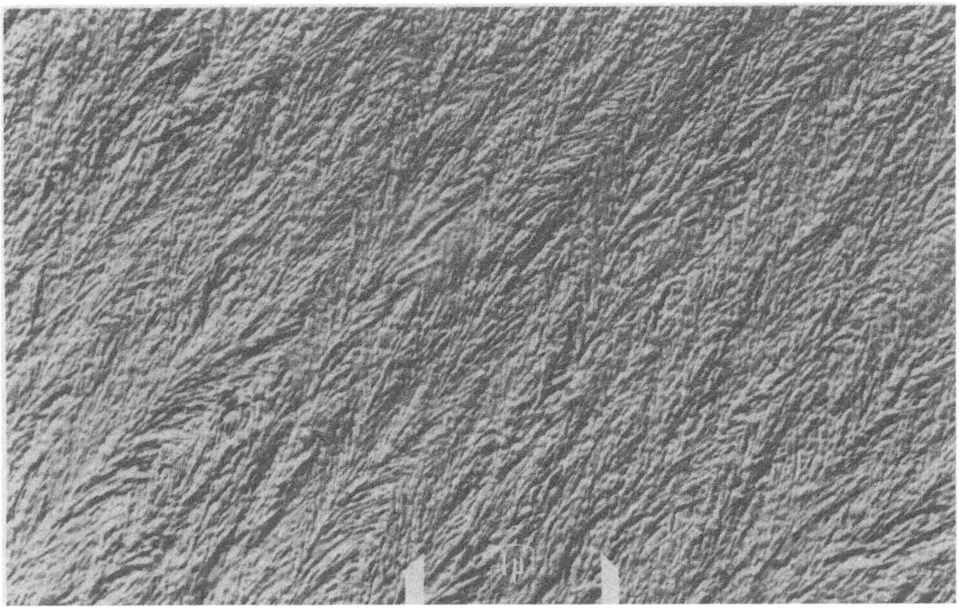

Figure 7 - Surface of a 19.6% Block Copolymer Cast From Chloroform (Mutual Solvent)

Figure 8 - Surface of 19.6% PS Block Copolymer Cast From Ethyl Benzene (Preferential to PS) One Step Pt-C Replica Electron Micrograph

EFFECTS OF SOLVENT EVAPORATION RATE

While the observations described up to this point involved cast samples in which solvent evaporation was as slow as possible, higher solvent evaporation rates resulted in quite different surface textures in several instances; we assume because equilibrium phase separation was not achieved. A striking example is illustrated in the next two figures.

Figures 9 and 10 are low magnification scanning electron micrographs of 39.6% PS films cast from nitromethane. While the solvent in the case of figure 9 was slowly evaporated, the film in Figure 10 was rapidly dried by placing the uncovered solution in a laboratory hood. The hood air currents caused rapid solvent evaporation resulting in the intriguing crater like ringed morphology. In this figure the "twin" crater structure represents two distinct spherulites as verified by optical microscopy. Notice that the ring patterns and spacings are almost identical. This is especially apparent at the interspherulite boundary where each ring runs into its identical twin in the adjacent spherulite. Since these rings are not equally or periodically spaced around the spherulite centers, we believe that the rings are caused by random fluctuations in solvent evaporation rates, perhaps caused by variations in air flow. Since the SEM can only detect topographical variations, these rings represent high and low regions on the film surface. It should be kept in mind that the ringed structure in this low magnification photograph is far greater in size than individual phase separated domains.

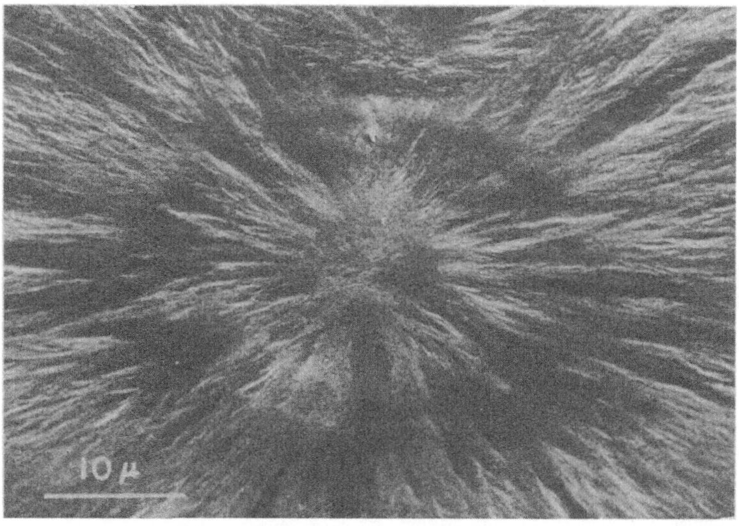

Figure 9 - Scanning Electron Micrograph of a 39.3% PS Block Copolymer Cast From Nitromethane Using the Slow Evaporation Procedure

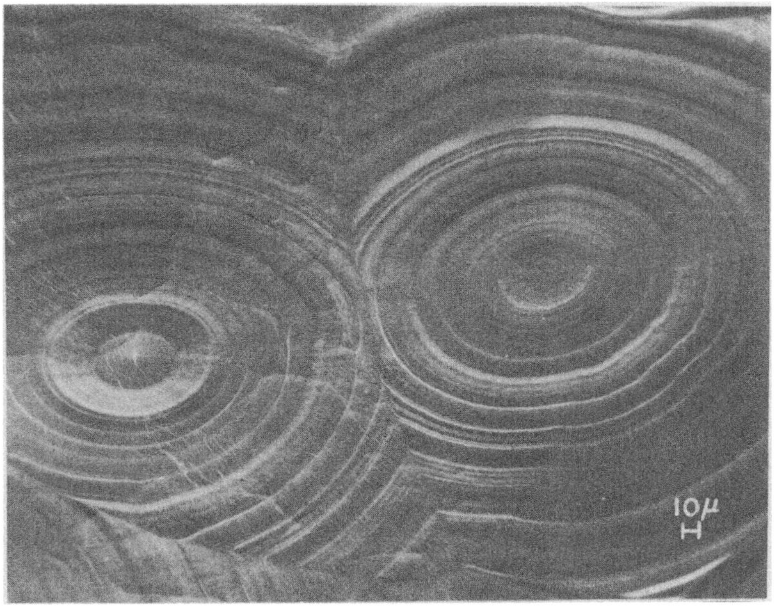

Figure 10 - Scanning Electron Micrograph of a 39.3% PS Block Copolymer Cast From Nitromethane Using a Rapid Evaporation Procedure

MELT CAST SPECIMENS

The final four figures illustrate the effect of thermal annealing on melt crystallized block copolymer surface morphologies. O'Malley, Crystal, and Erhardt[4] indicated that the melting points of the PEO phase and the glass transitions of the PS phase in these materials are independent of block composition and are approximately equal to the Tm's and Tg's of homopolymers whose molecular weights correspond to the respective block segment molecular weights. For all specimens used in this study, the PS glass transitions were 20 to 35°C above the PEO melting points. Thus, we would expect that the resultant surface morphologies of melt crystallized block copolymer would depend strongly upon whether or not the PS glass transition was exceeded during the melting cycle. For samples whose surfaces and matrices are dominated by PS, little change in morphologies would be anticipated. Unless the PS phase flows sufficiently, the phase separated structure would be frozen in by the glassy component. Our previous study[3] has confirmed that few changes in bulk morphology are realized under these conditions. By way of illustration, the final figures illustrate the effect of melting temperature on the 19.6% specimen cast from nitromethane. Figure 11 is a scanning electron micrograph of the specimen surface after heating to 75°C followed by PEO crystallization. Notice that the striking three dimensionality associated with PEO spherulitic crystallization is quite evident.

Comparison of figures 5 and 6 indicate that the surface roughness associated with spherulitic crystallization is still evident and perhaps is somewhat increased. Notice the high points at spherulite centers and low points at interspherulitic boundaries. We can assume that crystallization from the melt has proceeded with little interference from the glassy PS spherical clusters. Figure 12 is again a scanning electron micrograph of a nitromethane cast 19.6% PS sample, this time heated to 95°C, which is 4°C above the reported 91°C glass transition of the PS phase. Notice that the surface topography is considerably smoother than in Figure 11. Closer inspection as shown in the higher magnification scanning electron micrograph in figure 13 indicates that a smooth amorphous PS skin has formed over the surface. The skin presumably forms in the melt via coalescence of PS clusters, and, as the melt is air cooled to room temperature, the skin becomes glassy. Subsequent PEO crystallization below 65°C exerts a minor effect on the surface morphology as indicated by the apparent spherulitic texture. Finally, when the 19.6% PS sample cast from nitromethane is heated to 120°C, a smooth surface morphology is realized (Figure 14). We anticipate that as the PS phase is heated sufficiently above its Tg, its viscosity is sufficiently low to permit the formation of a complete PS skin which becomes glassy, upon cooling, before PEO crystallization.

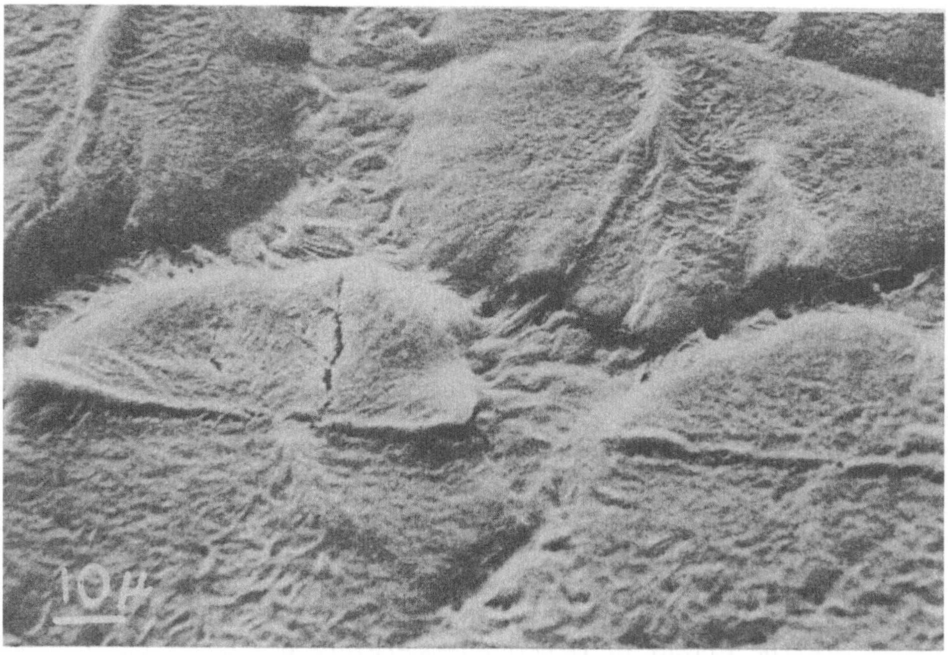

Figure 11 - Scanning Electron Micrograph Of A 19.6% PS Specimen Cast From Nitromethane That Has Been Heated To 75°C And Subsequently Air Cooled. The Bar Is 10u.

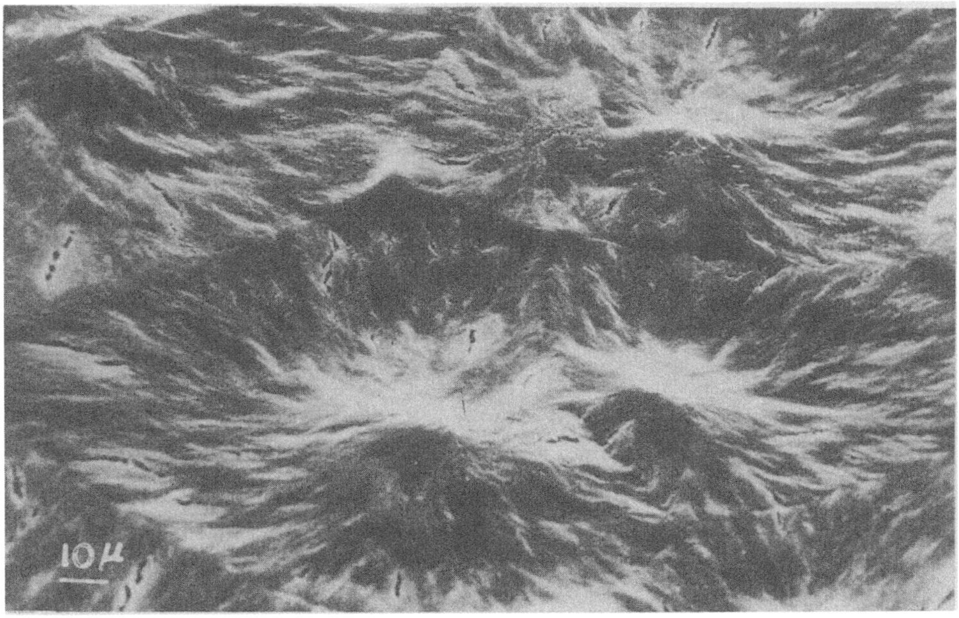

Figure 12 - Scanning Electron Micrograph of a 19.6% PS specimen Cast From Nitromethane That Has Been Heated To 95°C and Subsequently Air Cooled. The Bar is 10u.

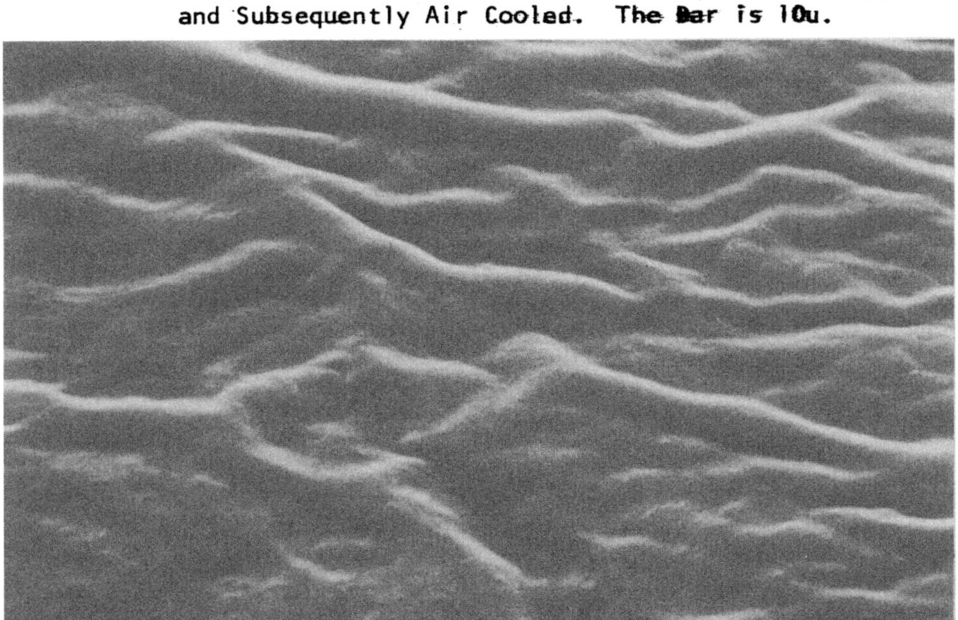

Figure 13 - Higher Magnification of Figure 11, a 19.6% PS Specimen Heated to 95°C. The Bar is 1u.

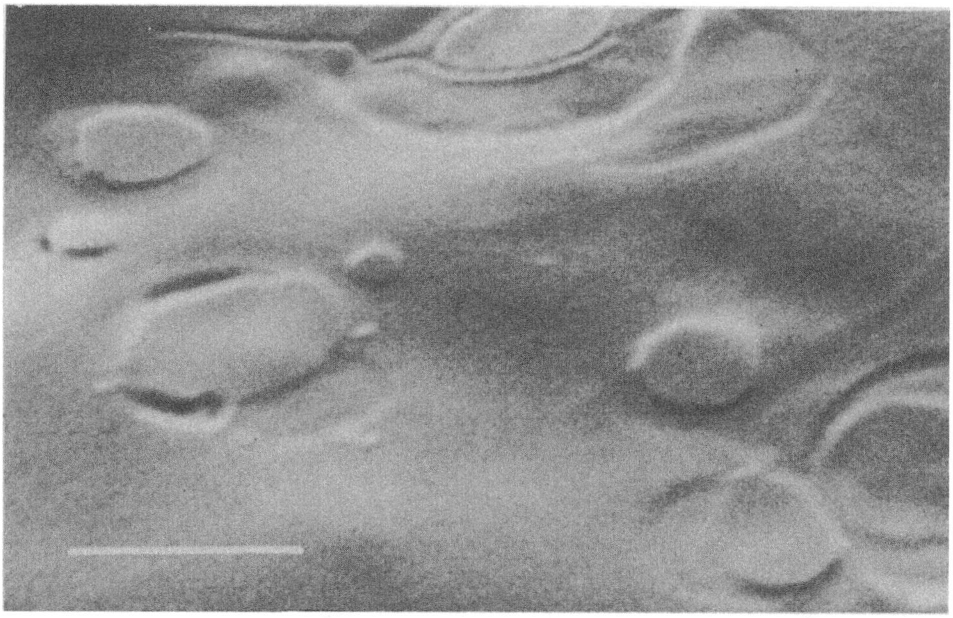

Figure 14 - Scanning Electron Micrograph of a 19.6% PS Specimen Cast From Nitromethane and Heated To 120°C and Subsequently Air Cooled. The Bar is 1u.

CONCLUSIONS

Variations in both casting solvent and block copolymer component concentrations lead to marked changes in both surface morphologies and relative compositions of each component at the surface. Variations in solvent evaporation rates, however, can also lead to more gross topographic nonuniformities. The surface morphologies of melt crystallized specimens are strongly dependent on composition as well as whether or not the PS glass transition is exceeded during melting. Heating above the PS Tg results in smooth surfaces;heating below the PS Tg results has little effect.

ACKNOWLEDGMENTS

The author would like to thank Mr. F. J. Walton for invaluable assistance in much of these studies, Dr. J. J. O'Malley for block copolymer synthesis, and Dr. J. M. Short for his enthusiasm and contributions to scanning electron micrographs.

REFERENCES

1. A. E. Skoulios, G. Tsoulandye, and E. Franta, J. Poly. Sci., C-4, 507 (1963).

2. T. Inoue, T. Soen, T. Hashimoto, and H. Kwai, IUPAC Macromolecular Symposium, Toronto, Canada, Sept. 1968.

3. R. G. Crystal, J. J. O'Malley, P. F. Erhardt, Polymer Preprints, 10, 2, 804 (1969); "Block Copolymers", S. L. Aggarwal Ed., Plenum Press, New York (1970) pp. 179

4. J. J. O'Malley, R. G. Crystal, and P. F. Erhardt, Polymer Preprints, 10, 2, 796 (1969); "Block Copolymers", S. L. Aggarwal Ed., Plenum Press, New York (1970) pp. 163

5. J. M. Short and R. G. Crystal, Applied Polymer Symposia, in press

INTERFACIAL MORPHOLOGY AND AUTOHESION OF A-B-A TRIBLOCK COPOLYMERS

D. H. Kaelble and E. H. Cirlin
Science Center, North American Rockwell Corporation
Thousand Oaks, California 91360

M. Shen
Dept. of Chem. Eng., University of California
Berkeley, California 94720

The tensile-recovery response, optical transparency, and autohesion properties of an A-B-A triblock copolymer where A is polystyrene and B is polybutadiene have been analyzed in terms of the interfacial mechanisms of reversible bonding. This study indicates that cavitation and void formation accompanies the characteristic initial yielding and cold drawing of this polymer system. Tensile stress-strain curves are described by a theoretical function that defines both the loss and regain of interfacial continuity between the continuous B phase and the discrete A phase domains. The study indicates the initially formed cavities act as light scattering centers. Under prolonged tensile stress these cavities coalesce to produce visible void structures. Both types of defects disappear after sufficient recovery time to regenerate the original optical and mechanical response of the undeformed material.

INTRODUCTION

Recently reported tensile studies of A-B-A triblock copolymer films, where A = polystyrene and B = polybutadiene or polyisoprene, has revealed a significant nonlinearity of viscoelastic response which is typified by initial yielding and cold drawing indicative of a "Mullins dewetting effect" (1-3). Thermally annealing these films apparently produces a complete recovery process (1). This paper re-examines this interesting internal adhesion phenomena and correlates the experimental results with the interfacial morphology of A-B-A block copolymers. The results of tensile studies, UV-visible spectral properties, and kinetics of auto-adhesive bonding by peeling experiments are discussed.

TABLE 1: Extension-recovery history represented in the curves of Figure 3 for Kraton 101.

Curve No.	Strain History	Unrecovered Strain	Initial area cm^2	B_o
1	Initial extension from undeformed state	0	0.0155	1.00
3	Second extension after 6 minute recovery time	0.158	0.0134	0.31
7	Third extension after 4320 minutes (3 days) recovery at 22°C	0.093	0.0142	0.68
9 (same as 3)	Fourth extension after 6 minutes recovery time	0.183	0.0131	0.31
13 (same as 1)	Fifth extension after 1 hr recovery at 66°C	0.045	0.0148	1.00

EXPERIMENTAL

Films of Kraton 101 (Shell Chemical Co.) were cast from benzene solution in a centrifugal casting device which produces films free of internal defects and displaying high surface and thickness uniformity (4). Films prepared for tensile experiments were cast on a TFE-Teflon film substrate which was later delaminated. Films prepared for peeling studies were reinforced by placing a nylon cloth over the TFE-Teflon film substrate. The nylon cloth becomes imbedded in the Kraton 101 film during centrifugal casting and forms an internal reinforcing member which prevents undesired tensile stretching during the peeling experiment. Film thickness for tensile testing was controlled at 0.0350±.0005 cm and peel testing was 0.0425±.0005 cm. Subsequent to centrifugal casting the films were further dried in vacuum for 6 weeks.

Tensile specimens were cut from the film by means of a microtensile die (ASTM Method D1708-66). These tensile specimens were clamped in specially modified Instron jaws that provide a well defined line clamp by compression between a 1/8" diameter cylindrical surface and a flat face. This line clamp acts at the inner extremity of the sample tabs so that the active gage length of 2.25 cm includes only the uniform 0.445 cm central width of the specimen plus the small fillets at each end of the central section. This special sample geometry and clamping arrangement insures uniform deformation of the sample and absence of slippage or failure at the clamp line. The tensile extension and recovery experiments were conducted at 22°C and a cross head speed of 1 cm/min and nominal extension rate of $\dot{\varepsilon} = 7.41 \cdot 10^{-3}$ sec^{-1}.

Peel specimens for autohesive bonding were die cut to 0.5 cm width by 7.8 cm length from the nylon reinforced film. The smooth surfaces formed at the polymer-air interface in centrifugal casting were mechanically bonded by compression between stainless steel rollers at 22°C. The effects of both dwell time at constant temperature as well as conditioning on a thermal gradient bar where temperature varies 6.7°C/cm have been studied and a portion of these results are included in this report. A 180 degree T-type peeling with crosshead rate of 0.1 cm/min and peeling rate of 0.05 cm/min at 22°C were applied to the bonded specimens.

TENSILE RESPONSE

The initial tensile yielding, cold drawing and subsequent hysteresis in recovery for benzene cast Kraton 101 are shown in the ascending and descending curves of nominal tensile stress S and strain ε in the upper part of Figure 1. The lower curve of Figure 1 illustrates the form of the complete curve of nominal stress versus strain from initial deformation to fracture where the nominal tensile strength is $S_b = 193.6$ Kg/cm^2 and $\varepsilon_b = 8.44$ with ultimate extension ratio $\lambda_b = 9.44$.

The data that form the full stress-strain curve in the lower curve of Figure 1 are replotted on bilogarithmic coordinates of true tensile stress $\sigma = S\lambda$ versus strain in Figure 2. The experimental curve of true stress versus strain shown in Figure 2 displays a smooth transition between two analytical functions:

$$\frac{\sigma}{\varepsilon}\left(\frac{kg}{cm^2}\right) = E_{,}(t) = 510 \tag{1}$$

$$\frac{\sigma}{\varepsilon}\left(\frac{kg}{cm^2}\right) = \frac{E_m(t)}{1-(\lambda/\lambda_m)^2} = \frac{31}{1-(\lambda/10.5)^2} \tag{2}$$

Equation (1) describes the initial stress-strain properties prior to elastic yielding at $\varepsilon \geqslant 0.02$. Equation (2) describes the stress-strain function from $\varepsilon = 2.0$ to $\varepsilon = \varepsilon_b = 8.44$. The general stress-strain relation that includes equations (1) and (2) as special cases has the following form:

$$\frac{\sigma}{\varepsilon}\left(\frac{kg}{cm^2}\right) = E_I B(t,\varepsilon) + \frac{E_m(t)}{1-(\lambda/\lambda_m)^2} \tag{3}$$

where $E_I = E_{,}(t) - E_m(t) = 479$ Kg/cm^2 and $B(t,\varepsilon)$ is the degree of bonding for the interfacial phase which intervenes the elastomeric B phase continuum and the spherical A phase along domains in Kraton

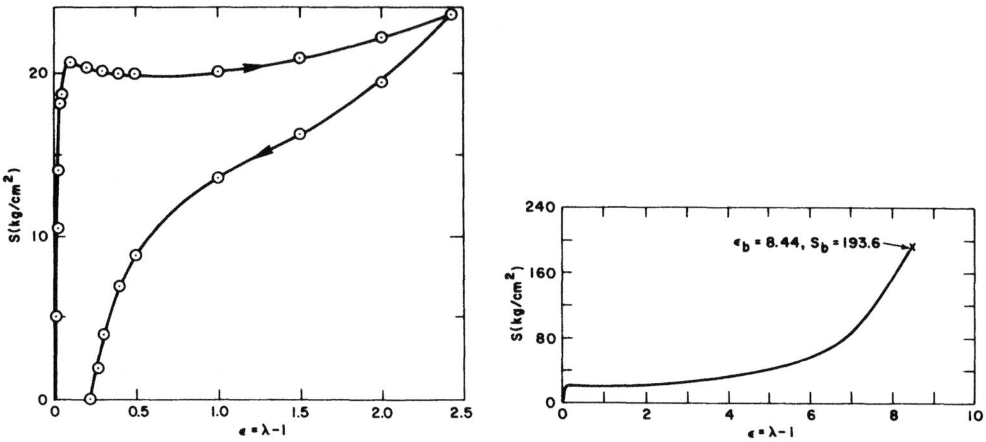

Fig. 1 Extension and recovery (upper curve) and extension to break (lower curve) for Kraton 101 at 22°C and constant strain rate showing nominal stess S vs extension $\varepsilon = \Delta L/L_o$.

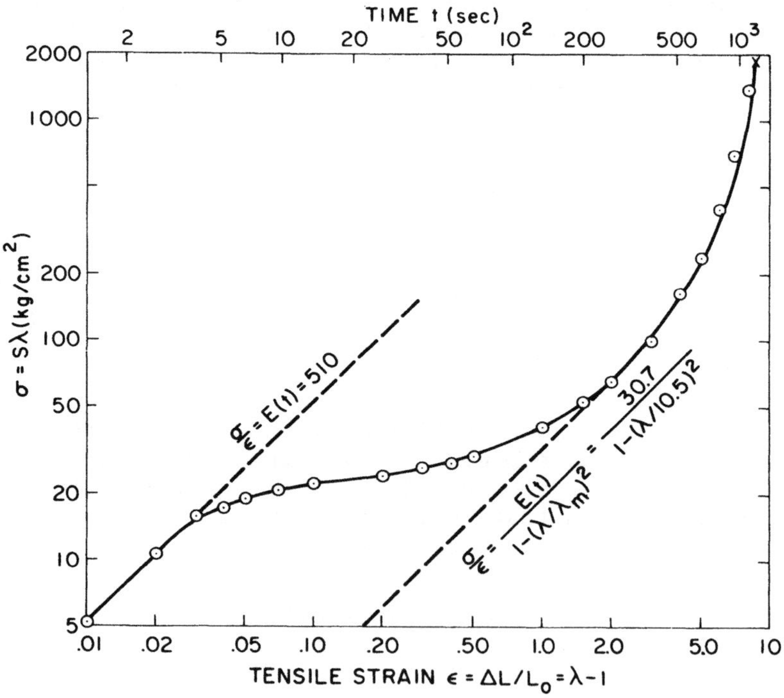

Fig. 2 True stress $\sigma = S\lambda$ vs extension ε for Kraton 101 at 22°C and constant strain rate.

101. The modulus E_I in equation (3) is defined as the rigidity contribution of the continuous interfacial phase that is obtained when $B(t,\varepsilon) = 1.0$. Equation (3) introduces a new hypothesis that defines the existence of a loss of interfacial continuity within the deformed block copolymer material where both time t and tensile strain ε interact to vary $B(t,\varepsilon)$ between the limits of zero and unity.

Figure 3 presents bilogarithmic plots of true stress versus tensile strain which represent a series of extension-recovery cycles of the type depicted in the upper curves of Figure 1. The relation of the curves of Figure 3 to the accumulated history of extension and recovery is summarized in Table 1. The lower curve in Figure 3 is a plot of equation (2) and is seen to define the convergence of the three experimental curves at $\varepsilon \geqslant 2.0$. The experimental curves of Figure 3 are referenced to the stress-strain origin of that particular cycle of extension. The unrecovered strain and new cross section area for each cycle of extension is noted in Table 1.

The analysis of these five conditions of strain history is developed by rearranging equation (3) in the following form:

$$B(t,\varepsilon) = \frac{1}{E_I}\left[\frac{\sigma(t)}{\varepsilon} - \frac{E_m}{1-(\lambda/\lambda_m)^2}\right] \qquad (4)$$

which expresses the time (or strain) dependence of stress $\sigma(t)$ in terms of the bonding parameter $B(t,\varepsilon)$. Figure 4 illustrates the experimental points and curves of $B(t,\varepsilon)$ provided by application of equation (4). The strain independent values of $B(t,\varepsilon)$ calculated for $\varepsilon \leqslant 0.02$ are denoted as B_o and tabulated in the last column of Table 1. The plotting symbols for the curves of Figure 3 and Figure 4 correlate equivalent stress-strain data.

Figure 5 replots the data of Figure 4 and shows that a single universal curve of $B(t,\varepsilon)/B_o$ versus tensile strain defines the five sets of strain histories represented by Table 1. An increase in strain rate from $\dot{\varepsilon} = 7.41 \cdot 10^{-3}$ sec. to $\dot{\varepsilon} = 1.48 \cdot 10^{-2}$, $3.70 \cdot 10^{-2}$, and $7.41 \cdot 10^{-2}$ at 22°C produced the expected slight changes in E_I, E_m, σ_b, and λ_b without changing the form of the $B(t,\varepsilon)/B_o$ versus tensile strain function shown in Figure 5.

OPTICAL PROPERTIES

We observed a consistent loss of transparency in our tensile test specimens when $B(t,\varepsilon)$ diminished from unity and a regain of transparency when $B(t,\varepsilon)$ returned to unity. This visual result correlates with the creation of internal voids and light scat-

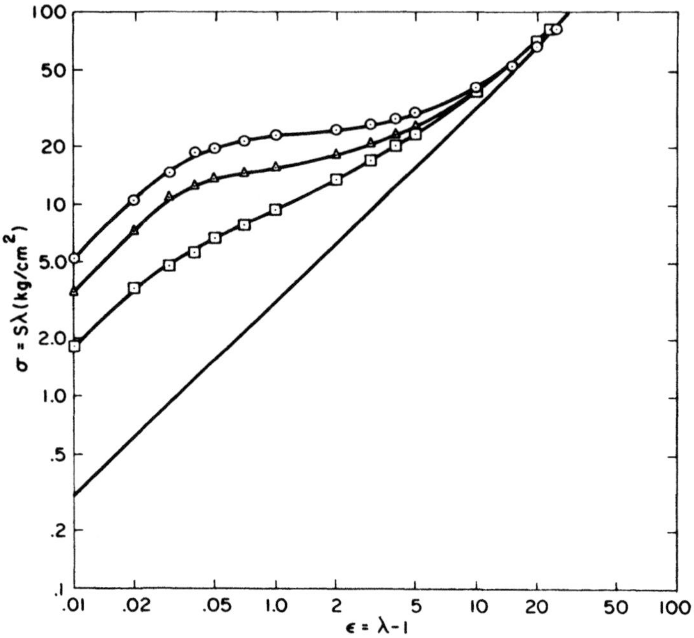

Fig. 3 True stress vs extension curves for Kraton 101 subjected to extension and recovery cycles shown in Table 1.

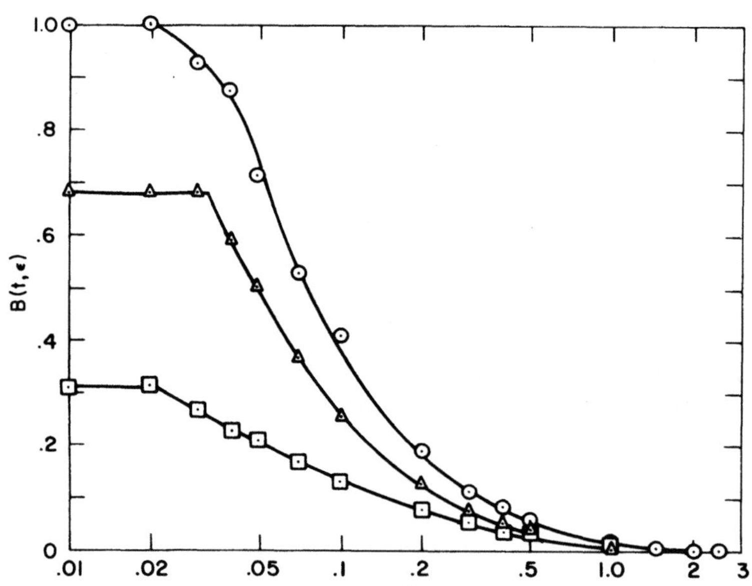

Fig. 4 Functions of the time-extensions functions for the degree of bonding $B(t,\varepsilon)$.

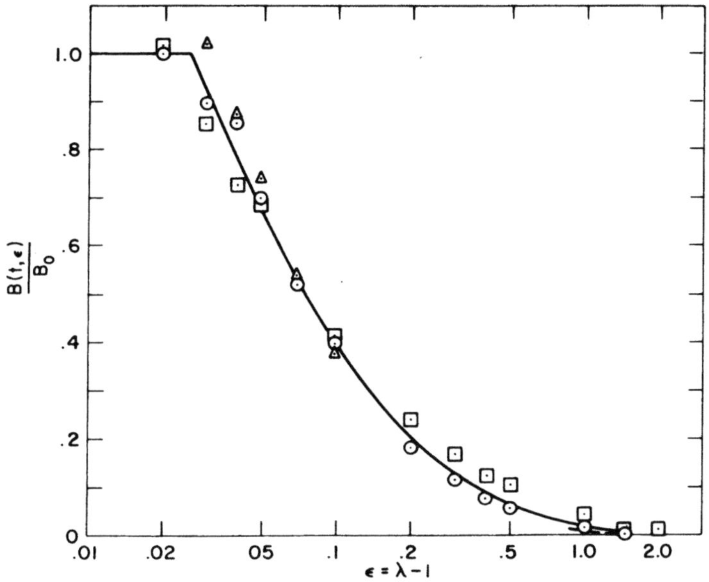

Fig. 5 Normalized function of $B(t,\varepsilon)/B_o$ vs extension for extension recovery histories described by Table 1.

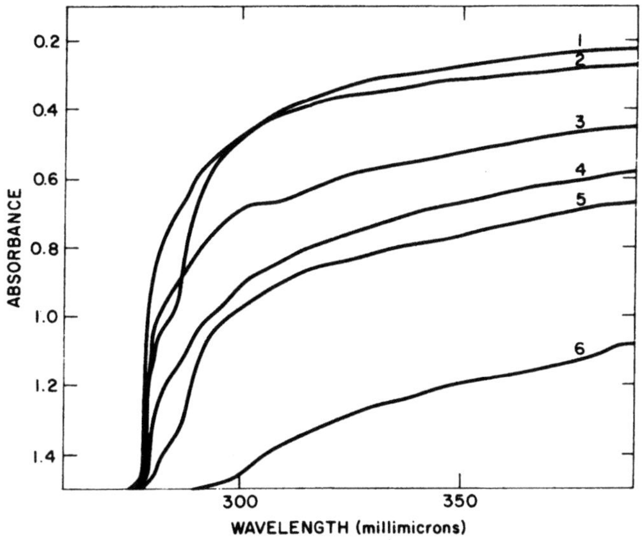

Fig. 6 Relative absorbence vs wave length: (1) undeformed film; (2) stretched to ε=2.5 (6 cycles) annealed 1 hour at 60°C stretched one more cycle and heated 1 hour at 90°C; (3) stretched to ε=2.5 (6 cycles) and annealed 1 hour at 60°C; (4) sample extended to break ε_b=8.4 and annealed 1 hour at 90°C; stretched to ε=2.5 (6 cycles) and not annealed; (6) stretched to break ε_b=8.4 and not annealed.

tering surfaces which exceed the wave length of light and is symptomatic of a cavitation process within the specimen. Yagii and Tscheogl had noted this same phenomena during the course of stress-relaxation studies on Kraton 102 (5). We employed a Perkin-Elmer Model 202 UV-visible spectrophotometer to analyze optical transmission properties at wave lengths from 200 to 750 millimicrons (nm). Absorbence curves at wave lengths from 250 to 390 nm are presented in Figure 6. All samples were highly absorptive at wave lengths shorter than 265 nm due to molecular absorption of the polystyrene A phase (6,7). In the visible region, from 300 to 750 nm, the relative absorbence between samples is independent of wave length while transparency increases monotonically with wave length.

The curves of Figure 6 clearly indicate that mechanical strain and recovery history can be correlated with sample transparency in the visible region. The undeformed film (curve 1) displays the greatest transparency while a sample deformed to fracture, $\varepsilon_b = 8.4$, shows the highest absorbence (curve 6). Heating this broken sample for 1 hour at 90°C (curve 4) does not produce a full recovery of initial transparency.

Extension cycling a sample to $\varepsilon = 2.5$ where the interfacial structure described by $B(t,\varepsilon)$ is destroyed but the polystyrene domains are not disturbed produces an initial loss of transparency (curve 5) which is partially recovered by 60°C annealing (curve 3) and almost completely recovered by 90°C annealing (curve 2).

The mechanism for loss of transparency on stretching and the resultant transition in mechanical properties becomes visibly evident if the sample is held extended at $\varepsilon_b = 2.5$ for 16 hours at 22°C. The left hand micrograph of Figure 7 shows agglomerated cavities as the chain like structures within the specimen and more sharply defined penny shaped cracks on the sample surfaces of the stretched specimen. This same specimen in the unstressed state display simple planes of defects which are the collapsed cavities and surface cracks under network constraints described by equation (2) which defines the residual elastic properties for the conditions $B(t,\varepsilon) = 0$. Upon initial stretching to $\varepsilon = 2.5$ and $B(t,\varepsilon) = 0$ visible light microscopy does not resolve the individual cavities, although the curves of Figure 6 do clearly resolve the change in sample transparency. The internal void structures shown in Figure 7 develop gradually by interaction of the primary cavities through a communal coalescence and growth process. On annealing overnight at 90°C the void structure illustrated in Figure 7 disappeared and the sample displayed its original transparency.

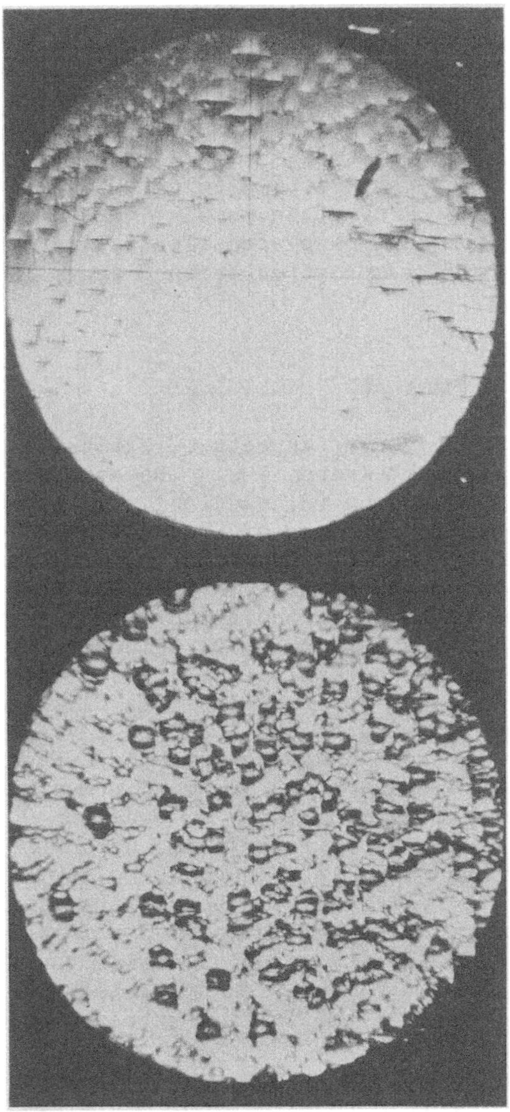

Fig. 7 Void structure in Kraton 101 film in stretched (left view) and retracted (right view) states after holding at $\varepsilon=2.5$ for 16 hours at 22°C. Extension axis horizontal and 16x magnification.

PEEL ADHESION

Figure 8 shows the effect of bonding temperature, using a dwell time of 15 minutes on a temperature gradient bar, on the peeling properties at 22°C and constant peeling rate of 0.05 cm/min. A test specimen is mechanically laminated at 22°C as described in the experimental section and the laminate annealed on the thermal gradient bar. Each point along the sample length thus experiences a different temperature-time history. Subsequent peeling advances the fracture line along the sample length and samples the effect of the varying annealing temperature. Figure 8 shows a rapidly increasing peeling force at temperatures below 60°C with a plateauing in peeling force above 60°C. This result is generally consistent with the recovery of B_o properties by tensile testing shown in Table 1 and the recovery of sample transparency shown in curves 5, 3, and 2 of Figure 6.

SUMMARY AND CONCLUSIONS

This study has examined three aspects of extension and recovery cycling by benzene cast Kraton 101. The analysis of tensile properties provides, in equations (3) and (4), a mathematical model for the entire stress-strain response of Kraton 101 at 22°C and strain rates $\dot{\varepsilon} = 7.41 \cdot 10^{-3}$ sec^{-1}. Equation (2) provides a reasonable approximation to the inverse Langevin function that describes the affine extension of crosslinked chains to a fully extended state defined by λ_m where E_m is the network modulus in the low extension region where Gaussian statistics apply (8, 9, 10). The applicability of Equation (2) to strains $\varepsilon \geqslant 2.5$ must indicate the essential stability of the styrene domains as the locus of physical crosslinking to strains $\varepsilon \simeq \varepsilon_b = 8.4$ characteristic of an approach to $\lambda \simeq 0.9\ \lambda_m$. The stress relaxation and recovery properties relating to the modulus E_I and the function $B(t,\varepsilon)$ of equation (4) appear to clearly correlate the processes of interfacial debonding and rebonding of the interfacial AB phase that intervenes the glassy A domains and the elastomeric B phase. This result is consistent with a new model for interfacial morphology of symmetrical A-B-A triblock polymer typical of Kraton 101 (10, 11) and correlates with the fully reversible character of the $B(t,\varepsilon)$ function by varied time and temperature histories for recovery of the initial stress-strain properties.

Changes in optical transparency and substantial recovery of initial transparency subsequent to extension to $\varepsilon = 2.5$ are indicative of the growth and desinence of cavities of dimensions sufficiently large to act as light scattering centers. Continued strain at $\varepsilon = 2.5$ for extended times shows the visible void structure generated

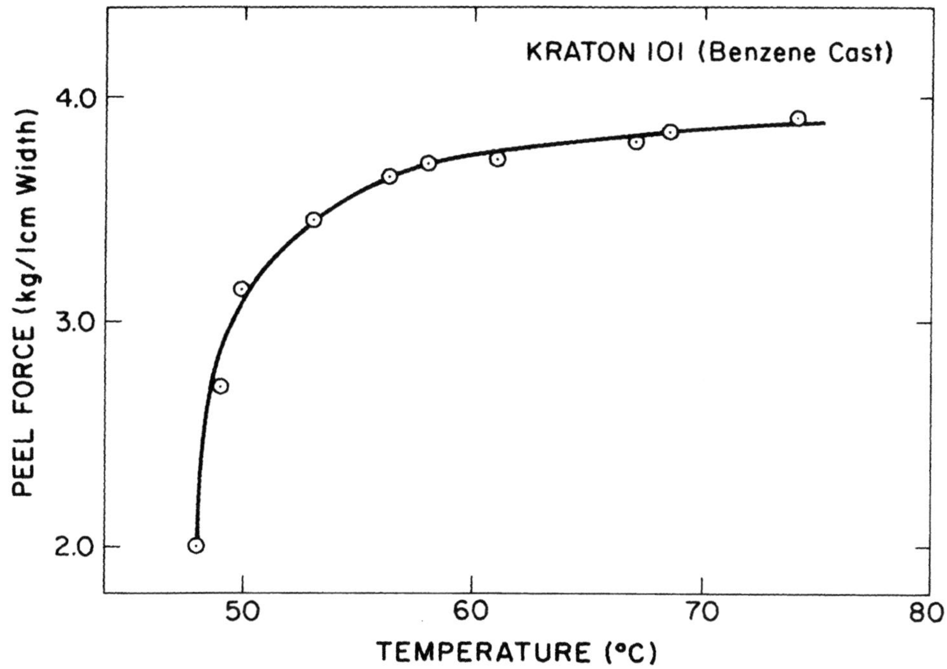

Fig. 8 Peeling force at 22°C vs bonding temperature on a temperature gradient bar (15 minute dwell time) for auto-adhered Kraton 101 laminates.

by cavity coalescence as shown in Figure 7. The form of the curve for dependence of 22°C peeling force upon the temperature-time history of autohesive bonding shown by Figure 8, is consistent with other aspects of this study.

REFERENCES

1. J. F. Beecher, L. Marker, R. D. Bradford and S. L. Aggarwal, J. Poly-Sci., Part C, No. 26, 117 (1969).

2. D. M. Brunwin, E. Fischer, and J. F. Henderson, Ibid., 135 (1969).

3. L. Mullins, J. Rubber Res., 16, 275 (1947).

4. D. H. Kaelble, J. Appl. Poly. Sci. 9, 1209 (1965).

5. K. Yagii and N. W. Tscheogl, Second Annual Report, Contract Number F04611-67-C-0057, AFRPL-TR-69-180, California Institute of Technology, August 1969, pp. V-9.

6. E. J. Meehan, J. Poly. Sci., 1, 175 (1946).

7. J. G. Calvert and J. N. Pitts, Jr., Photochemistry, Wiley, New York, 1966, p. 254.

8. L. R. G. Treloar, Physics of Rubber Elasticity, 2nd Edition, Clarendon Press, Oxford 1958.

9. R. F. Landel and R. F. Fedors, "Fracture Processes in Polymeric Solids" (Editor: B. Rosen) Chap. 3-B, Wiley-Interscience, New York (1964).

10. D. H. Kaelble, Physical Chemistry of Adhesion, Chapter 11 Wiley, New York (1970) in production.

11. D. H. Kaelble, "Interfacial Morphology and Mechanical Properties of A-B-A Triblock Copolymers", Presented before Society of Rheology, Pasadena, Feb. 1970, to be published in Trans. Soc. Rheol.

INFLUENCE OF MORPHOLOGY ON THE VISCOELASTIC BEHAVIOR OF A SOLVENT-CAST S-B-S BLOCK COPOLYMER

M. Shen
Dept. of Chem. Eng., University of California
Berkeley, California 94720

E. H. Cirlin and D. H. Kaelble
Science Center, North American Rockwell Corporation
Thousand Oaks, California 91360

It has recently been demonstrated in the literature that styrene block copolymers give rise to different morphological structures depending on the condition of sample preparation. Thus polymers cast from benzene will have a butadiene-continuous matrix in which are embedded polystyrene regions. On the other hand, methyl ethyl ketone/tetrahydrofuran mixture will render the sample styrene-continuous. In this work we have carried out stress-relation measurements on Kraton 101 over an extended temperature range (-150°C to 100°C). Viscoelastic master curves were constructed that cover 25 decades of time (in seconds). It is found that the MEK/THF cast samples possess considerably higher modulus than the benzene cast samples above the glass transition temperature. The shift factors utilized to effect time-temperature superposition do not follow the classical Williams-Landel-Ferry equation. Those for the benzene-cast block copolymer can be fitted to a modified WLF equation for a multiphase system. But those for the MEK/THF-cast ones cannot be adequately described by either of these equations. In addition, the presence of plasticizer is found to affect the viscoelastic behavior of the benzene-cast block copolymers in an unique manner. A model based on the possible morphological structures of these samples is introduced to explain the observed phenomena.

INTRODUCTION

The unique properties possessed by polymers consisting of blocks of thermodynamically incompatible segments are now known to be attributable to the presence of domains caused by microphase separation.[1-7] Electron photo-micrographs have demon-

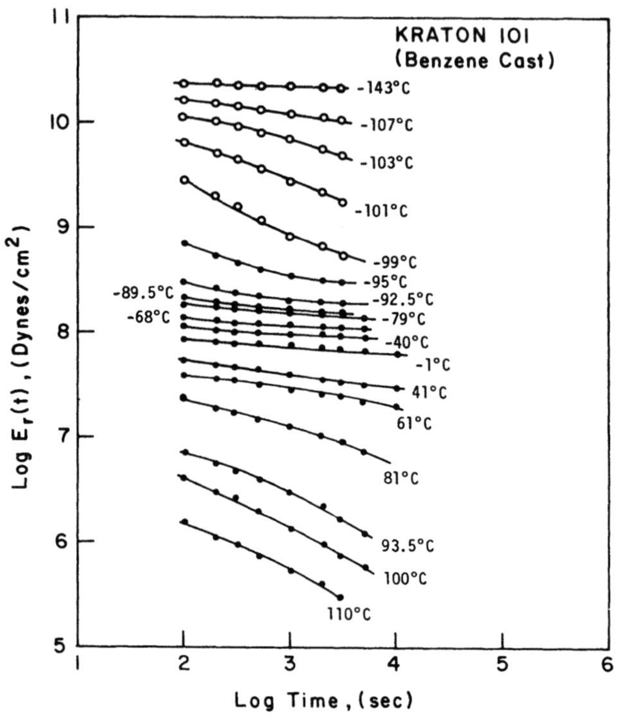

Figure 1: Modulus-time isotherms for benzene-cast Kraton 101, between -143°C and 110°C. Open circles represent data obtained by the fluxural mode, closed circles bv the tensile mode.

strated not only the presence of these domains, but the myriads of possible morphological structures of these domains.[8-12] That these different morphological structures evince different mechanical behavior in the block copolymer has already been reported.[13,14] Stress-strain data of Beecher, et al.,[13] corroborated by electron microscopic evidence, showed that styrene-butadiene-styrene block copolymer cast from benzene/heptane mixture shows a butadiene-continuous phase, and is essentially rubbery. On the other hand, the same polymer cast from a methyl ethyl ketone/tetrahydrofuran mixture is styrene-continuous and exhibits some yielding. In this work, we shall present viscoelastic data on these two kinds of samples of S-B-S polymer in an effort to study the effect of these supramolecular structures.

EXPERIMENTAL

Two samples of styrene-butadiene-styrene (S-B-S) block copolymers were studied. These are Kraton 101 and Thermolastic

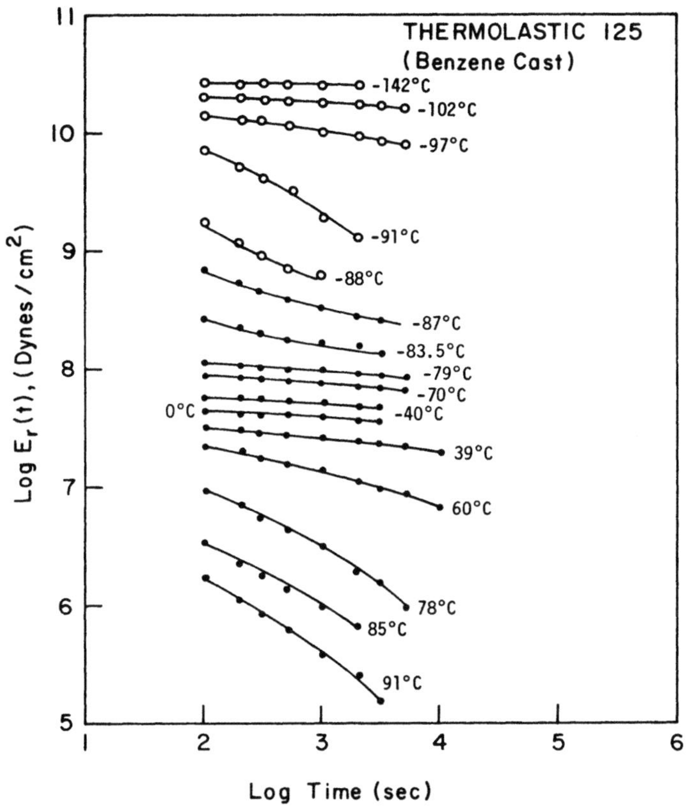

Figure 2: Modulus-time isotherms for benzene-cast Thermolastic 125 between -142°C and 91°C.

125 manufactured by the Shell Chemical Company. Kraton 101 is a pure S-B-S block copolymer, and contains no other additive than an antioxidant. The number average molecular weight of the polybutadiene portion was found [13,15] to be 76,000 and that of the polystyrene portion 11,000. From infrared analysis, Beecher, et al.[13] showed that the polybutadiene chain consists of 51% trans 1,4, 41% cis-1, 4 and 8% of the vinyl structure. Thermolastic 125 is also an S-B-S block copolymer, but contains some plasticizer. Refluxing in boiling methanol for three days removed 21% of an extractable material, which is presumably the plasticizer. The number average molecular weight of the polybutadiene portion was shown[16] to be 47% cis, 19% trans and 4% vinyl.

Both Kraton 101 and thermolastic 125 were cast into sheets approximately 0.25 cm thick by slowly evaporating from a benzene solution. In addition, Kraton 101 was also cast from a tetrahydrofuran solution containing 10% methyl ethyl ketone. Samples

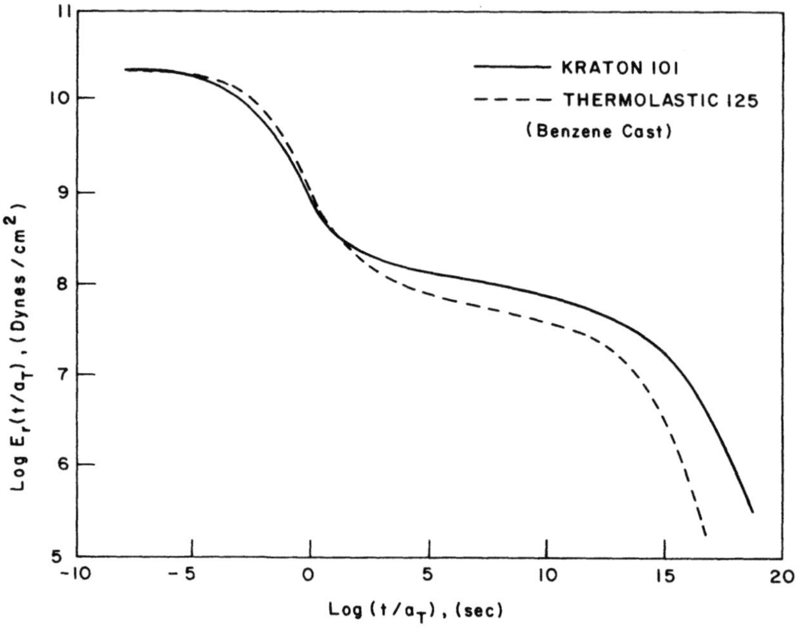

Figure 3: Viscoelastic master curves for Kraton 101 (solid curve) and Thermolastic 125 (broken curve). Both samples were cast from a benzene solution. Reference temperatures are their respective inflection temperatures (see Table 1).

were conditioned in a vacuum oven at 60°C for at least one month prior to use. Stress-relaxation measurements were carried out on an automatic tensile testing machine described previously.[17] Relaxation measurements covered temperatures from -140° to +110°C and times from 10^2 to 10^4 sec. For values of modulus less than 10^9 dynes/cm^2, strain was applied by uniaxial extension. If moduli were greater than this value, a flexural mode[18] of deformation was employed for more accurate results. For the benzene cast samples, strains (ε) of less than 10% was used. Since it has been demonstrated by Beecher, et al.[15] that yielding occurs in THF/MEK cast samples of Kraton 101 for strains greater than 3%, of all measurements for these samples were made at $\varepsilon<0.02$.

Dynamic mechanical experiments were performed on the THF/MEK cast Kraton 101 sample. Benzene-cast samples were not measured, because they are not sufficiently rigid for use in the acoustic spectrometer.[19,20] A sample of the dimension 0.5 x 0.5 x 6.5 cm^3 was clamped at the nodal points. An acoustic frequency at ~2kc was fed through one end of the sample by means of an electromagnetic transducer, and the damped signal detected on the other

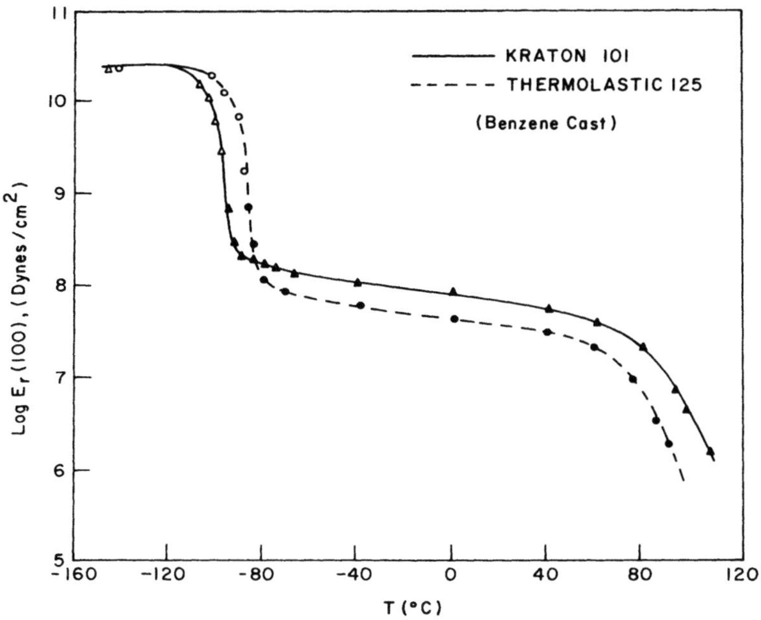

Figure 4: Isochronal modulus-temperature curves for the benzene-cast Kraton 101 (solid curve) and Thermolastic 125 (broken curve). Reference time is 100 seconds. Open and closed data points have the same significance as given in Figure 1.

end by a similar transducer. From the resonant frequency, the storage modulus E' can be readily obtained.[19,20]

RESULTS AND DISCUSSION

1. Effect of Plasticization

In Figures 1 and 2 are shown the modulus-time isotherms for benzene-cast Kraton 101 and Thermolastic 125. The open circles represent data taken by the flexural mode of the stress-relaxometer, and the closed ones represent the tensile mode. The agreement between data taken by these two methods appears to be quite good, and curves are readily superposable. By using the well-known time-temperature superposition principle,[21,22] viscoelastic master curves are obtained. These are given in Figure 3. Reference temperatures used for these master curves are the inflection temperatures of the respective polymers.

By definition the inflection temperature,[21] T_i, is the temperature at which the modulus of the polymer, measured at a

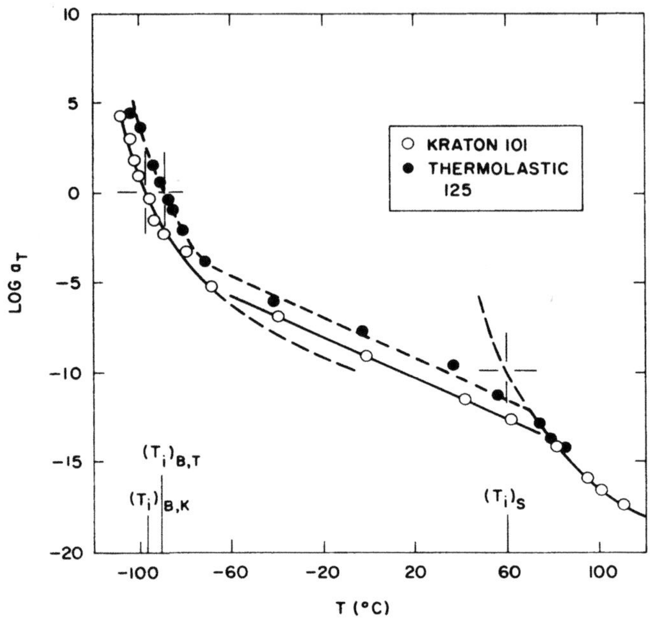

Figure 5: Log a_T vs. temperature plot for benzene cast Kraton 101 (open circles) and Thermolastic 125 (closed circles). Curves are calculated from the modified WLF equation to obtain best fit. $(T_i)_{B,T}$ is the inflection temperature of the polybutadiene phase of Thermolastic 125, $(T_i)_{B,K}$ that of the Kraton 101, and $(T_i)s$ refers to the polystyrene phase.

fixed time scale, reaches 10^9 dynes/cm^2. On the basis of the data given in Figures 1 and 2, we obtain the modulus-temperature curves for the benzene-cast Kraton 101 and Thermolastic 125 (Figure 4). Here we have taken 100 seconds as the reference time, rather than the usual 10 seconds.[21] In addition to T_i, a number of other characteristic viscoelastic parameters can be obtained from Figures 3 and 4. These are the glassy modulus E_1, rubbery modulus E_2, the negative slope of the master curve in the transition region ($n = - d \log E_r/d \log t$) and the negative slope of the modulus-temperature curve in the transition region ($s = -d \log E_r/dT$). These quantities are given in Table 1. For the sake of comparison, we have also included in Table 1 the characteristic viscoelastic parameters for polystyrene, polybutadiene and a styrene (25%) - butadiene (75%) random copolymer (SBR).[23]

DATA SHEET

Table 1. Characteristic Viscoelastic Parameters of Homopolymers and Copolymers of Butadiene and Styrene

Polymer	$E_1 \times 10^{-10}$ (dyn/cm^2)	$E_2 \times 10^{-7}$ (dyn/cm^2)	T_i (°C)	n	p	np	s
Kraton 101 (THF/MEK cast)	2.8	75.0	- 96	0.20	0.67	0.14	0.18
Kraton 101 (Benzene cast)	3.7	7.9	- 96	0.50	0.36	0.18	0.15
Thermolastic 125 (Benzene cast)	3.5	7.7	- 86	0.53	0.30	0.16	0.16
Polystyrene*	2.00	0.34	93.1	0.87	0.26	0.22	0.21
Polybutadiene*	2.00	1.07	-106.2	-	-	-	0.19
SBR (75% butadiene - 25% styrene)*	1.74	2.76	- 50.8	0.71	0.34	0.24	-

* From ref. 23

While glassy moduli of the block copolymers are close to those of homopolymers and random copolymers, there is a substantial increase in rubbery moduli of the former. We should note, however, that E_2 for the block copolymers above T_i changes rapidly with time and temperature. The numbers quoted in Table 1 are intermediate average values for the sake of discussion. The increased modulus in these block copolymers is believed to be the direct consequence of the presence of the domain structure.[24,25] The rigid polystyrene domains act on both cross link sites and fillers to enhance the properties of the rubbery polybutadiene phase. The interpretation is consistent with the electron microscopic evidence[13] that Kraton 101 cast from a benzene-heptane mixture has a polybutadiene-continuous matrix into which spherical domains of polystyrene are interspersed. As long as the glass transition temperature of polystyrene is not exceeded, the above observed phenomenon will persist.

The inflection temperatures observed in Figure 4 is believed to be associated with the onset of glass transition of the polybutadiene phase. The value of -96°C for Kraton 101 is lower than the dilatometrically determined T_g of -85°C for polybutadiene,[26] but is higher than the reported T_i of -106°C for the same rubber.[23] In general, inclusion of plasticizers in a polymer lowers its glass transition temperature.[26] It is thus puzzling to observe that for Thermolastic 125, which contains more than 20% by weight of a plasticizer, T_i = -80°C. Here the increased value of inflection temperature has in fact an "antiplasticization effect." Another unique behavior of Thermolastic 125 that is not in line with the "normal" plasticized polymers is its slopes s and n in the transition region. Usually the slopes broaden (s and n decrease in value) as plasticizers are introduced. For instance, for unplasticized polyvinyl chloride s = 0.25 and n = 0.59. Incorporation of 30% dictyl phthalate rendered the values of s and n to be lowered to 0.05 and 0.29 respectively.[27] In Table 1 we note that there is little difference in the values of s and n between Thermolastic 125 and its unplasticized analog - Kraton 101.

A possible interpretation of these phenomena is as follows. The plasticizer in Thermolastic 125 may have acted in such a way that it permits some of the polystyrene segments to "invade" the polybutadiene matrix. This small amount of rigid segments could have raised the glass transition temperature of the "impure" polybutadiene phase, even though part of this effect may be offset by the plasticizer. We note in Table 1 that incorporation of 25% styrene in polybutadiene as a random copolymer raised its T_i by 56°C. Thus an increase of 10°C in T_i for Thermolastic 125 over that of Kraton 101 is not inconceivable. Since values of s and n are essentially unaffected by the introduction of plasticizer into the block copolymer, it is inferred that most

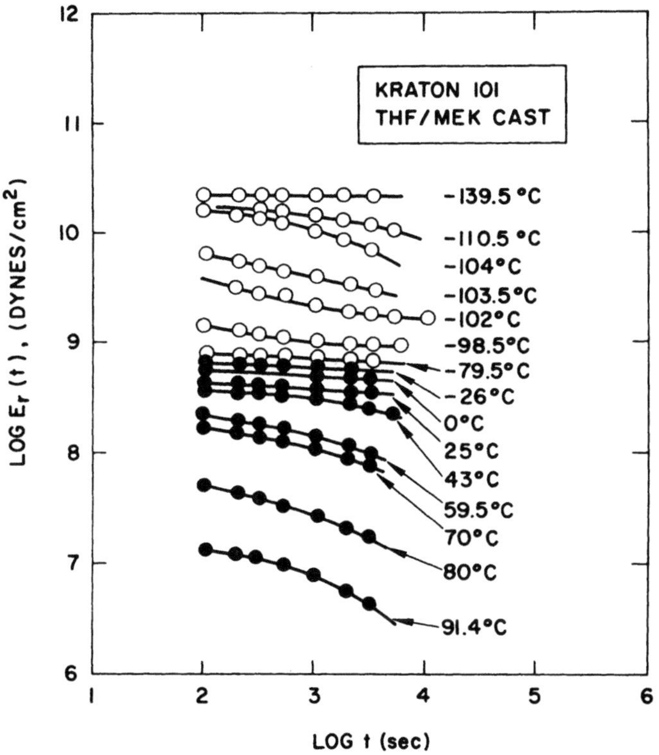

Figure 6: Modulus-time isotherms for Kraton 101 cast from a 10:1 mixture of tetrahydrofuran and methyl ethyl ketone.

of the plasticizer must be present in the polystyrene regions. This would be consistent with the observation that rubbery flow commences at a lower temperature for Thermolastic 125 than for Kraton 101. Here the plasticizer not only has lowered the glass transition temperature, but also shortened the maximum relaxation times of the polymers (Figures 3 and 4).

Although no difficulty is encountered in constructing the master curves from the modulus-time isotherms, the shift factors a_T employed in the superposition procedure do not follow the classic WLF equation[22]:

$$\log a_T = \frac{-16.14\ (T-T_i)}{56 + T - T_i} \qquad (1)$$

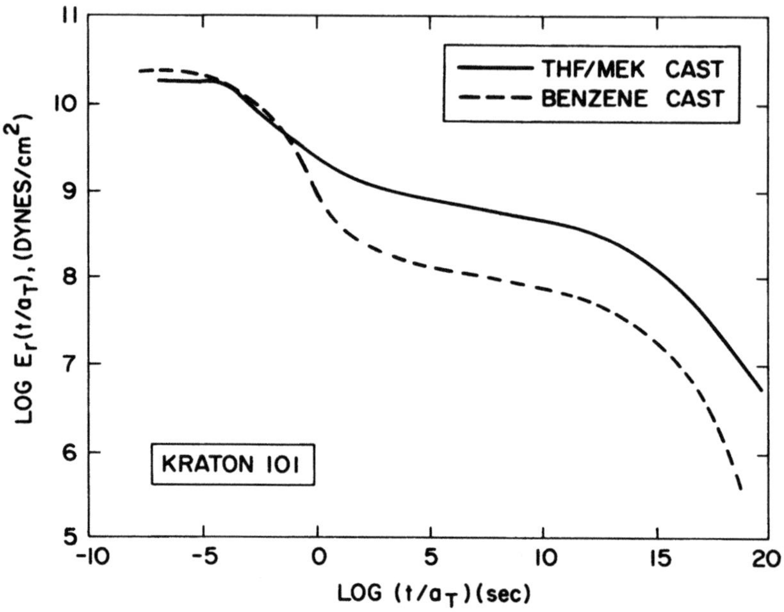

Figure 7: Viscoelastic master curves for Kraton 101 cast from a 10:1 mixture of tetrahydrofuran and methyl ethyl ketone (solid curve) and that from benzene. Reference temperatures are their respective inflection temperatures (see Table 1).

The basis of the WLF equation is the concept of an isofractional free volume (f_g) below the glass transition temperature (T_g) and then a constant thermal expansion coefficient (β_f) for fractional free volume above T_g. Since S-B-S block copolymers are recognized to possess two phases, the "effective" fractional free volume will depend upon the phase which contributes the molecular mechanisms for stress relaxation response. At temperatures below -50°C and above 80°C the experimental time-temperature shift factors, a_T, are well defined by eq. 1 (Figure 5) for the respective T_i's of polybutadiene and polystyrene. Intervening these low and high temperature regions of simple WLF response the data points of Figure 5 show a broad intermediate temperature range of constant slope, where $d \log a_T/dT = -.0578$, and equation (1) might be presumed inapplicable. Similar linear segments for the log a_T versus T curve have been reported by both Beecher and coworkers,[13] and also Smith and Dickie.[15] Figure 5 indicates our interpretation[28] of this broad central region of response as being defined by

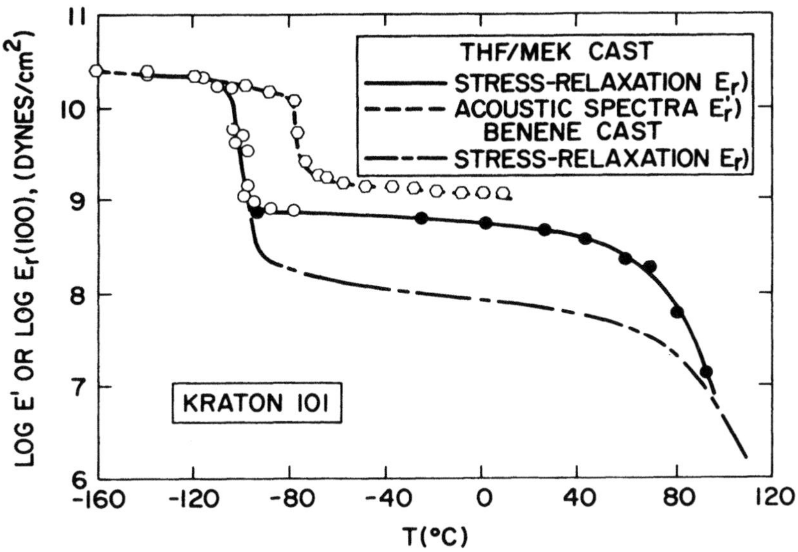

Figure 8: Isochronal modulus-temperature curves for Kraton 101 cast from a 10:1 tetrahydrofuran and methyl ethyl ketone mixture (circles). Data for the benzene-cast sample are included for the sake of comparison. Hexagons represent dynamic storage moduli for the THF/MEK-cast sample determined at ~ 2kc. Significance of the open and closed circles are the same as in Figure 1.

$$\frac{d \log a_T}{dT} = \frac{-c_1 c_2}{2.303\,(c_2 + T - T_i)^2} \qquad (2)$$

for the constant condition $T-T_i = 66°C$. If we regard a_T to be proportional to exp (H/RT), where R is the ideal gas constant, then it can be readily shown that H, the activation energy for viscoelastic relaxation times in this region, is 24 kcal/mole.

At temperatures between -60°C and +80°C, the molecular mechanisms for stress relaxation are being contributed by an interfacial phase that intervenes the polybutadiene continuum and the spherical polystyrene domains. This interfacial phase can be visualized as a series of spherical shells enclosing each of the pure polystyrene domains and characterized by a fairly sharp

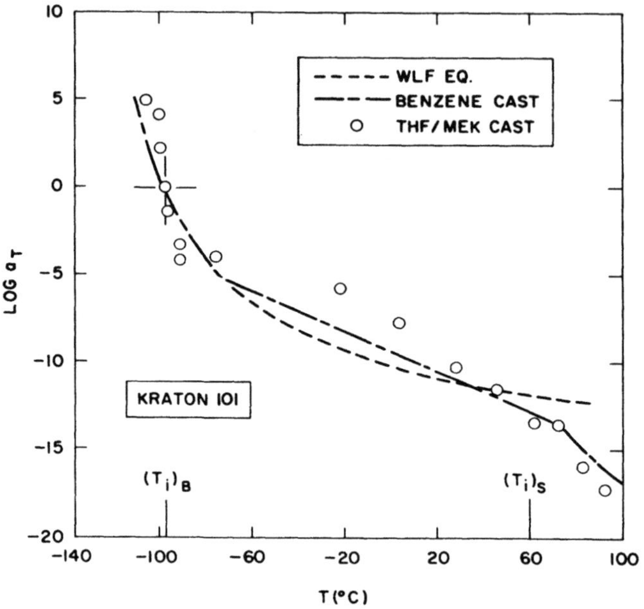

Figure 9: Log a_T vs. temperature plot for Kraton 101 cast from a 10:1 mixture of tetrahydrofuran and methyl ethyl ketone (circles). The classical WLF curve for benzene-cast sample are included for the sake of comparison. $(T_i)_B$ is the inflection temperature of the polybutadiene phase, and $(T_i)_B$ that of the polystyrene phase.

concentration gradient between shells of mixed polybutadiene and polystyrene segments. The outermost shell of the interfacial phase displays the lowest T_i value and lowest concentration of polystyrene segments. Both T_i and the volume fraction of polystyrene segments increase rapidly in succeeding interior shells. The constant value of $T - T_i = 65°C$ indicates that stress relaxation is occurring in succeeding interior shells of higher $T_i \cong T_g$ as the test temperature is raised above -60°C.

This interpretation of the viscoelastic response of S-B-S triblock polymers would indicate that the interfacial phase plays a dominant role in mechanical response at ambient temperatures. The three-phase model is consistent with the several theoretical analyses for the thermodynamics of mixing between polystyrene and polybutadiene segments in S-B-S block copolymers.[6,7]

It is of interest to note that in a recent work by Lim, Carmichael and Tschoegl,[29] the following equation was suggested for an S-B-S block copolymer

$$\log a_T = -\frac{c_1(T-T_1)}{c_2 + T-T_1}\, h(T) + \frac{\Delta H}{2.303R}\left(\frac{1}{T} - \frac{1}{T_o}\right) h(T-T_o) \qquad (3)$$

where h(T) is a unit step function, and C_1, C_2, T_1, T_o are empirically determined parameters. These workers found excellent agreement for their viscoelastic data between -70° to + 80°C for Kraton 102 cast from benzene and from cyclohexane. Eq. 3, however, cannot be used to fit our data on Kraton 101 and Thermolastic 125. The discrepancy could be due to possible difference in morphological structures. Kraton 102 has a molecular weight of 9500 for the polystyrene portion and 37,500 for the polybutadiene portion, being substantially lower than the samples used in this study.

An additional viscoelastic parameter that can be obtained from our data is the negative slope p of the log a_T vs. T plot for temperatures close to T_1. According to Tobolsky,[21] s ≈ np. This approximate relation is obeyed for our block copolymers, as shown in Table 1.

2. Effect of Casting Solvent

Fifteen modulus-time isotherms between -139.5°C and 91.4°C were taken for Kraton 101 cast from a 10:1 tetrahydrofuran/methyl ethyl ketone solution (Figure 6). These isotherms were superposed to yield a viscoelastic master curve in the usual manner. Figure 7 shows this curve, along with the master curve of the benzene-cast Kraton 101. It is evident that the viscoelastic property of the same block copolymer has been drastically modified by the use of different casting solvents.

It has been demonstrated that a THF/MEK-cast S-B-S block copolymer has a styrene-continuous structure in which the polybutadiene portion is present as discontinuous domains.[13] The above electron microscopic evidence is corroborated by the observation that yielding occurs above a strain of 3% which is characteristic of the glassy polystyrene. The enhanced rubbery modulus, being nearly an order of magnitude greater than the benzene-cast sample (Table 1), is consistent with the above-mentioned morphological structure. The transition region slope n is broader, possibly due to the presence of greater number of interfacial regions. However, Figure 8 indicates that the slopes s are not significantly affected. In addition, flow appears to commence around the same temperature for both the benzene-cast and THF/MEK-cast Kraton 101. The latter behavior is to be expected, since above the glass transition temperature of polystyrene flow of the polymer will begin regardless of the structure.

The log a_T - T plot for this sample, shown in Figure 9, cannot be described by either the classical or the modified WLF equation. Here it is possible that there is a distribution of sub-structures in the interfacial region, which respond to the temperature in a more complex manner than that in the benzene-cast sample. Thus one might surmise that the role of the supramolecular structure is in general more complicated than the benzene-cast Kraton 101. We should mention that Cooper, Estes and Tobolsky[30] recently also found that the log a_T-T data for a segmented polymer Estane is not amenable to either of the WLF analysis.

In Figure 8 is also shown the modulus-temperature curve for THF/MEK-cast Kraton 101, determined by acoustic spectrometer at ~ 2kc. The shape of this curve is similar to the one determined at a time scale of 100 sec. As one would expect, decrease in time scale or increase in frequency increases both the transition temperature and the rubbery modulus. If we redefine T_i as the temperature at which the inflection in the transition region slope occurs, then it is about 40°C higher than that obtained from the $E_r(100)$ curve. This is in agreement with the data in Figure 9 that a shift of 5 decades of time (10^{-3} to 10^{-2}) is equivalent to a 40°C increase in temperature.

REFERENCES

1. D. J. Meier, J. Polymer Sci., Part C, 26, 81 (1969).

2. S. Krause, J. Polymer Sci., Part A-2, 7, 249 (1969).

3. H. Ohnuma, T. Kotaka and H. Inagki, Polymer, 10, 501 (1969).

4. T. Inoue, T. Soen, T. hashimoto and H. Kawai, J. Polymer Sci., Part A-2, 7, 1283 (1969).

5. U. Bianchi, E. Pedemonte and A. Turturro, Polymer Letters, 7, 785 (1969); polymer, 11, 1268 (1970).

6. D. F. Leary and M. C. Williams, Polymer Letters, in press.

7. D. H. Kaelble, "Interfacial Morphology and Mechanical Properties of A-B-A Triblock Polymers," Presented Soc. Rheology, Pasadena (Feb. 1970).

8. H. Hendus, K. H. Illers and E. Ropte, Kolloid-Z. u.Z.f. Polymere, 216, 110 (1967).

9. E. B. Bradford and E. Vazo, J. Polymer Sci., Part A-1, 6, (1968).

10. E. Fisher, J. Macromol. Sci., Part A, 2, 1285 (1968).

11. M. Matsuo, T. Veno, H. Horino, S. Chujyo and H. Asai, Polymer, 9, 425 (1968).

12. T. Inoue, T. Soen, T. Hashimoto and H. Kawai, Polymer Preprints, 10, 538 (1969).

13. J. F. Beecher, L. Marker, R. D. Bradford and S. L. Aggarwal, J. Polymer Sci., Part C, 26, 117 (1969).

14. G. A. Harpell and C. E. Wilkes, Polymer Preprints, 10, 678 (1969).

15. T. L. Smith and R. A. Dickie, J. Polymer Sci., Part C, 26, 117 (1969).

16. N. H. Canter, J. Polymer Sci., Part A-2, 6, 155 (1968).

17. M. Shen, D. A. McQuarrie and J. L. Jackson, J. Appl. Phys., 38, 791 (1967).

18. ASTM Designation D790-66.

19. H. Schlein and M. Shen, Rev. Sci. Intr., 40, 587 (1969).

20. M. Shen, J. D. Strong and H. Schlein, J. Macromol-Sci., Part A, 3, 1315 (1969).

21. A. V. Tobolsky, "Properties and Structure of Polymers", Wiley, New York, 1960.

22. J. D. Ferry, "Viscoelastic Properties of Polymers," 2nd ed., Wiley, New York, 1970.

23. A. V. Tobolsky and M. Takahashi, J. Appl. Polymer Sci., 1, 1341 (1963).

24. G. Holden, E. T. Bishop and N. R. Legge, J. Polymer Sci., Part C, 26, 37 (1969).

25. R. F. Fedors, J. Polymer Sci., Part C, 26, 189 (1969).

26. M. Shen and A. Eisenberg, Rubber Chem. and Tech., 43, 95 (1970).

27. M. Shen and A. V. Tobolsky, Adv. Chem. Sen., 48, 118 (1965).

28. M. Shen and D. H. Kaelble, Polymer Letters, 8, 149 (1970).

29. C. K. Lim, L. T. Carmichael and N. W. Tschoegl, to be published.

30. S. L. Cooper, G. M. Estes and A. V. Tobolosky, Polymer Preprints, 10, 686 (1969).

INDEX

GPSR Compliance

The European Union's (EU) General Product Safety Regulation (GPSR) is a set of rules that requires consumer products to be safe and our obligations to ensure this.

If you have any concerns about our products, you can contact us on ProductSafety@springernature.com

In case Publisher is established outside the EU, the EU authorized representative is:

Springer Nature Customer Service Center GmbH
Europaplatz 3
69115 Heidelberg, Germany

Batch number: 10684222

Printed by Printforce, the Netherlands